Carbon Filaments and Nanotubes: Common Origins, Differing Applications?

NATO Science Series

A Series presenting the results of activities sponsored by the NATO Science Committee. The Series is published by IOS Press and Kluwer Academic Publishers, in conjunction with the NATO Scientific Affairs Division.

A.	Life Sciences	IOS Press
B.	Physics	Kluwer Academic Publishers
C.	Mathematical and Physical Sciences	Kluwer Academic Publishers
D.	Behavioural and Social Sciences	Kluwer Academic Publishers
E.	Applied Sciences	Kluwer Academic Publishers
F.	Computer and Systems Sciences	IOS Press

1.	Disarmament Technologies	Kluwer Academic Publishers
2.	Environmental Security	Kluwer Academic Publishers
3.	High Technology	Kluwer Academic Publishers
4.	Science and Technology Policy	IOS Press
5.	Computer Networking	IOS Press

NATO-PCO-DATA BASE

The NATO Science Series continues the series of books published formerly in the NATO ASI Series. An electronic index to the NATO ASI Series provides full bibliographical references (with keywords and/or abstracts) to more than 50000 contributions from international scientists published in all sections of the NATO ASI Series.
Access to the NATO-PCO-DATA BASE is possible via CD-ROM "NATO-PCO-DATA BASE" with user-friendly retrieval software in English, French and German (WTV GmbH and DATAWARE Technologies Inc. 1989).

The CD-ROM of the NATO ASI Series can be ordered from: PCO, Overijse, Belgium

Carbon Filaments and Nanotubes: Common Origins, Differing Applications?

edited by

L.P. Biró
Research Institute for Technical Physics and Materials Science,
Laboratory for Nanostructure Research,
Budapest, Hungary

C.A. Bernardo
Universidade do Minho,
Campus de Azurém,
Guimarães, Portugal

G.G. Tibbetts
General Motors R&D Center,
Physics and Physical Chemistry Department,
Warren, U.S.A.

and

Ph. Lambin
Facultés Universitaires Notre Dame de la Paix,
Physics Department,
Namur, Belgium

Kluwer Academic Publishers

Dordrecht / Boston / London

Published in cooperation with NATO Scientific Affairs Division

Proceedings of the NATO Advanced Study Institute on
Carbon Filaments and Nanotubes: Common Origins, Differing Applications?
Budapest, Hungary
19–30 June 2000

A C.I.P. Catalogue record for this book is available from the Library of Congress.

ISBN 0-7923-6907-6 (HB)
ISBN 0-7923-6908-4 (PB)

Published by Kluwer Academic Publishers,
P.O. Box 17, 3300 AA Dordrecht, The Netherlands.

Sold and distributed in North, Central and South America
by Kluwer Academic Publishers,
101 Philip Drive, Norwell, MA 02061, U.S.A.

In all other countries, sold and distributed
by Kluwer Academic Publishers,
P.O. Box 322, 3300 AH Dordrecht, The Netherlands.

Printed on acid-free paper

TABLE OF CONTENTS

Roundtable discussions

PREFACE

Intriguing recent discoveries in the behavior of low dimensional materials have captivated the attention of physicists, chemists, biochemists, and engineers. Nano-scale objects in particular seem to offer attractive prospects for novel properties and revolutionary applications. Despite numerous investigations into the fundamental science of such materials, new mysteries seem to arise as rapidly as old questions are answered.

The discovery of carbon nanotubes by Iijima a decade ago marked the opening of a new and important chapter in nano-scale materials science. Carbon nanotubes are unique because they combine one-dimensional solid-state physics arising from their macroscopic length with molecular physics due to their nanoscopic diameter. The scientific community soon learned that researchers who had been producing carbon filaments had been unknowingly growing nanotubes years before Iijima's publication. Thus, deep and meaningful connections have linked these two fields from the outset. However, the linkage between the two scientific communities: scientists working on carbon nanotubes, and scientists working on carbon filaments has not been sufficiently close. The flow of ideas, concepts and information seemed to require improvement.

Achieving closer relationships and improvements in communications between these two communities seemed like a worthwhile goal, motivating us to organize a scientific meeting presenting both research fields in parallel. We aimed to review in a somewhat tutorial manner the most important achievements of the last few years, concluding at the forefront of present research. Our intention was to create a framework that will generate a synergistic reformulation of our accumulated knowledge and experience, leading to new research efforts that would utilize perspective and wisdom from both fields.

Our effort to implement this goal was to organize in June 2000, in Budapest, Hungary, a NATO Advanced Study Institute. This book presents a selection of the lectures given by experts which provide broad perspective on the state-of-the-art of the various disciplines of carbon nanotube and vapor grown carbon fiber (VGCF) research and on the directions for future research. While the snapshot archived in the following pages is necessarily ephemeral in view of the rapid advances in carbon naostructures research, we hope that the perspectives offered here can positively effect both the carbon nanotube and VGCF fields.

We thank all the authors who responded to the demanding task of promptly preparing high quality manuscripts. We gratefully acknowledge the generous support of the NATO Scientific and Environmental Affairs Division, the Hungarian Academy of Sciences, the Hungarian Scientific Research Fund, and the Research and Development Division of the Ministry of Education, Hungary. Further thanks are due to our Institutes, Universities, and Companies, since they made the Advanced Study Institute and this resulting volume possible. We also express our gratitude to all those who have contributed to the ASI and to Kluwer for the efficient publication of these contributions.

L. P. Biró	**C. A. Bernardo**	**G. G. Tibbetts**	**Ph. Lambin**
Budapest	**Guimarães**	*Warren, Michigan*	*Namur*
Hungary	*Portugal*	*USA*	*Belgium*

Budapest, Guimarães, Warren and Namur, December 2000

increasing recent discoveries in the behavior of low dimensional materials, have

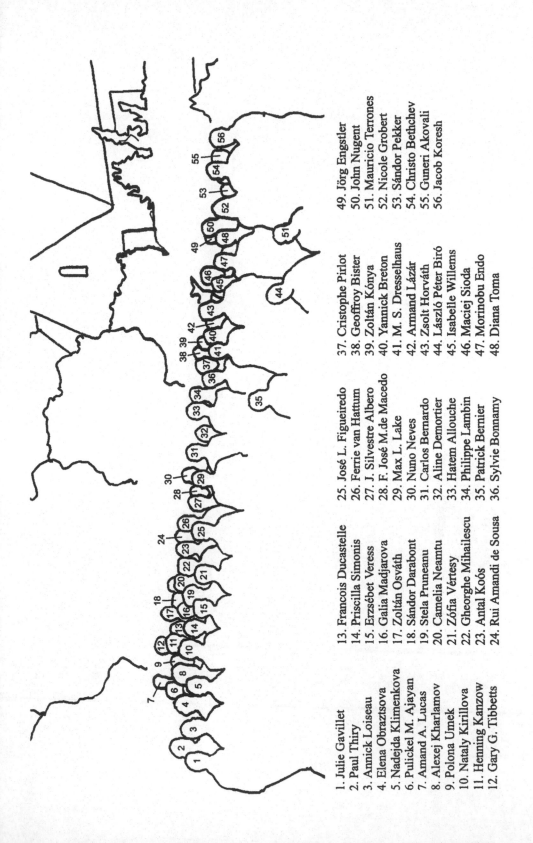

1. Julie Gavillet
2. Paul Thiry
3. Annick Loiseau
4. Elena Obraztsova
5. Nadejda Klimenkova
6. Pulickel M. Ajayan
7. Amand A. Lucas
8. Alexej Kharlamov
9. Polona Umek
10. Nataly Kirillova
11. Henning Kanzow
12. Gary G. Tibbetts
13. Francois Ducastelle
14. Priscilla Simonis
15. Erzsébet Veress
16. Galia Madjarova
17. Zoltán Osváth
18. Sándor Darabont
19. Stela Pruneanu
20. Camelia Neamtu
21. Zófia Vértesy
22. Gheorghe Mihailescu
23. Antal Koós
24. Rui Amandi de Sousa
25. José L. Figueiredo
26. Ferrie van Hattum
27. J. Silvestre Albero
28. F. José M.de Macedo
29. Max L. Lake
30. Nuno Neves
31. Carlos Bernardo
32. Aline Demortier
33. Hatem Allouche
34. Philippe Lambin
35. Patrick Bernier
36. Sylvie Bonnamy
37. Cristophe Pirlot
38. Geoffroy Bister
39. Zoltán Kónya
40. Yannick Breton
41. M. S. Dresselhaus
42. Armand Lázár
43. Zsolt Horváth
44. László Péter Biró
45. Isabelle Willems
46. Maciej Sioda
47. Morinobu Endo
48. Diana Toma
49. Jörg Engstler
50. John Nugent
51. Mauricio Terrones
52. Nicole Grobert
53. Sándor Pekker
54. Christo Bettichev
55. Guneri Akovali
56. Jacob Koresh

ORGANIZING AND SCIENTIFIC COMMITTEE

AND

EDITORS

Co-Directors

Dr. L. P. Biró
Research Institute for Technical
Physics and Materials Science
Laboratory for Nanostructure Research
Budapest, HUNGARY

Prof. C. A. Bernardo
Universidade do Minho
Campus de Azurém
Guimarães, PORTUGAL

Members

Dr. G. G. Tibbetts
General Motors R&D Center
Physics and Physical Chemistry
Department
Warren, USA

Prof. Ph. Lambin
Facultés Universitaires Notre Dame de
la Paix,
Physics Department
Namur, BELGIUM

LIST OF CONTRIBUTORS

Lecturers

Prof. P. M. Ajayan
Department of Materials Science &
Engineering
Rensselaer Polytechnic Institute Troy,
New York, USA

Prof. C. A. Bernardo
Universidade do Minho
Campus de Azurém
Guimarães, PORTUGAL

Dr. P. Bernier
Groupe de Dynamique des Phases
Condensees
Universite de Montpellier II,
Montpellier, FRANCE

Dr. L. P. Biró
Laboratory for Nanostructure Research
Research Institute for Technical
Physics and Materials Science
Budapest, HUNGARY

Dr. J-C. Charlier
Université Catholique de Louvain
Unité de Physico-Chimie et de
Physique des Matériaux
Louvain-la-Neuve, BELGIUM

Prof. D. D. L. Chung
Department of Mechanical and
Aerospace Engineering
State University of New York at
Buffalo
Buffalo, USA

Prof. M. S. Dresselhaus
Department of Electrical Engineering
and Computer Science
and Department of Physics
Massachusetts Institute of Technology
Cambridge, Massachusetts, USA

Prof. M. Endo
Shinshu University
Faculty of Engineering
Wakasato, Nagano-shi, JAPAN

Prof. J. L. Figueiredo
Faculdade de Engenharia - DEQ
Rua dos Bragas
Porto, PORTUGAL

Dr. A. Fonseca
Facultés Universitaires Notre Dame de
la Paix,
Physics Department
Namur, BELGIUM

Dr. Z. Kónya
Szegedi Egyetem
Alkalmazott és Környezeti Kémiai Tsz.
Szeged, HUNGARY

Mr. M. L. Lake
Applied Sciences, Inc.
Cedarville, USA

Prof. Ph. Lambin
Facultés Universitaires Notre Dame de
la Paix,
Physics Department
Namur, BELGIUM

Dr. Annick Loiseau
Laboratoire d'Etudes des
Microstructures
Unite Mixte ONERA-CNRS
Châtillon, FRANCE

Prof. A. A. Lucas
Facultés Universitaires Notre Dame de
la Paix,
Physics Department
Namur, BELGIUM

Dr. Mauricio Terrones
Max-Planck-Institut für
Metallforschung
Stuttgart, GERMANY

Dr. G. G. Tibbetts
General Motors R&D Center

Physics and Physical Chemistry
Department
Warren, USA

Dr. F. W. J. van Hattum
Centre for Lightweight Structures
(CLS) TUD-TNO
Delft, THE NETHERLANDS

Notes taken at the Roundtable discussions by:

Ms. Nicole Grobert
University of Sussex
Fullerene Science Centre
Brighton , UK

Mrs. Camelia Neamtu
National Inst. for Research and Dev.
for Isotopic and Molec. Technologies
Cluj- Napoca, ROMANIA

Dr. F. W. J. van Hattum
Centre for Lightweight Structures (CLS) TUD-TNO
Delft, THE NETHERLANDS

Dr. Diana Toma
University of Applied Science
Gelsenkirchen
Gelsenkirchen, GERMANY

Mr. John Nugent
Rensselaer Polytechnic Institute
Troy, USA

INTRODUCTION

Carbon fibers grown from carbon filaments were observed for the first time more than a century ago[1]. The study of these objects, with typical diameters in the range of 100 nm to 10 μm, accelerated after the invention of the electron microscope. By the mid 1980's, continuous production of vapor grown carbon fibers (VGCF) was achieved. In contrast, carbon nanotubes – members of the fullerene family – were only discovered in 1991. However, due to the remarkable properties of these nanotubes, combining one dimensional solid-state like characteristics with molecular dimensions,[2] nanotube research has undergone explosive growth in the past decade. Recent results have convincingly demonstrated that at the core of each vapor-grown carbon fiber is a carbon nanotube[3]. In other words, there is a deep and intrinsic connection between these two varieties of filamentous carbon.

Carbon nanotubes are quasi one-dimensional tubular structures that exist in two classes: single-walled and multi-walled. Single-walled nanotubes consist of one graphene sheet rolled up in the form of a cylinder with a diameter of 1 to 2 nm. Multi-walled nanotubes are composed of several such concentric graphene cylinders. As discussed during the ASI, the maximum external diameter of a multi-walled carbon nanotube is around 30 nm. Below that value, the circumferences of the layers are small enough to give rise to sizable quantum size effects, with the consequence that the nanotube may be either metallic or semiconducting, depending on how the honeycomb structure is wrapped. This unique electronic property is one of the main differences from VGCF, which have a larger diameter and a more complex structure. Quantum effects have been detected in metallic nanotubes, intramolecular junctions, and nanotube field-effect transistors, and these devices may well lead the way to nanoelectronic applications.

The book starts with an introduction to carbon filaments and nanotubes. Part I includes a review on the achievements of the past and the barriers that still face researchers searching for applications for VGCF and filaments. The first chapter outlines how various manufacturers are now bringing VGCF materials to American and Japanese markets in areas that exploit their electrical conductivity and electrochemical activity and concludes that while significant problems to full-scale commercial production still remain, they are being solved rapidly. The second chapter introduces the structure and the most important properties of carbon nanotubes. It describes the quantum effects that have been detected in carbon nanotubes and model devices produced from nanotubes. The introduction includes a third chapter, describing the applications of VGCF in the Japanese market, highlighting their use in batteries now in production in that country. This chapter also discusses the similarities and differences between NT and VGCF, including production methods and some recent applications of nanotubes as electronic components.

---

[1] Vapor-grown carbon fiber research and applications: achievements and barriers, by *Gary G. Tibbetts*

[2] Electronic properties of carbon nanotubes and applications, by *Mildred S. Dresselhaus*

[3] From vapor-grown carbon fibers (VGCFs) to carbon nanotubes, by *M. Endo et al.*
Optimizing growth conditions for carbon filaments and vapor-grown carbon fibers, by *J. L. Figueiredo and Ph. Serp*

The body of the book is divided in two further parts. Part II deals with the production, mainly via the catalytic route, of both vapor grown carbon fibers and nanotubes. Part III describes their properties, characterization, and applications.

Part II has 9 chapters, covering the production of VGCF and NT, including the more exotic tubular $B_xC_yN_z$ architectures. The enhancement of fiber activity by gasification and surface modification and the modeling of NT growth from first principles are also presented in Part II, which ends with the description of large-scale production of VGCF. Different techniques exist for the production of NT, but it now seems that the catalytic method, in which carbon-containing gases are thermally decomposed on catalytic nanoparticles, is the one most likely to lead to mass production. This method is very similar to the one used for making the filaments that grow into VGCF. Accordingly, the catalytic route is emphasized in this part of the book. However, the growth rate of nanotubes is smaller than that of the fibers and the pyrolysis stage used to thicken the filaments must be scrupulously avoided. As mentioned during the Institute, high-pressure disproportionation of carbon monoxide is a recently developed and promising variant of the catalytic route for single-wall nanotubes that could simplify the purification process. The more detailed conditions used for optimizing the growth of filaments described in this part of the book can provide ideas that might be useful for developing more dynamic NT growth processes.

Part III has 12 chapters, starting with the theoretical background necessary for detailed structural analysis by transmission electron microscopy (TEM) and diffraction. The physical characterization of nanotubes requires special techniques. Due to their small diameters, TEM is a mandatory tool, but scanning tunneling microscopy (STM) and atomic force microscopy are also very helpful. With single-walled nanotubes, the ultimate characterization consists of the two wrapping indexes that define the helical structure. As shown by theoretical modeling and computer simulation[4], the corrected STM data may provide such structural information.

The greater perfection of the atomic structure of nanotubes compared with VGCF leads to better mechanical properties. Single-walled nanotubes have the highest tensile modulus, which means high stiffness, yet they remain extremely flexible. This unique combination of properties makes them ideal tips for nanoprobe microscopes. On the other hand, the structural perfection of NT is a disadvantage in composite materials, because of the resulting poor interactions with the matrix. Unless defects are intentionally introduced, or functional groups grafted to the surface of nanotubes, VGCF may be superior for reinforcement. Some of the chapters deal with the mechanical properties of VGCF and VCCF composites with different matrices, emphasizing the progress made in developing high performance composites in the past few years. The importance of rheological studies to optimize the production of polymer matrix composites is also highlighted[5].

This part of the book finishes with an overview of the applications of VGCF and nanotubes. A general conclusion for both carbon materials is that specific applications have to be found which can justify their industrial usage despite their relatively high present price. Several chapters emphasize how the electrical conductivity of VGCF may

---

[4] Interpretation of the STM images of carbon nanotubes, by *Ph. Lambin, V. Meunier*

[5] The role of rheology in the processing of vapor grown carbon fiber/thermoplastic composites, by *C. A. Bernardo et al.*

lead to conducting composites at relatively low filler volume fractions. Electromagnetic interference shielding, electromagnetic reflection and surface electrical conduction seem to promise applications for the near future. For NT, field emission devices will probably be the first to reach the market if the production costs of nanotubes can be cut. Nanotubes may also lead to electromechanical devices such as actuators that convert charging effects into mechanical deformations. In the long term, NT may provide a way to fabricate tools for manipulation at the nanoscopic scale. However, nanoelectronic applications will remain uncertain as long as no real breakthrough appears to solve the assembly problem.

The book ends with a brief description of the conclusions of three roundtables that took place during the Institute, focusing on the similarities and differences between the growth of filaments and nanotubes, the structure of these materials, and their near and long term applications.

L. P. Biró **C. A. Bernardo** **G. G. Tibbetts** **Ph. Lambin**
Budapest *Guimarães* *Warren, Michigan* *Namur*
Hungary *Portugal* *USA* *Belgium*

Budapest, Guimarães, Warren and Namur, December 2000

VAPOR-GROWN CARBON FIBER RESEARCH AND APPLICATIONS: ACHIEVEMENTS AND BARRIERS

GARY G. TIBBETTS
General Motors Research & Development Center
30500 Mound Rd., Warren, MI, 48090-90555, USA

Abstract

Vapor-Grown Carbon Fibers (VGCF) are now being produced in moderate volume by several companies around the world, making larger quantities of these materials available for experiment and manufacturing. These fibers have been used to fabricate thermoset and thermoplastic composites with improved tensile strength, tensile modulus, and electrical conductivity. Even so, barriers to more general application of these fibers will have to be surmounted. VGCFs are, difficult to permeate with resins, and to a certain extent, fragile. Moreover, the fibers as produced have a low density, making them difficult to handle and compound. Nagging worries continue that the fibers will present a lung hazard if they are dispersed in air. Present composites have shown properties far below optimum, because they are made of nearly randomly oriented fibers. To achieve really optimal properties, composites with reasonably well oriented fibers must be fabricated.

1. Introduction

The earliest reference to vapor-grown carbon fibers is an 1889 patent by Hughes and Chambers [1] describing the growth of "hair-like carbon filaments" from a mixture of hydrogen and methane in an iron crucible. These structures, apparently being chemical vapor deposited thickened carbon filaments about 20 μm in diameter, were thick enough to be observed by the naked eye.

It was not until after the invention of the electron microscope that Davis, Slawson, and Rigby [2] were able to observe nanometer-sized in diameter carbon filaments for the first time. Their sample was simply furnace soot. Carbon filaments are generally under 100 nm in diameter, while VGCF may range from slightly larger than this to 10 μm in diameter.

This discovery of filamentous carbon inspired a multitude of observational papers on carbon filaments [3] Meanwhile, a parallel literature on vapor-grown carbon fibers grew [4] without clear understanding that VGCF were simply carbon filaments thickened by vapor-deposited carbon.

With the development of conventional rayon and PAN-based carbon fibers of 7-10 μm in diameter, researchers in VGCF recognized that their materials could be

1

L.P. Biró et al. (eds.), Carbon Filaments and Nanotubes: Common Origins, Differing Applications, 1–9.
© 2001 *Kluwer Academic Publishers. Printed in the Netherlands.*

useful as a reinforcement. Efforts to produce VGCF of approximately 10 μm in diameter with a length longer than 1 mm were published by Koyama and Endo in Japan [5], Benissad et al [6] in France, and Tibbetts [7] in the USA. Despite the inexpensive feedstocks used by these researchers, the long periods required and sparse product produced by these methods proved to be impractical for most applications.

By the mid 1980's, Koyama and Endo [8], Hatano, Ohsaki, and Arakawa [9], and others sought to utilize a continuous process to produce smaller diameter VGCF with a higher yield. In this process, nanometer sized catalyst particles are created in the gas phase and move through a tubular reactor while they grow filaments, thicken with a small CVD layer, and are collected as they exit.

2. VGCF Manufacture and Achievements

2.1 HYPERION CATALYSIS - FIBRILS

In the USA, Hyperion Catalysis (Fibrils.com) pursued an innovative implementation scheme. They devised multi-component metallic catalysts [10], for example Fe-Mo and maintained a fine dispersion by supporting the catalysts with fumed alumina to avoid sintering. These particles could thus be utilized to grow fibers at relatively low temperature, about 600°C, from a more reactive hydrocarbon, such as ethylene. The manufacturers state that proper control of the catalyst and growth conditions allows the fabrication of a multi-walled nanotube of 10 nm diameter comprised of 16 concentric carbon layers whose graphene planes are parallel with the longitudinal axis of the "fibril" (Figure 1).

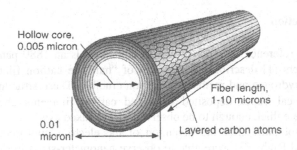

Figure 1. Diagram of a Hyperion fibril.

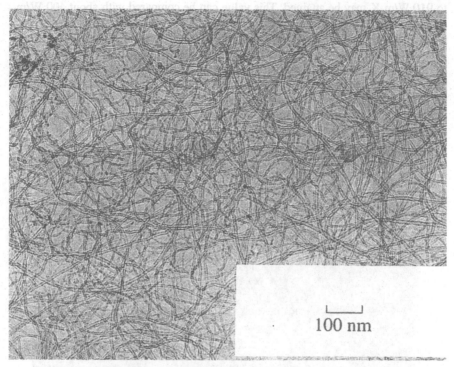

100 nm

Figure 2. SEM of Hyperion fibrils (Courtesy of Hyperion Catalysis, Inc.)

Hyperion and its development partners vigorously sought to utilize fibrils. They investigated applications in polymer and rubber reinforcement, conducting polymeric composites, Li-ion battery electrodes, and supercapacitors. Because of the difficulty of coupling these very graphitic fibrils to polymers and the strong competition of fiberglass, reinforcement applications have come disappointingly slowly. However, because of the low fibril diameter, good conductivity can be achieved with small fractional additions of fibrils to thermoplastics [11]. This conductivity is a useful property because it allows the manufacturer to avoid putting on a conductive primer before the customary electrostatic painting. Moreover, these low diameter fibers do not degrade the surface finish of the plastics, and excellent parts can be fabricated from homogeneous mixtures.

2.2 APPLIED SCIENCES, INC. – PYROGRAF II VGCF

Applied Sciences's (apsci.com) first marketable products utilized 5 μm macroscopic VGCF produced from methane in a tubular furnace with a 2 hour CVD step at 1100°C to achieve the large diameter. ASI found that these fibers could be laid up in a mold to give longitudinal fiber orientation, converted to a porous preform with a furfuryl alcohol binder, infiltrated with carbon, and graphitized to 3000°C. Even though these materials are quite expensive, applications in spacecraft and electronics are feasible as thermal conductivities up

4

to 910 W/m-K may be attained. This value can be compared with about 400 W/m-K for the much denser metallic copper.

ASI licensed technology from General Motors for producing 200 nm fibers using a continuous process [12] and developed further innovations to cut costs and to scale up production. Applied Sciences is now producing PYROGRAF III VGCF, a fiber averaging 200 nm in diameter (Figure 3) at the rate of about 35,000 kg/year. The material now costs $55/kg and up for different grades of fiber, but management hopes that equipment coming on stream shortly can dramatically cut these costs.

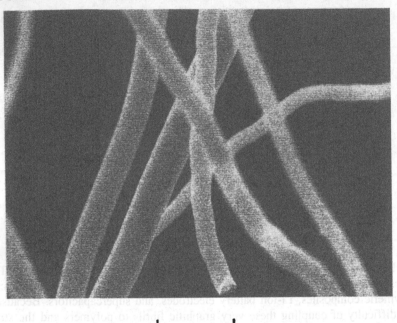

Figure 3. SEM of PYROGRAF™ III VGCF

Working with General Motors and a team supported by a United States NIST ATP award, ASI has researched the surface treatment of PYROGRAF III to make composites having good mechanical properties, developing over twenty different surface treatments and morphologies.

2.3. SHOWA DENKO KK; NIKKISO CO. LTD - GRASKER FIBERS

Because Japan does not have sufficient natural gas reserves, Japanese researchers have worked with more easily imported hydrocarbon liquids. They have devised a method of making straight VGCF by spraying a liquid benzene-catalyst mixture into a flow of waste gas from the making of steel [13] (this Linz-Donawitz gas is comprised of CO, CO_2, N_2, and H_2).

Special heaters and ring inlets prevent convection and encourage the growth of straight fibers. Showa Denko operates a pilot plant that provides 40,000 kg of fibers yearly to Li-ion battery electrode manufacturers. This plant utilizes a horizontal reactor to grow 200 nm in diameter fibers.

Nikkiso has successfully utilized 'Grasker whiskers' of various diameters (Figure 4) in thermoset plastics to reduce friction and to improve flex strength, compressive strength, wear resistance, electrical conductivity, and microwave and vibration absorption.

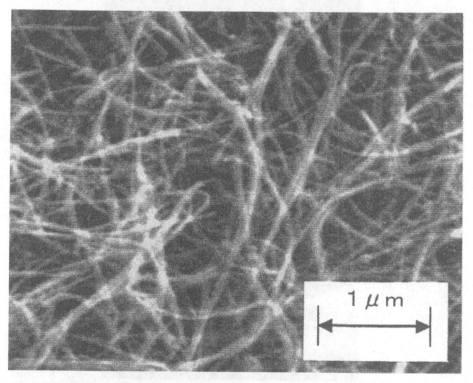

Figure 4. SEM view of Nikkiso fibers. (Courtesy of Nikkiso, Ltd.)

2.4. SUBSTRATE GROWTH

Abundant and uniform fibers sprouting form a planar substrate will enable the application of VGCF in flat panel displays and field emission devices. They key problem to be solved is providing a surface densely seeded with nm sized catalyst particles which do not aggregate despite heating to growth temperature. In pursuing the goal of growing uniform, straight, vertically aligned fibers on a glass substrate, researchers have developed some useful and distinctly different growth techniques.

In 1996 Li *et al.* [14] produced a silica networked surface with pores containing iron catalyst particles which could grow uniform straight fibers. In 1998, Ren *et al.* [15] utilized a sputtered thin film of Ni deposited on a glass substrate to catalyze the growth of a dense layer of uniformly long fibers in acetylene/ammonia plasma at 666°C (Figure 5). The Ni layer was first pre-treated in an ammonia plasma; the temperature of 666°C is low enough to allow the use of glass substrates.

6

Fan *et al.* [16] used Fe films deposited on a porous Si substrate to catalyze the growth of parallel and vertical fibers in ethylene at 700°C without using plasma activation. Choi *et al.* [17] demonstrated a flat panel display incorporating nanotubular field emitters in 1998.

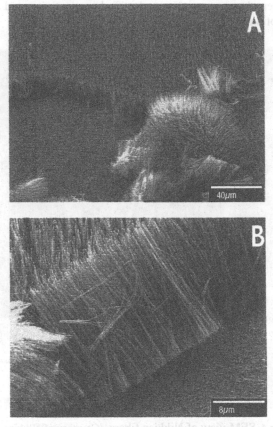

Figure 5. **A.** Nanofibers grown on a glass substrate. **B.** Enlarged view of A along the peeled edge showing uniformity of growth. (Courtesy of Z. F. Ren, Department of Physics, Boston University, and Science magazine)

3. Problem Areas

3.1. SURFACE TREATMENTS

Before fiberglass or conventional carbon fibers found suitable applications as reinforcements, considerable research had to be performed on altering the surfaces to bond well with each matrix material. VGCF has required similar extensive development. This work is described in detail by M. L. Lake in the present volume.

3.2. INFILTRATION

It is difficult to form void-free composites with thin fibers because of the high viscous dynamic drag impeding the infiltration of polymer between the fibers. A one-dimensional solution to Darcy's law [18] which describes a viscous fluid of viscosity μ, applied pressure P_o, and duration of infiltration τ permeating a volume fraction of fibers V, with the permeability of the preform or mass of fibers being K, will give an infiltration depth d of

$$d = \sqrt{\frac{2KP_0}{\mu(1-V)}}.$$

(1)

For fibrous materials of radius r, the equation

$$K = \frac{2r^2\sqrt{2}}{9V}(1-\sqrt{\frac{4V}{\pi}})$$

(2)

has been shown to represent the permeability fairly well [18]. The resulting dependence on r^2 means that K for 0.1 μm fibers is only 10^{-4} times the value for conventional 10 μm fibers. This extremely small value of K leads to very slow infiltration rates.

What is to be done about this? Dispersing and mixing the fibers before infiltration, infiltrating at high pressures, using less viscous polymers, and infiltrating for long periods will all help. Our most practical solution has been to decrease the average size of the clumps of fibers, making complete infiltration of the clumps easier. Significantly, equation (1) yields a one-dimensional nylon infiltration depth of 150 μm for 0.1 μm in radius PYROGRAF fibers, implying that the maximum diameter clump that may be infiltrated under these conditions is about 300 μm. This is in good agreement with the clump size measured with SEM for fibers carefully ball milled for optimal composite properties.

For the thinner NT's, this problem will be more difficult than with VGCF, as the infiltration depth decreases with fiber diameter. For example, for SWNT's with a diameter of 1.4 nm, the infiltration depth under the above conditions will be 1 μm!

3.3. DEBULKING

Conventional carbon fibers, both pitch and PAN, are spun and wound on a spinneret in fiber form. By contrast, VGCF is produced from 10^{12} catalyst particles simultaneously. Moreover, during the CVD thickening process the masses of entangled filaments become cemented together by the carbon coating, rendering any prospect of 'combing' the fibers into a more tractable form extremely difficult.

In the growth of PYROGRAF fibers by Applied Sciences, well-formed clumps of long fibers coming out of the reactor have an apparent bulk density less than 0.001 g/cc. It is easy to compress this material to about 0.05 g/cc, but further compression can be destructive to the fibers and diminish the mechanical or electrical properties of the resulting composites.

Converting this substance to a dense material which may be conveniently poured into the hopper of an injection-molding machine is a challenge which the industry must solve before many applications can be realized.

3.4. ORIENTATION

The excellent longitudinal mechanical properties of conventional graphite fibers are a direct consequence of the anisotropy of the graphite lattice. Composites formed with randomly oriented fibers have considerably lower mechanical properties than those in which the fibers are one- or two-dimensionally oriented [19]. Composites that have been fabricated from VGCF have achieved some degree of anisotropy, primarily due to the intrinsic tendency of injection and compression molding to align fibers. The effect of orientation on the properties of VGCF-thermoplastic composites will be discussed in more detail in other chapters of this book.

Applications engineers who wish to exploit the impressive properties claimed for nanotubes will have to devise clever schemes to fabricate composites reinforced with long, well oriented nanotubes.

3.5. HAZARDS

Since both nanotubes and VGCF are submicron fibers, it is vital that both manufacturers and end users of these products avoid lung damage to workers and consumers.

Several assays of the air quality at Applied Sciences have demonstrated that the tendency of VGCF to form clumps keeps individual fibers from dispersion in the atmosphere. Only in the unfiltered exhaust of the growth reactors was the fiber concentration larger than the EPA asbestos limit of 0.1 fiber/ cc [20]. However, vigilance must be maintained after any mechanical operation that might break loose individual fibers and disperse them in the atmosphere.

4. Acknowledgements

I would like to thank Robert Hoch of Hyperion Catalysis, Carlton Crothers of Mitsui America, and Z. F. Ren for kindly providing figures and information. Thanks to the NIST ATP (cooperative agreement # 70NANB5H1173) for their generous support of VGCF research. I would also like to thank J. J. McHugh, J. C. Finegan, and C. Kwag, my colleagues at the General Motors R&D Center. Finally, many thanks to R. L. Alig, M. L. Lake, D. G. Glasgow, and the staff of Applied Sciences, Inc.

5. References

1. Hughes, T. V., and Chambers, C. R. (June 18, 1889) Manufacture of carbon 'filaments', US Patent 405,480.
2. Davies W. R., Slawson, R. J., and Rigby, G. R. (1953) An unusual form of carbon, *Nature* **171**, 756.
3. A good review is: Baker, R. T. K., and Harris, P. S. (1978) The formation of filamentous carbon, in P. L. Walker, and P. A. Thrower (eds.), *Chemistry and Physics of Carbon*, Vol. 14, p. 83-165. Dekker, New York (1978).
4. Hillert, M., and Lang, N. (1958) The structure of graphite filaments, *Zeit. Krist.* **111**, 24-34

5. Koyama, T. and Endo, M. (1973), Structure and growth processes of Vapor-Grown Carbon Fibers (in Japanese), *O. Buturi* **42**, 690.
6. Benissad, F., Gadelle, P., Coulon, M., and Bonnetain, L. (1988) Formation de fibres de carbone a partir du methane: I Croissance catalytique et epaississement pyrolytique, *Carbon* **26**, 61-69.
7. Tibbetts, G. G. (1985) Lengths of carbon fibers grown from iron catalyst particles in natural gas, *J. Cryst. Growth* **73**, 431-438.
8. Koyama, T., and Endo, M. (October 22, 1983) Method for manufacturing carbon fibers by a vapor phase process, Japanese Patent No. 1982-58,966.
9. Hatano, M., Ohsaki, T., and Arakawa, K. (1985) Graphite whiskers by new process and their composites, Advancing Technology in Materials and Processes, *Science of Advanced Materials and Processes*, National SAMPE Symposium 30, 1467-1476.
10. Mandville, W., and Truesdale, L. K. (1990) Carbon fibrils and a catalytic vapor growth method for producing fibrils, International Patent Application WO90/07023.
11. Miller, B. (1996) Tiny graphite tubes create high-efficiency conducting plastics, *Plastics World*, September, 73-77.
12. Tibbetts, G. G., Bernardo, C. A., Gorkiewicz, D. W., and Alig, R. L. (1994) Role of sulphur in the production of carbon fibers in the vapor phase, *Carbon* **32**, 569-576.
13. Ishioka, M., Okada, T., and Matsubara, K. (1093) Preparation of Vapor-Grown Carbon Fibers in straight form by floating catalyst method, *Carbon* **31**, 123-127.
14. Li, W. Z., Xie, S. S., Qian, L. X., Chang, B. H., Zou, B. S., Zhou, W. Y., Zhao, R. A., and Wang, G. (1996) Large-scale synthesis of aligned Carbon Nanotubes, *Science* **274**, 1701-1703.
15. Ren, Z. F., Huang, Z. P., Xu, J. W., Wang, J. H., Bush, P., Siegal, M. P., and Provencio, P. N. (1998) Synthesis of large arrays of well-aligned Carbon Nanotubes on glass, *Science* **282**, 1105-1107.
16. Fan, S., Chapline, M. G., Franklin, N. R., Tombler, T. W., Cassell, A. M., and Dai, H. (1999) Self-oriented arrays of Carbon Nanotubes and their field emission properties, *Science* **283**, 512-514.
17. Choi, W. B., Chung, D. S., Kang, J. H., Kim, H. Y., Jin, Y. W., Han, I. T., Lee, Y. H., Jung, J. E., Lee, N. S., Park, G. S., and Kim, J. M. (1999) Fully sealed, high brightness Carbon Nanotube field emission display, *Appl. Phys. Lett.* **75**, 3129-3131
18. Mortensen, A., Masur, L. J., Cornie, J. A., and Fleming, M. C. (1989), Infiltration of fibrous preforms by a pure metal: Part I. Theory, *Metall. Trans. A* **20A**, 2535-2547.
19. Tibbetts, G. G., and McHugh, J. J. (1999) Mechanical properties of Vapor-Grown Carbon Fiber composites with thermoplastic matrices, *J. Mater. Res.* **14**, 2871-2880
20. Code of Federal Regulations 29-CFR, Standard No. 1910.1001.

ELECTRONIC PROPERTIES OF CARBON NANOTUBES AND APPLICATIONS

MILDRED S. DRESSELHAUS
Massachusetts Institute of Technology
Cambridge, MA 02139, USA

Abstract.

A brief overview will be given of the remarkable structural and electronic properties of carbon nanotubes, which are tiny structures of molecular dimensions in the form of hollow cylinders with about 20 carbon atoms around the circumference of the cylinders and microns in length. Unusual properties follow as a consequence of quantum mechanical phenomena associated with this one-dimensional system. The unique electronic properties of these carbon nanotubes are that they can be either semiconducting or metallic depending on their geometry. From this, stem other remarkable and unique properties, as observed in their vibrational spectra and in their strength and stiffness. Though less than a decade since their discovery, carbon nanotubes are already finding practical applications based on their unique properties.

1. Introduction

In these notes on the electronic properties of carbon nanotubes and applications, we start by reviewing the basic structural notation for carbon nanotubes that is used to describe the electronic properties. Then the electronic properties of a 2D graphene sheet are reviewed, from which the electronic structure of the nanotubes are derived. This is followed by a discussion of the remarkable properties of the 1D density of electronic states. The experimental study of the 1D density of states behavior by scanning tunneling microscopy/spectroscopy, resonant Raman scattering and by optical reflectivity techniques is then reviewed through application of the unique electronic properties of these techniques. Application to transport properties and hydrogen storage in nanotubes is also briefly discussed.

11

L.P. Biró et al. (eds.), Carbon Filaments and Nanotubes: Common Origins, Differing Applications, 11–49.
© 2001 *Kluwer Academic Publishers. Printed in the Netherlands.*

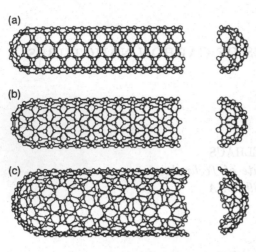

Figure 1. Schematic models for single-wall carbon nanotubes with the nanotube axis normal to the chiral vector which, in turn, is along: (a) the $\theta = 30°$ direction [an "armchair" (n, n) nanotube], (b) the $\theta = 0°$ direction [a "zigzag" $(n, 0)$ nanotube], and (c) a general θ direction, such as $\overrightarrow{OB}$ (see Fig. 2), with $0 < \theta < 30°$ [a "chiral" (n, m) nanotube]. The actual nanotubes shown here correspond to (n, m) values of: (a) (5, 5), (b) (9, 0), and (c) (10, 5). The nanotube axis for the (5,5) nanotube has 5-fold rotation symmetry, while that for the (9,0) nanotube has 3-fold rotation symmetry [1].

2. Structure of Carbon Nanotubes

The structure of carbon nanotubes (see Fig. 1) has been explored by high resolution TEM and STM techniques, yielding direct confirmation that the nanotubes are cylinders derived from the honeycomb lattice (graphene sheet). The structure of a single-wall carbon nanotube is conveniently explained in terms of its 1D (one-dimensional) unit cell as shown in Fig. 1(a). The three basic categories of carbon nanotubes, the armchair, zigzag, and chiral nanotubes are shown in Fig. 1, and their structures are explained as follows.

The circumference of any carbon nanotube is expressed in terms of the chiral vector $\vec{C}_h = n\hat{a}_1 + m\hat{a}_2$ which connects two crystallographically equivalent sites on a 2D graphene sheet [see Fig. 1(a)] [2]. The construction in Fig. 1(a) shows the chiral angle θ between $\vec{C}_h$ and the "zigzag" direction ($\theta = 0$) and shows the unit vectors $\hat{a}_1$ and $\hat{a}_2$ of the hexagonal honeycomb lattice, while the armchair nanotube corresponds to a chiral angle of $\theta = 30°$ [Figs. 1(a) and 2]. An ensemble of chiral vectors specified by pairs of integers (n, m) denoting the vector $\vec{C}_h = n\hat{a}_1 + m\hat{a}_2$ is given in Fig. 1(a) [4]. The intersection of OB with the first lattice point determines the fundamental 1D translation vector $\vec{T}$ and hence the unit cell of the 1D lattice [Fig. 1(a)].

The cylinder connecting the two hemispherical caps of the carbon nanotube is formed by superimposing the two ends of the vector $\vec{C}_h$ and the cylinder joint is made along the two lines OB and AB' in Fig. 1(a). The lines OB and AB' are both perpendicular to the vector $\vec{C}_h$ at each end of

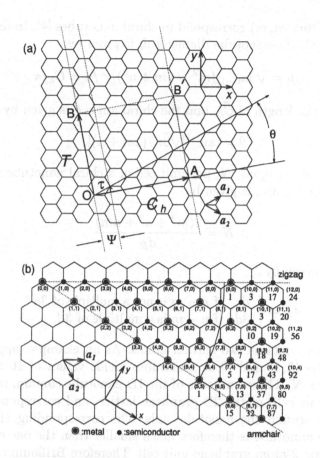

Figure 2. (a) The chiral vector $\overrightarrow{OA}$ or $\vec{C}_h = n\hat{a}_1 + m\hat{a}_2$ is defined on the honeycomb lattice of carbon atoms by unit vectors $\hat{a}_1$ and $\hat{a}_2$ and the chiral angle θ with respect to the zigzag axis. Along the zigzag axis $\theta = 0°$. Also shown are the lattice vector $\overrightarrow{OB} = \vec{T}$ of the 1D nanotube unit cell and the rotation angle ψ and the translation τ which constitute the basic symmetry operation $R = (\psi|\tau)$ for the carbon nanotube. The diagram is constructed for $(n, m) = (4, 2)$ [2]. (b) Possible vectors specified by the pairs of integers (n, m) for general carbon nanotubes, including zigzag, armchair, and chiral nanotubes. Below each pair of integers (n, m) is listed the number of distinct caps that can be joined continuously to the carbon nanotube denoted by (n, m) [2]. The encircled dots denote metallic nanotubes while the small dots are for semiconducting nanotubes [3].

$\vec{C}_h$ [2]. The chiral nanotube, thus generated has no distortion of bond angles other than distortions caused by the cylindrical curvature of the nanotube. Differences in chiral angle θ and in the nanotube diameter d_t give rise to differences in the properties of the various carbon nanotubes.

In the (n, m) notation for $\vec{C}_h = n\hat{a}_1 + m\hat{a}_2$, the vectors $(n, 0)$ or $(0, m)$ denote zigzag nanotubes and the vectors (n, n) denote armchair nanotubes.

All other vectors (n, m) correspond to chiral nanotubes [4]. In terms of the integers (n, m), the nanotube diameter d_t is given by

$$d_t = \sqrt{3}a_{C-C}(m^2 + mn + n^2)^{1/2}/\pi = C_h/\pi \tag{1}$$

where C_h is the length of $\vec{C}_h$, and the chiral angle θ is given by

$$\theta = \tan^{-1}[\sqrt{3}n/(2m + n)]. \tag{2}$$

The number of hexagons, N, per unit cell of a chiral nanotube is specified by integers (n, m) and is given by

$$N = \frac{2(m^2 + n^2 + nm)}{d_R} \tag{3}$$

where d_R is given by

$$d_R = \begin{cases} d & \text{if } n - m \text{ is not a multiple of } 3d \\ 3d & \text{if } n - m \text{ is a multiple of } 3d. \end{cases} \tag{4}$$

Each hexagon contains two carbon atoms. As an example, application of Eq. (3) to the $(5, 5)$ and $(9, 0)$ nanotubes yields values of 10 and 18, respectively, for N. These unit cells of the 1D nanotube contain, respectively, 10 and 18 unit cells of the 2D graphene lattice, and correspond to a ring of hexagons around the nanotube axis. The corresponding 1D Brillouin zone for the nanotube is therefore much smaller than the one corresponding to a single 2-atom graphene unit cell. Therefore Brillouin zone-folding techniques have been commonly used to obtain approximate electron and phonon dispersion relations for carbon nanotubes with specific symmetry (n, m).

Measurements of the nanotube diameter d_t are conveniently made by using STM (scanning tunneling microscopy) and transmission electron microscope (TEM) techniques. Measurements of the chiral angle θ have been made using high resolution TEM [5], and θ is normally defined by taking $\theta = 0°$ and $\theta = 30°$ (modulo, $2\pi/6$), for zigzag and armchair nanotubes, respectively. While the ability to measure the diameter d_t and chiral angle θ of individual single-wall nanotubes has been demonstrated, it remains a major challenge to determine d_t and θ for specific nanotubes used for an actual physical property measurement, i.e., resistivity, Raman scattering, magnetic susceptibility, etc.

In the multi-wall carbon nanotubes, the measured interlayer distance is 0.34 nm [6], comparable to the value of 0.344 nm in turbostratic carbons. In turbostratic graphite, the carbon atoms in each graphene layer are arranged as in crystalline graphite, but there is no site correlation between the

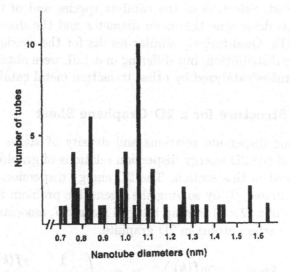

Figure 3. Histogram of the single-wall nanotube diameter distribution for Fe-catalyzed nanotubes. Note the small magnitude of the average nanotube diameter, and the narrow distribution of the nanotube diameters [5].

carbon atom locations in turbostratic graphite. Likewise for a nanotube, where the graphene sheets are arranged in concentric cylinders, only limited site correlation between carbon atoms in adjacent layers is possible. Typical lengths of the arc-grown multi-wall nanotubes are $\sim 1\,\mu$m, giving rise to an aspect ratio (length to diameter ratio) of 10^2 to 10^3. Because of their small diameter, involving only a small number of carbon atoms, and because of their large aspect ratio, carbon nanotubes are classified as 1D carbon systems. Most of the theoretical work on carbon nanotubes has been carried out on single-wall nanotubes (see Fig. 1) and has emphasized their 1D properties. Single-wall nanotubes, just like the multi-wall nanotubes (and also conventional vapor grown carbon fibers), have hollow cores along the axis of the nanotube.

The diameter distribution of single-wall carbon nanotubes is of great interest for both theoretical and experimental reasons, since theoretical studies indicate that the physical properties of carbon nanotubes are strongly dependent on nanotube diameter. Early results for the diameter distribution of Fe-catalyzed single-wall nanotubes (Fig. 3) show a diameter range between 0.7 nm and 1.6 nm, with the largest peak in the distribution at 1.05 nm, and with a smaller peak at 0.85 nm [5]. In general, the smallest reported diameter for a single-wall carbon nanotube is 0.7 nm [5], the same as the diameter of the C_{60} molecule (0.71 nm) [2]. Recently, smaller diameter nanotubes down to a 0.4 nm diameter (corresponding to a C_{20} cap) have been reported. Selection of the catalyst species and of the detailed growth conditions determine the mean diameter and the diameter distribution for SWNTs. Qualitatively similar results for the average diameter and the diameter distribution, but differing in detail, were obtained for the single-wall nanotubes catalyzed by other transition metal catalysts.

3. Electronic Structure for a 2D Graphene Sheet

The 1D electronic dispersion relations and density of states is obtained by zone folding of the 2D energy dispersion relations of graphite π bands, which are reviewed in this section. The 2D energy dispersion relations of graphite are calculated [3] by solving the eigenvalue problem for a (2×2) Hamiltonian $\mathcal{H}$ and a (2×2) overlap integral matrix $\mathcal{S}$, associated with the two inequivalent carbon atoms in 2D graphite,

$$\mathcal{H} = \begin{pmatrix} \varepsilon_{2p} & -\gamma_0 f(k) \\ -\gamma_0 f(k)^* & \varepsilon_{2p} \end{pmatrix} \text{ and } \mathcal{S} = \begin{pmatrix} 1 & sf(k) \\ sf(k)^* & 1 \end{pmatrix}, \quad (5)$$

where ε_{2p} is the site energy of the $2p$ atomic orbital and

$$f(k) = e^{ik_x a/\sqrt{3}} + 2e^{-ik_x a/2\sqrt{3}} \cos \frac{k_y a}{2}, \quad (6)$$

where $a = |\mathbf{a}_1| = |\mathbf{a}_2| = \sqrt{3} a_{C-C}$. Solution of the secular equation $\det(\mathcal{H} - E\mathcal{S}) = 0$ implied by Eq. (5) leads to the eigenvalues

$$E_{g2D}^{\pm}(\vec{k}) = \frac{\varepsilon_{2p} \pm \gamma_0 w(\vec{k})}{1 \mp sw(\vec{k})} \quad (7)$$

for the C-C transfer energy $\gamma_0 > 0$, where s denotes the overlap of the electronic wave function on adjacent sites, and E^+ and E^- correspond to the π^* and the π energy bands, respectively. Here we conventionally use γ_0 as a positive value. The function $w(\vec{k})$ in Eq. (7) is given by

$$w(\vec{k}) = \sqrt{|f(\vec{k})|^2} = \sqrt{1 + 4\cos \frac{\sqrt{3}k_x a}{2} \cos \frac{k_y a}{2} + 4\cos^2 \frac{k_y a}{2}} \quad (8)$$

leading to the 2D dispersion relations $E_{g2D}(k_x, k_y)$ for electrons on a graphene sheet in the limit $s = 0$ and $\varepsilon_{2p} = 0$ [7]

$$E_{g2D}(k_x, k_y) = \pm \gamma_0 \left\{ 1 + 4\cos\left(\frac{\sqrt{3}k_x a_0}{2}\right) \cos\left(\frac{k_y a_0}{2}\right) + 4\cos^2\left(\frac{k_y a_0}{2}\right) \right\}^{1/2} \quad (9)$$

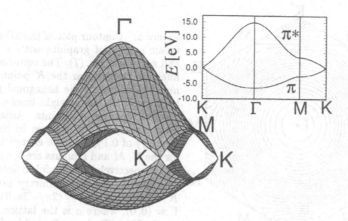

Figure 4. The energy dispersion relations for 2D graphite with $\gamma_0 = 3.013$ eV, $s = 0.129$ and $\varepsilon_{2p} = 0$ in Eq. (7) are shown throughout the whole region of the Brillouin zone. The inset shows the energy dispersion along the high symmetry lines between the Γ, M, and K points. The valence π band (lower part) and the conduction π^* band (upper part) are degenerate at the K points in the hexagonal Brillouin zone which corresponds to the Fermi energy [3]

where $a_0 = 1.421 \times \sqrt{3}$ Å is the lattice constant for a 2D graphene sheet and γ_0 is the nearest-neighbor C–C energy overlap integral [8]. A set of 1D energy dispersion relations is obtained from Eq. (9) by considering the small number of allowed wave vectors in the circumferential direction of each (n, m) nanotube.

In Fig. 4 we plot the electronic energy dispersion relations for 2D graphite [Eq. (7)] as a function of the two-dimensional wave vector $\vec{k}$ in the hexagonal Brillouin zone in which we adopt the parameters $\gamma_0 = 3.013$ eV, $s = 0.129$ and $\varepsilon_{2p} = 0$ so as to fit both the first principles calculation of the energy bands of 2D turbostratic graphite [9, 10]. The corresponding energy contour plot of the 2D energy bands of graphite with $s = 0$ and $\varepsilon_{2p} = 0$ is shown in Fig. 5. The Fermi energy corresponds to $E = 0$ at the K points. In this limit the valence and conduction bands become symmetric with respect to each other.

Near the K-point at the corner of the hexagonal Brillouin zone of graphite, $w(\vec{k})$ in Eq. (7) has a linear dependence on $k \equiv |\vec{k}|$ as measured from the K point:

$$w(\vec{k}) = \frac{\sqrt{3}}{2}ka + \dots, \quad \text{for } ka \ll 1. \tag{10}$$

Thus, the expansion of Eq. (7) for small k yields

$$E_{g2D}^{\pm}(\vec{k}) = \varepsilon_{2p} \pm (\gamma_0 - s\varepsilon_{2p})w(\vec{k}) + \dots, \tag{11}$$

18

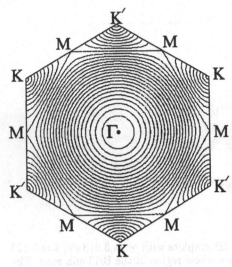

Figure 5. Contour plot of the 2D electronic energy of graphite with $s = 0$ and $\varepsilon_{2p} = 0$ in Eq. (7). The equi-energy lines are circles near the K point and near the center of the hexagonal Brillouin zone, but are straight lines which connect nearest M points. Adjacent lines correspond to changes in height (energy) of $0.1\gamma_0$ and the energy value for the K, M and Γ points are 0, γ_0 and $3\gamma_0$, respectively. It is useful to note the coordinates of high symmetry points: $K = (0, 4\pi/3a)$, $M = (2\pi/\sqrt{3}a, 0)$ and $\Gamma = (0, 0)$, where a is the lattice constant of the 2D sheet of graphite [11]

so that in this approximation, the valence and conduction bands are symmetric near the K point, independent of the value of s. When we adopt $\varepsilon_{2p} = 0$ and take $s = 0$ for Eq. (7), and assume a linear k approximation for $w(k)$, we get the linear dispersion relations for graphite given by [7, 12]

$$E(k) = \pm\frac{\sqrt{3}}{2}\gamma_0 ka = \pm\frac{3}{2}\gamma_0 ka_{\text{C-C}}. \tag{12}$$

If the physical phenomena under consideration only involve small k vectors, it is convenient to use Eq. (12) for interpreting experimental results relevant to such phenomena. Figure 5 shows the regular behavior close to points K and K' in the Brillouin zone. Very different behavior is observed near the M point in the Brillouin zone.

4. Dispersion Relations for Single Wall Carbon Nanotubes

Single-wall carbon nanotubes are interesting examples of a one-dimensional periodic structure along the axis of the nanotube. Confinement in the radial direction is provided by the monolayer thickness of the nanotubes. In the circumferential direction, periodic boundary conditions apply to the enlarged unit cell that is formed in real space. The dispersion relations for the single wall nanotube are obtained by the zone folding in reciprocal space of the dispersion relations for a graphene sheet (see §3) into the Brillouin zone for the 1D nanotube. Zone-folding arguments have been used to obtain 1D dispersion relations for electrons [4, 13, 14] and phonons [15, 16] in single-wall carbon nanotubes.

Calculation of the dispersion relations for carbon nanotubes shows remarkable electronic properties, namely that for small diameter graphene nanotubes, about 1/3 of the nanotubes are metallic and 2/3 are semiconducting, depending on their fiber diameter d_t and chiral angle θ. These unusual properties follow from the simplest arguments based on the tight binding approximation and zone folding of the 2D dispersion relations of a graphene sheet.

The 1D energy dispersion relations of a SWNT are derived from Eq. (9) by use of periodic boundary conditions:

$$E_\mu^\pm(k) = E_{g2D}^\pm\left(k\frac{\mathbf{K}_2}{|\mathbf{K}_2|} + \mu\mathbf{K}_1\right),$$
$$\left(-\frac{\pi}{T} < k < \frac{\pi}{T}, \text{ and } \mu = 1, \cdots, N\right), \tag{13}$$

where T is the magnitude of the translational vector $\mathbf{T}$ (see Fig. 2), k is a 1D wave vector along the nanotube axis, and N denotes the number of hexagons of the graphite honeycomb lattice that lie within the nanotube unit cell (see Fig. 2). It should be noted that solutions of the dispersion relations for $E_{g2D}^\pm$ that have so far been used in Eq. (13) employ a simplified version of the tight binding approximation which assumes that $s = 0$ and $\varepsilon_{2p} = 0$ for the graphene sheet. Once detailed experiments regarding the band structure and level filling are available, the simplifications now being used ($s = 0$ and $\varepsilon_{2p} = 0$) can be relaxed for carbon nanotubes as they are for graphite. T and N in Eq. (13) are given, respectively, by

$$T = \frac{\sqrt{3}C_h}{d_R} = \frac{\sqrt{3}\pi d_t}{d_R}, \text{ and } N = \frac{2(n^2 + m^2 + nm)}{d_R}. \tag{14}$$

where d_R is the greatest common divisor of $(2n + m)$ and $(2m + n)$ for a (n, m) nanotube [3, 17]. Furthermore, the reciprocal lattice vectors $\mathbf{K}_1$ and $\mathbf{K}_2$ denote, respectively, a discrete unit wave vector along the circumferential direction, and a reciprocal lattice vector along the nanotube axis direction (see Fig. 6). For a (n, m) nanotube $\mathbf{K}_1$ and $\mathbf{K}_2$ are given by

$$\begin{aligned}\mathbf{K}_1 &= \{(2n + m)\mathbf{b}_1 + (2m + n)\mathbf{b}_2\}/Nd_R \\ \mathbf{K}_2 &= (m\mathbf{b}_1 - n\mathbf{b}_2)/N,\end{aligned} \tag{15}$$

where $\mathbf{b}_1$ and $\mathbf{b}_2$ are the reciprocal lattice vectors of 2D graphite and are given in x, y coordinates by

$$\mathbf{b}_1 = \left(\frac{1}{\sqrt{3}}, 1\right)\frac{2\pi}{a}, \quad \mathbf{b}_2 = \left(\frac{1}{\sqrt{3}}, -1\right)\frac{2\pi}{a}. \tag{16}$$

(a)

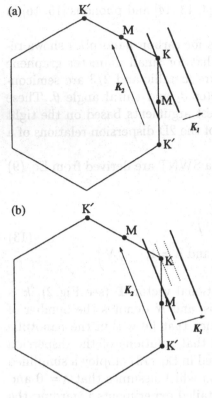

(b)

Figure 6. The wave vector k for one-dimensional carbon nanotubes is shown in the two-dimensional Brillouin zone of graphite (hexagon) as bold lines for (a) metallic and (b) semiconducting carbon nanotubes. In the direction of $\mathbf{K}_1$, discrete k values are obtained by periodic boundary conditions for the circumferential direction of the carbon nanotubes, while in the direction of the $\mathbf{K}_2$ vector, continuous k vectors are shown in the one-dimensional Brillouin zone. (a) For *metallic* nanotubes, the bold line intersects a K point (corner of the hexagon) at the Fermi energy of graphite. (b) For the *semiconducting* nanotubes, the K point always appears one-third of the distance between two bold lines. It is noted that only a few of the N bold lines are shown near the indicated K point. For each bold line, there is an energy minimum (or maximum) in the valence and conduction energy subbands, giving rise to the energy differences $E_{pp}(d_t)$

The periodic boundary condition for a carbon nanotube (n, m) gives N discrete k values in the circumferential direction. The N pairs of energy dispersion curves given by Eq. (13) correspond to the cross sections of the two-dimensional energy dispersion surface shown in Fig. 4, where cuts are made on the lines of $k\mathbf{K}_2/|\mathbf{K}_2| + \mu\mathbf{K}_1$. In Fig. 6 several cutting lines near one of the K points are shown. The separation between two adjacent lines and the length of the cutting lines are given by the $\mathbf{K}_1$ and $\mathbf{K}_2$ vectors of Eq. (15), respectively, whose lengths are given by

$$|\mathbf{K}_1| = \frac{2}{d_t}, \quad \text{and} \quad |\mathbf{K}_2| = \frac{2\pi}{T} = \frac{2d_R}{\sqrt{3}d_t}. \tag{17}$$

If, for a particular (n, m) nanotube, the cutting line passes through a K point of the 2D Brillouin zone [Fig. 6 (a)], where the π and π^* energy bands of two-dimensional graphite are degenerate (Fig. 4) by symmetry, then the one-dimensional energy bands have a zero energy gap. Since the degenerate point corresponds to the Fermi energy, and the density of states are finite as shown below, single wall nanotubes (SWNTs) with a zero band gap are *metallic*. When the K point is located between two cutting lines, then the K point is always located in a position one-third of the distance between two adjacent $\mathbf{K_1}$ lines [Fig. 6 (b)] [17] and this situation gives rise to a *semiconducting* nanotube with a finite energy gap. The rule for a carbon nanotube to be metallic or semiconducting is:

$$n - m = 3q \qquad \text{metallic}$$
$$n - m \neq 3q \qquad \text{semiconducting,}$$

(18)

where q is an integer [3, 4, 9, 18].

Thus the electronic structure can be either metallic or semiconducting depending on the choice of (n, m), although there is no difference in the local chemical bonding between the carbon atoms in the nanotubes, and no doping impurities are present [4]. These surprising results can be understood on the basis of the electronic structure of a graphene sheet which is found to be a zero gap semiconductor [10] with bonding and antibonding π bands that are degenerate at the K-point (zone corner) of the hexagonal 2D Brillouin zone (see Fig. 8). The periodic boundary conditions for the 1D carbon nanotubes of small diameter permit only a few wave vectors to exist in the circumferential direction, and these satisfy the relation $n\lambda = \pi d_t$, where $\lambda = 2\pi/k$ and n is an integer and λ is a wavelength. Metallic conduction occurs when one of the allowed wave vectors k passes through the K-point of the 2D Brillouin zone, where the valence and conduction bands are degenerate because of the symmetry of the 2D graphene lattice (see Fig. 8).

As the nanotube diameter increases, more wave vectors become allowed for the circumferential direction, the nanotubes become more two-dimensional and the semiconducting band gap disappears, as is illustrated in Fig. 7 which shows the semiconducting band gap to be proportional to the reciprocal diameter $1/d_t$. At a nanotube diameter of $d_t \sim 3\,\text{nm}$ (Fig. 7), the bandgap becomes comparable to thermal energies at room temperature, showing that small diameter nanotubes are needed to observe 1D quantum effects.

It is useful for pedagogic purposes to give explicit expressions for the dispersion relations for the case of armchair and zigzag nanotubes, for which the real space rectangular unit cells can be chosen in accordance with Fig. 8. The area of each real space unit cell for both the armchair and zigzag nanotubes contains two hexagons or four carbon atoms. The area of the cylindrical ring around the nanotube (and defined by the vectors $\vec{T}$ and $\vec{C}_h$) is n times larger than the real space unit cells shown in Fig. 8. Also shown in this figure are the real space unit cells for a 2D graphene layer

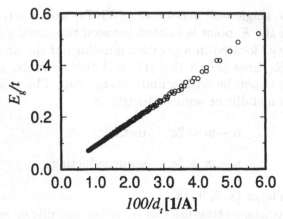

Figure 7. The energy gap E_g for a general chiral carbon nanotube as a function of 100 Å$/d_t$, where d_t is the nanotube diameter in Å [19].

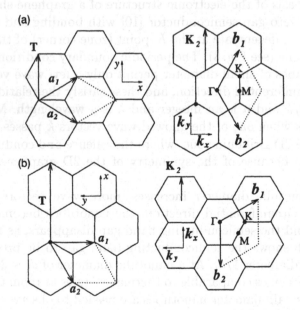

Figure 8. Real space unit cell and Brillouin zone for (a) armchair and (b) zigzag nanotubes (dotted lines). Those for a 2D graphene sheet (dashed lines) are shown for comparison [4].

and the corresponding unit cells in reciprocal space for armchair and zigzag nanotubes. The appropriate periodic boundary conditions used to obtain the energy eigenvalues for the (n, n) armchair nanotube define the small number of allowed wave vectors $k_{x,q}$ in the circumferential direction

$$n\sqrt{3}a_0 k_{x,q} = q2\pi, \quad q = 1, \ldots, n. \tag{19}$$

Substitution of the discrete allowed values for $k_{x,q}$ given by Eq. (19) into Eq. (9) yields the energy dispersion relations $E_q^a(k)$ for the (n, n) armchair nanotube [9]

$$E_q^a(k) = \pm\gamma_0 \left\{1 \pm 4\cos\left(\frac{q\pi}{n}\right)\cos\left(\frac{ka_0}{2}\right) + 4\cos^2\left(\frac{ka_0}{2}\right)\right\}^{1/2},$$
$$(-\pi < ka_0 < \pi), \quad (q = 1, \ldots, n) \tag{20}$$

in which the superscript a refers to the armchair nanotube, k is a one-dimensional vector along the nanotube axis, and n refers to the armchair index, i.e., $(n, m) \equiv (n, n)$ [20].

The energy bands for the $\mathbf{C}_h = (n, 0)$ zigzag nanotube $E_q^z(k)$ can be obtained likewise from Eq. (9) by writing the periodic boundary condition on k_y as:

$$nk_{y,q}a_0 = 2\pi q, \quad (q = 1, \ldots, 2n), \tag{21}$$

to yield the 1D dispersion relations for the $4n$ states for the $(n, 0)$ zigzag nanotube (denoted by the superscript z)

$$E_q^z(k) = \pm\gamma_0 \left\{1 \pm 4\cos\left(\frac{\sqrt{3}ka_0}{2}\right)\cos\left(\frac{q\pi}{n}\right) + 4\cos^2\left(\frac{q\pi}{n}\right)\right\}^{1/2},$$
$$\left(-\frac{\pi}{\sqrt{3}} < ka_0 < \frac{\pi}{\sqrt{3}}\right), \quad (q = 1, \ldots, 2n). \tag{22}$$

The resulting calculated 1D dispersion relations $E_q^z(k)$ for the $(9, 0)$ and $(10, 0)$ zigzag nanotubes are shown in Figs. 9(b) and (c), respectively, while Fig. 9(a) shows the corresponding dispersion relations for the (5,5) armchair nanotube.

The resulting 1D dispersion relations for zigzag and chiral carbon nanotubes are plotted in Fig. 9, which shows that the condition for metallic nanotubes $|n - m| = 3q$ is satisfied for the (5,5) and (9,0) nanotubes, but that the (10, 0) nanotube satisfies $|n - m| \neq 3q$ and is therefore semiconducting with a bandgap that is large compared to kT. The calculations show that all armchair (n, n) nanotubes are metallic [Figure 9(a)], but only 1/3 of the possible zigzag nanotubes are metallic [Fig. 9(b)] [4, 13, 14]. Likewise only 1/3 of the chiral nanotubes are metallic [see Fig. 1(b)]. A plot of the energy gap E_g for the semiconducting tubes as a function of nanotube diameter is shown in Fig. 7, reflecting the $1/d_t$ dependence of E_g.

Calculation of the electronic structure for two concentric nanotubes shows that pairs of coaxial metal-semiconductor or semiconductor-metal

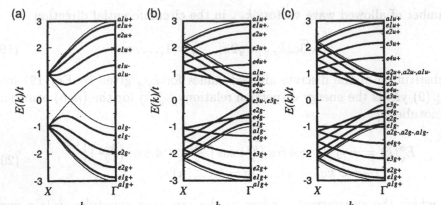

Figure 9. One-dimensional energy dispersion relations for (a) armchair (5, 5) nanotubes, (b) zigzag (9, 0) nanotubes, and (c) zigzag (10, 0) nanotubes, with the Fermi level placed at $E = 0$ [21].

nanotubes are stable. The stacking of coaxial single wall nanotubes, one within another gives rise to multi-wall carbon nanotubes which are routinely synthesized in the laboratory [22, 23]. Since nanotubes can be either semiconducting or metallic, one could imagine designing an electronic device. For example, two concentric carbon nanotubes, with a smaller diameter metallic inner nanotube surrounded by a larger diameter semiconducting (or insulating) outer nanotube, form a shielded cable of nanometer size. This concept could be extended to the design of tubular metal-semiconductor all-carbon devices without introducing any doping impurities [4].

5. 1D Density of Electronic States

The 1D density of states (DOS) in units of states/C-atom/eV is calculated from the dispersion relations $E_\mu^\pm(k)$ by

$$D(E) = \frac{T}{2\pi N} \sum_{\pm} \sum_{\mu=1}^{N} \int \frac{1}{\left| \frac{dE_\mu^\pm(k)}{dk} \right|} \delta(E_\mu^\pm(k) - E)dE, \qquad (23)$$

where the summation is taken over the N conduction (+) and valence (−) 1D bands and T is the length of the lattice vector in the direction of the tube axis. Since the energy dispersion near the Fermi energy [Eq. (12)] is linear, the density of states of metallic nanotubes is constant at the Fermi energy:

$$D(E_{\mathrm{F}}) = a/(2\pi^2 \gamma_0 d_t), \qquad (24)$$

and is inversely proportional to the diameter of the nanotube. It is noted that we always have two cutting lines (1D energy bands) at the two equivalent symmetry points K and K' in the 2D Brillouin zone in Fig. 5. The integrated value of $D(E)$ for the energy region of $E_\mu(k)$ is 2 for any (n, m) nanotube, which includes the plus and minus signs of E_{g2D} and the spin degeneracy.

It is clear from Eq. (23) that the density of states becomes large when the energy dispersion relation becomes flat as a function of k, yielding one-dimensional van Hove singularities (vHs) in the DOS, which are known to be proportional to $(E^2 - E_0^2)^{-1/2}$, and therefore singular, at both the energy minima and maxima $(\pm E_0)$ of the dispersion relations for carbon nanotubes. Shown in Figure 10 are the 1D density of states predicted for a metallic (9,0) zigzag nanotube and for a semiconducting (10,0) zigzag nanotube. These density of states curves are important for determining many solid state properties of carbon nanotubes, such as the spectra observed by scanning tunneling spectroscopy (STS) [24, 25, 26, 27, 28], optical absorption [29, 30, 31], and resonant Raman spectroscopy technique [32, 33, 34, 35, 36].

The one-dimensional van Hove singularities of single wall nanotubes near the Fermi energy come from the energy dispersion along the bold lines in Fig. 6 near the K point of the Brillouin zone of 2D graphite. Within the linear k approximation for the energy dispersion relations of graphite given by Eq. (12), the energy contour as shown in Fig. 5 around the K point is circular, and thus the energy minima of the 1D energy dispersion relations are located at the closest positions to the K point. Using the small k approximation of Eq. (12), the energy differences $E_{11}^M(d_t)$ and $E_{11}^S(d_t)$ for metallic and semiconducting nanotubes between the highest-lying valence band singularity and the lowest-lying conduction band singularity in the 1D electronic density of states curves are expressed by substituting for k the values of $|\mathbf{K}_1|$ of Eq. (17) for metallic nanotubes and of $|\mathbf{K}_1/3|$ and $|2\mathbf{K}_1/3|$ for semiconducting nanotubes, respectively [37, 38]. We therefore can write the relations

$$E_{11}^M(d_t) \simeq 6a_{C-C}\gamma_0/d_t \quad \text{and} \quad E_{11}^S(d_t) \simeq 2a_{C-C}\gamma_0/d_t \qquad (25)$$

which are approximately valid near the K point in the Brillouin zone. When we use the number p $(p = 1, 2, \ldots)$ to denote the order of the valence π and conduction π^* energy bands symmetrically located with respect to the Fermi energy, optical transitions $E_{pp'}$ from the p-th valence band to the p'-th conduction band occur in accordance with the selection rules of $\delta p = 0$ and $\delta p = \pm 1$ for parallel and perpendicular polarizations of the electric field with respect to the nanotube axis, respectively [30]. However, in the case of perpendicular polarization, the optical transition is suppressed

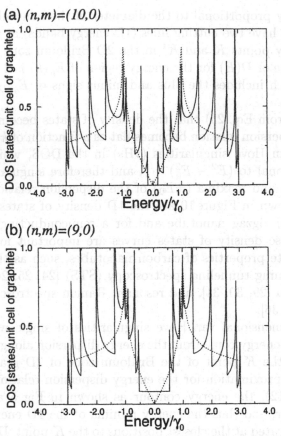

(a) (n,m)=(10,0)

(b) (n,m)=(9,0)

Figure 10. Electronic 1D density of states per unit cell of a 2D graphene sheet for two zigzag nanotubes: (a) the $(9,0)$ nanotube which has metallic behavior, (b) the $(10,0)$ nanotube which has semiconducting behavior. Also shown in the figure is the density of states for the 2D graphene sheet [22].

by the depolarization effect [30], and thus optical absorption only occurs for $\delta p = 0$. For mixed samples containing both metallic and semiconducting carbon nanotubes with similar diameters, optical transitions may appear with the following energies, starting from the lowest energy, $E_{11}^S(d_t)$, $2E_{11}^S(d_t)$, $E_{11}^M(d_t)$, $4E_{11}^S(d_t)$, ...

In Fig. 11, values for both $E_{pp}^S(d_t)$ and $E_{pp}^M(d_t)$ at the K point are plotted as a function of nanotube diameter d_t for all chiral angles [11, 29, 39]. The width of the $E_{ii}(d_t)$ curves depends on trigonal warping effects [11], which become more pronounced as the tube diameter and energy increase. Zigzag tubes show the maximum trigonal warping effects and armchair nanotubes show the least. This plot is very useful for interpreting the optical absorption spectra for carbon nanotubes and for determining the resonant

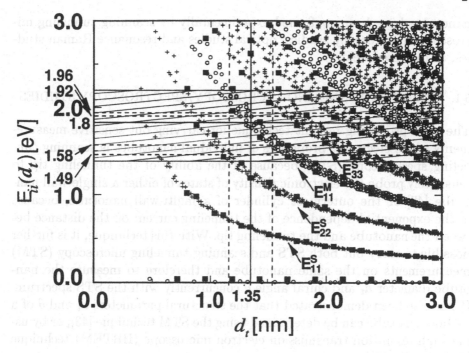

Figure 11. Calculation [39, 40, 41] of the energy separations $E_{ii}(d_t)$ for all (n, m) values vs nanotube diameter in the range $0.7 < d_t < 3.0$ nm using $\gamma_0 = 2.9$ eV. Semiconducting and metallic nanotubes are indicated by crosses and open circles, respectively, and the four lowest energy transitions are labelled by $E_{11}^S(d_t)$, $E_{22}^S(d_t)$, $E_{11}^M(d_t)$, and $E_{33}^S(d_t)$, where S and M, respectively, refer to semiconducting and metallic nanotubes. The filled squares denote zigzag tubes. The vertical lines denote $d_t = 1.35 \pm 0.20$ nm for a particular single wall carbon nanotube sample [11]. The horizontal lines denote laser energies used to probe the optical transition energies.

energy in the resonant Raman spectra corresponding to a particular nanotube diameter.

6. Experimental Determination of the Density of States

Experimental measurements to test the remarkable theoretical predictions of the electronic structure of carbon nanotubes can be considered to be an application of the electronic structure. Such experiments are difficult to carry out because of: (1) the strong dependence of the predicted electronic properties on nanotube diameter and chirality, and (2) the experimental difficulties in making electronic or optical measurements on individual single-wall nanotubes that have also been characterized with regard to diameter and chiral angle (d_t and θ). Despite these difficulties, the remarkable electronic structure and 1D electronic density of states for carbon nanotubes have been confirmed experimentally by scanning tunneling microscopy/spectroscopy studies, optical studies and resonance Raman studies, as discussed below.

6.1. SCANNING TUNNELING MICROSCOPY/SPECTROSCOPY STUDIES

The most promising present technique for carrying out sensitive measurements of the electronic properties of individual nanotubes is scanning tunneling spectroscopy (STS) because of the ability of the tunneling tip to sensitively probe the electronic density of states of either a single-wall nanotube [42] or the outermost cylinder of a multi-wall nanotube, because of the exponential dependence of the tunneling current on the distance between the nanotube and the tunneling tip. With this technique, it is further possible to carry out both STS and scanning tunneling microscopy (STM) measurements on the same nanotube and therefore to measure the nanotube diameter d_t and chiral angle θ concurrently with the STS spectrum. It has also been demonstrated that the structural parameters d_t and θ of a carbon nanotube can be determined using the STM technique [43], or by using high-resolution transmission electron microscope (HRTEM) technique [5, 6, 44, 45, 46].

Since measurements of dI/dV in the scanning tunneling spectroscopy (STS) mode of a scanning tunneling microscope yields a signal which is proportional to the 1D density of states, the STS technique has become a powerful tool for studying the electronic structure of both metallic and semiconducting single wall carbon nanotubes [24, 25] and in particular in identifying the peaks in the experimental dI/dV curves with singularities in the electronic density of states. The top 4 traces in Fig. 12 show that the bandgap or energy separation is $E_{11}^S(d_t) \simeq 0.6\,\mathrm{eV}$ for the indicated semiconducting nanotubes, while the lower 3 traces show energy separations of $E_{11}^M(d_t) \simeq 1.8\,\mathrm{eV}$ for metallic nanotubes. The combined STM/STS studies allow identification of the observed density of states with specific (n, m) carbon nanotubes, and the results can thus be identified with specific metallic or semiconducting nanotubes [24, 47]. These experimental results are consistent with: (1) about 2/3 of the nanotubes being semiconducting, and 1/3 being metallic; (2) the density of states exhibiting van Hove singularities, characteristic of the expectations for 1D systems (see Fig. 10); and (3) energy gaps for the semiconducting nanotubes that are proportional to $1/d_t$. Using the approximate relation $E_g \simeq 2\gamma_0 a_{\mathrm{C-C}}/d_t$ for the band gap of single wall nanotubes, the nearest neighbor overlap energy γ_0 (or the transfer integral of a tight binding model) can be found. The STS experiments furthermore confirm that the density of electronic states near the Fermi level is zero for semiconducting nanotubes, and non-zero for metallic nanotubes

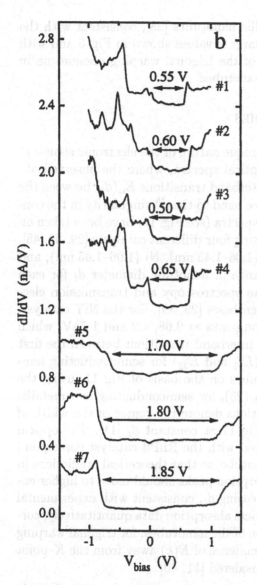

Figure 12. Derivative of the current-voltage (dI/dV) curves obtained by scanning tunneling spectroscopy on various isolated single wall carbon nanotubes with diameters near 1.4 nm. Nanotubes #1–4 are semiconducting and #5–7 are metallic [24].

[24]. Thus the combined STM/STS studies provided the first comprehensive experimental verification of the previously predicted electronic properties of carbon nanotubes and their unique 1D density of states features [24, 25].

Subsequent atomic resolution STM/STS studies coupled with tight-binding calculations [28] were able to make a detailed identification of the features in the dI/dV vs. voltage spectrum of an individual (13,7) metallic carbon nanotube with specific interband transitions. Comparison to the tight-binding calculations led to the first experimental demonstration of the trigonal warping effect in metallic nanotubes [28], consistent with the behavior of the energy contours at large k values shown in Fig. 5 and with a systematic theoretical study [11] of the trigonal warping phenomena in both metallic and semiconducting nanotubes.

6.2. OPTICAL ABSORPTION STUDIES

A second method for studying the unique nature of the electronic structure of carbon nanotubes comes from optical spectra, where the observed absorption peaks are identified with interband transitions $E_{ii}(d_t)$ between the ith van Hove singularity in the valence band to the ith singularity in the conduction band. Optical transmission spectra (see Fig. 13) have been taken on single wall nanotubes synthesized using four different catalysts [29, 39, 48], namely NiY (1.24–1.58 nm), NiCo (1.06–1.45 nm), Ni (1.06–1.45 nm), and RhPd (0.68–1.00 nm), where the range in nanotube diameter d_t for each catalyst, as characterized by Raman spectroscopy and transmission electron microscopy, is indicated in parentheses [29, 39]. For the NiY catalyst, Fig. 13 shows three strong absorption peaks at 0.68, 1.2 and 1.7 eV, which can, respectively, be identified with interband transitions between the first and second van Hove singularities (E_{11}^S and E_{22}^S) for semiconducting nanotubes and E_{11}^M for metallic nanotubes on the basis of Fig. 11, using the approximate relations given by Eqs. (25), for semiconducting and metallic nanotubes. These approximate relations denote the center of the width of the $E_{ii}(d_t)$ transition energies in Fig. 11 at constant d_t [11]. The optical spectra for the nanotubes synthesized with the RhPd catalyst correspond to a very much smaller diameter nanotube, so that theoretical predictions in Fig. 11 suggest that the optical absorption peaks should move to higher energies as $E_{ii}(d_t)$ increases with increasing d_t, consistent with experimental observations. In interpreting the optical absorption data quantitatively, corrections for the diameter distribution of the nanotubes, for trigonal warping effects, and for the non-linear k dependence of $E(k)$ away from the K-point in the Brillouin zone need to be considered [11, 40].

Optical absorption measurements on bromine and cesium doped single wall nanotubes ($1.24 < d_t < 1.58$ nm) [31] showed an upshift of E_F on doping the nanotubes with Cs and a downshift of E_F on doping with Br_2, as shown in Fig. 14, due to charge transfer effects between the dopants and the nanotubes [31]. When the doping concentration x in M_xC, (M = Cs, Br) is less than 0.005, the first peak at 0.68 eV decreases continuously in intensity with increasing x, without changing the intensity of the second and the third peaks. In subsequent doping in the range $0.005 < x < 0.04$, the two absorption peaks at 0.68 eV and 1.2 eV decrease in intensity. At the high doping level shown in Fig. 14(b), the small peak at 1.8 eV de-

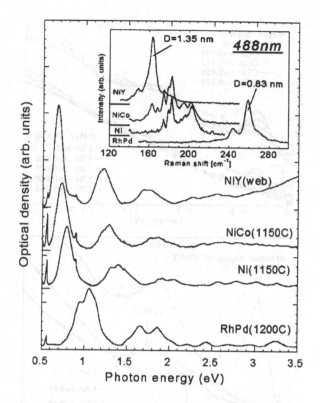

Figure 13. Optical absorption spectra taken for single wall nanotubes synthesized using four different catalysts at the indicated growth temperatures [29, 39], namely NiY (1.24–1.58 nm), NiCo (1.06–1.45 nm), Ni (1.06–1.45 nm), and RhPd (0.68–1.00 nm). Absorption peaks at 0.55 eV and 0.9 eV are due to absorption by the quartz substrate [49]. The inset shows the corresponding radial breathing mode Raman spectra obtained at 488 nm laser excitation with the same 4 catalysts. These radial breathing mode Raman spectra are used along with transmission electron microscopy to identify the average diameter and the diameter distribution of the single wall nanotube samples given in the caption

creases to a low level and new bands appear at 1.07 eV and 1.30 eV for $CBr_{0.15}$ and $CCs_{0.10}$, respectively. These doping-induced absorption peaks are tentatively identified with conduction to conduction inter-subband transitions and from valence to valence inter-subband transitions, respectively, for donor and acceptor type carbon nanotubes. The difference between the peak positions 1.07 eV and 1.30 eV for acceptor and donor type nanotubes [50], respectively, is consistent with the expected magnitude of the asymmetry between the π and π^* bands. This asymmetry is not yet established for carbon nanotubes and is implicitly neglected in present tight binding

32

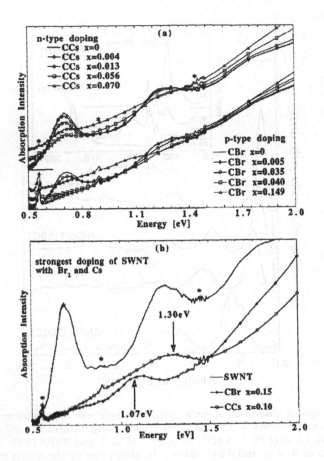

Figure 14. (a) Optical absorption spectra for Cs and Br doped single wall nanotube (SWNT) samples for CCs$_x$ and CBr$_x$ for various stoichiometries x. The entire set of spectra for the CCs$_x$ samples is offset for clarity with a short line on the left indicating the 0 level. The '*' in the figure indicates features coming from the quartz substrate and from spectrometer noise. (b) The absorption spectra for CCs$_x$ and CBr$_x$ for the nearly saturated doping regime [31]

models by taking the overlap integral $s = 0$ (see §3). However, the detailed assignments for the inter-subband transitions which are responsible for the doping-induced peaks are not clear within the rigid band model.

From observation of this reduction in the optical absorption intensity upon doping, the shift of E_F with Br$_2$ doping could be monitored, showing that at full Br$_2$ saturation concentration, E_F had decreased sufficiently so that optical transitions from an occupied state below the Fermi level to an unoccupied state above the Fermi level could only occur for $E_{laser} \geq$

2 eV, because of the emptying of the initial valence states. Thus, although optical absorption spectra for undoped and Br and Cs doped nanotubes are consistent with the framework of the predicted 1D electronic structure, the optical studies have not yet provided a stringent test for detailed theoretical predictions.

6.3. RESONANCE RAMAN SCATTERING STUDIES

The resonance Raman effect in single wall carbon nanotubes is unique in its own right and provides detailed information about the 1D electronic density of states through the strong electron-phonon coupling that occurs because of the van Hove singularities that appear in the density of states. Prominent in the Raman spectrum for single wall carbon nanotubes (Fig. 15) [32] are a number of modes near $1580\,cm^{-1}$, which are associated with graphite carbon-carbon in-plane G-band Raman feature, that exhibits a very weak dependence on nanotube diameter, and a strong mode at $\sim186\,cm^{-1}$, which is identified with a totally symmetric (A_{1g}) radial breathing mode and is strongly dependent on the nanotube diameter, as shown in Fig. 16.

Quantum effects are observed in the Raman spectrum through the resonance Raman enhancement effect which is seen experimentally by measuring the Raman spectra at a number of laser excitation energies, as shown in Fig. 15. Resonant enhancement in the Raman scattering intensity from carbon nanotubes occurs when the laser excitation frequency corresponds, as shown in Fig. 17, to a transition between the sharp features in the one-dimensional electronic density of states of the carbon nanotube, for either the incident or scattered photons [51].

Since the energies of these sharp features in the density of states are strongly dependent on the nanotube diameter, a change in the laser excitation energy for the Raman effect brings into resonance a carbon nanotube with a different diameter. However, armchair nanotubes with different diameters have different vibrational frequencies for the A_{1g} breathing mode (see Fig. 16). By examining the Raman spectra in Fig. 15, we see large differences in the vibrational frequency of the strong A_{1g} mode for different laser excitation energies, consistent with a resonant Raman effect involving nanotubes of different diameters. The dependence of the Raman intensity of the A_{1g} mode on laser excitation energy (see Fig. 15) is also consistent with a resonant Raman scattering mechanism. These resonant quantum effects lend strong credibility to the 1D aspects of the electronic and phonon structure of single-wall carbon nanotubes, and provide a very clear confirmation for the theoretical predictions about the singularities in the 1D electronic density of states.

In the resonance Raman effect, a large Raman scattering intensity is

34

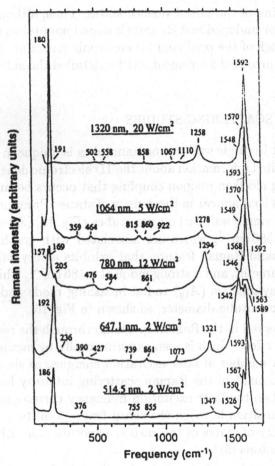

Raman Intensity (arbitrary units)

180

1320 nm, 20 W/cm² 1592

191 502 558 858 1067 1110 1258 1570
1548

1593
1570
1064 nm, 5 W/cm² 1549
359 464 815 860 922 1278

157 169 1294 1568 1592
205 780 nm, 12 W/cm² 1546
476 584 861

192 1542 1563
1589

236 647.1 nm, 2 W/cm² 1321
390 427 739 861 1073 1593

186 1567
514.5 nm, 2 W/cm² 1550
376 755 855 1347 1526

500 1000 1500
Frequency (cm⁻¹)

Figure 15. Experimental room temperature Raman spectra for purified single-wall carbon nanotubes excited at five different laser excitation wavelengths. The laser wavelength and power density for each spectrum are indicated, as are the vibrational frequencies (in cm^{-1}) [32]. The equivalent photon energies for the laser excitation are: 1320 nm → 0.94 eV; 1064 nm → 1.17 eV; 780 nm → 1.58 eV; 647.1 nm → 1.92 eV; 514.5 nm → 2.41 eV.

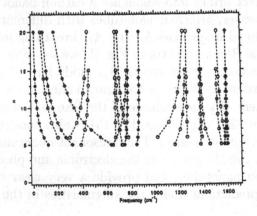

Figure 16. The armchair index n vs mode frequency for the Raman-active modes of single-wall armchair (n,n) carbon nanotubes [32]. The nanotube diameter can be found from n using the relation $\omega_{RBM} \sim 224/d_t \, cm^{-1}$ where d_t is given by Eq. (1) and where $n = m$ and $a_{C-C} = 0.142$ nm.

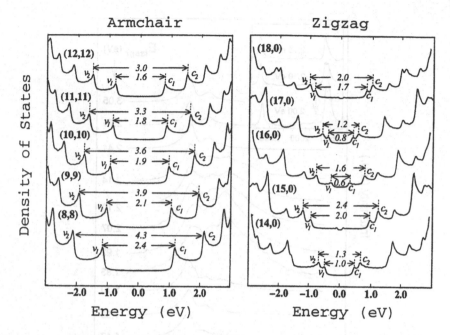

Figure 17. Electronic 1D density of states (DOS) calculated with a tight binding model for (8,8), (9,9), (10,10), (11,11), and (12,12) armchair nanotubes and for (14,0), (15,0), (16,0), (17,0), and (18,0) zigzag nanotubes and assuming a nearest neighbor carbon-carbon interaction energy $\gamma_0 = 3.0\,\text{eV}$ [52]. Wavevector conserving optical transitions can occur between mirror image singularities in the 1D density of states, i.e. $v_1 \to c_1$ and $v_2 \to c_2$, etc., and these optical transitions are given in the figure in units of eV. These interband transitions are denoted in the text by E_{11}, E_{22}, etc. and are responsible for the resonant Raman effect [52].

observed when either the incident or the scattered light is in resonance with calcelectronic transitions between van Hove singulatities in the valence and conduction bands $E_{pp}(d_t)$ for a given nanotube (n, m) [29, 32, 34, 35, 36]. In general, the size of the optical excitation beam is at least $1\,\mu\text{m}$ in diameter, so that many nanotubes with a large variety of (n, m) values (that is, nanotube diameters and chiralities) are excited by the optical beam simultaneously, as is also the case for the optical absorption measurements discussed in §6.2.

Most studies of the tangential G-band for single wall carbon nanotubes have focussed primarily on the dependence of the spectra on laser excitation energy, along with the spectral dependence on nanotube diameter. In Fig. 18 are plotted the resonance Raman spectra for a single wall nanotube sample with a diameter distribution $d_t = 1.37 \pm 0.20\,\text{nm}$ showing that a different spectrum (associated with metallic nanotubes) is found for $0.94 \leq E_{\text{laser}} \leq 3.05\,\text{eV}$) using different laser excitation energies ($1.8 \leq E_{\text{laser}} \leq 2.2\,\text{eV}$,

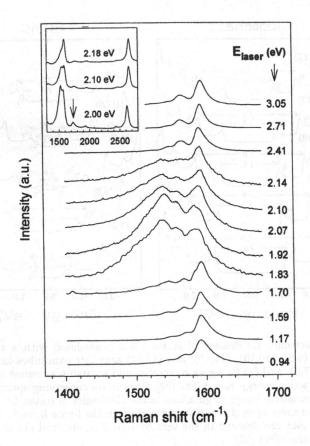

Figure 18. Raman spectra of the tangential *G*-band modes of carbon nanotubes with diameters in the range $d_t = 1.37 \pm 0.20$ nm, obtained with several different laser lines. The inset shows low resolution Raman spectra between 1300 and 2800 cm^{-1} in the range of laser energies 2.00–2.18 eV where the contributions from metallic nanotubes are dominant in the Raman spectra [35].

while the Raman spectra for $E_{laser} \leq 1.7$ eV and for $E_{laser} \geq 2.4$ eV are identified predominantly with semiconducting nanotubes. The range of E_{laser} which is resonant with metallic nanotubes, is called the "metallic window". This metallic window for the resonance Raman scattering effect is consistent with the optical density of the third peak as a function of laser excitation energy, as shown in Fig. 13, and with interband transition diagram shown in Fig. 11.

The lineshape of the tangential *G*-band for semiconducting nanotubes is fit by two sets of Lorentzian components, each set identified by polarization studies [53, 54] with one A_1, one E_1 and one E_2 symmetry component. One set is associated with displacements along the axis, and the other set with displacements in the circumferential direction. The Raman spectrum associated with the metallic nanotubes, on the other hand, is described in terms of two components both with A_1 symmetry, one of which is a Lorentzian component with displacements along the nanotube axis, and the other, with displacements in the circumferential direction, exhibiting a Breit–Wigner–Fano lineshape, where the A_{1g} phonon is coupled to the surface plasmon. This coupling to the surface plasmon is induced by the curvature of the nanotube which gives rise to some out-of-plane admixture of phonon states which couple to the plasmon.

Since the phonon energies of the G-band are large (0.2 eV), so that the resonant condition for the metallic energy window is generally different according to the energy difference for the incoming and scattered photon, thereby resonantly exciting different nanotubes. Furthermore, the resonant laser energies for phonon-emitted Stokes and phonon-absorbed anti-Stokes Raman spectra (discussed below) are different from each other by twice the energy of the corresponding phonon. Thus when a laser energy is selected, carbon nanotubes with different diameters d_t are resonant with the scattered photon for the radial breathing mode phonon and with the G-band phonon and also different nanotubes are excited in the Stokes spectra as compared to the anti-Stokes Raman spectra, as described below in more detail.

So far, almost all of the resonance Raman scattering experiments have been carried out on the Stokes spectra. The metallic window is determined experimentally as the range of E_{laser} over which the characteristic Raman spectrum for metallic nanotubes (Fig. 18) is seen, for which the most intense Raman component is near 1540 cm^{-1} [35]. Since there is essentially no Raman scattering intensity for semiconducting nanotubes at this phonon frequency, the intensity I_{1540} provides a convenient measure for the metallic window. The normalized intensity of the dominant Lorentzian component for metallic nanotubes $\tilde{I}_{1540}$ (normalized to a reference line) has a dependence on E_{laser} given by [35, 55]

$$
\begin{aligned}
\tilde{I}_{1540}(d_0) = & \sum_{d_t} A \exp\left[\frac{-(d_t - d_0)^2}{\Delta d_t^2/4}\right] \\
& \times \left[(E_{11}^M(d_t) - E_{\text{laser}})^2 + \Gamma_e^2/4\right]^{-1} \\
& \times \left[(E_{11}^M(d_t) - E_{\text{laser}} \pm E_{\text{phonon}})^2 + \Gamma_e^2/4\right]^{-1},
\end{aligned}
\tag{26}
$$

where d_0 and Δd_t are, respectively, the mean diameter and the width of the Gaussian distribution of nanotube diameters within the SWNT sample, E_{phonon} is the average energy (0.197 eV) of the tangential phonons and the $+$ ($-$) sign in Eq. (26) refers to the Stokes (anti-Stokes) process where a phonon is emitted (absorbed), Γ_e is a damping factor that is introduced to avoid a divergence of the resonant denominator, and the sum in Eq. (26)

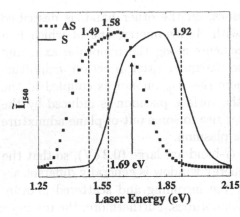

Figure 19. Metallic window for carbon nanotubes with diameter of $d_t = 1.49 \pm 0.20$ nm for the Stokes (solid line) and anti-Stokes (square points) processes plotted in terms of the normalized intensity of the phonon component at $1540\,\mathrm{cm}^{-1}$ for metallic nanotubes vs the laser excitation energy for the Stokes and the anti-Stokes scattering processes calculated from Eq. (26) [55]. The crossing between the Stokes and anti-Stokes curves is denoted by the vertical arrow, and provides a sensitive determination of the electronic energy γ_0 (2.94 $\pm$ 0.05 eV) [40, 55]

is carried out over the nanotube diameter distribution. Equation (26) indicates that the normalized intensity for the Stokes process $\tilde{I}^S_{1540}(d_0)$ is large when either the incident laser energy is equal to $E^M_{11}(d_t)$ or when the scattered laser energy is equal to $E^M_{11}(d_t)$ and likewise for the anti-Stokes process. This feature is an expression of the resonance Raman effect and relates to the van Hove singularities in the 1D electronic density of states. Since the phonon energy is on the same order of magnitude as the width of the metallic window for nanotubes with diameters d_t, the Stokes and the anti-Stokes processes can be observed at different resonant laser energies in the resonant Raman experiment. The dependence of the normalized intensity $\tilde{I}_{1540}(d_0)$ for the actual single wall nanotube (SWNT) sample on E_{laser} is primarily sensitive [34, 35, 36] to the energy difference $E^M_{11}(d_t)$ for the various d_t values in the sample, and the resulting normalized intensity $\tilde{I}_{1540}(d_0)$ is obtained by summing over d_t.

In Fig. 19 we present a plot of the expected integrated intensities $\tilde{I}_{1540}(d_0)$ for the resonance Raman process for metallic nanotubes for both the Stokes (solid curve) and anti-Stokes (square points) processes. This figure is used to distinguishes 4 regimes for observation of the Raman spectra for Stokes and anti-Stokes processes shown in Fig. 20: (1) the semiconducting regime (2.19 eV), for which both the Stokes and anti-Stokes spectra receive contributions from semiconducting nanotubes, (2) the metallic regime (1.58 eV), where metallic nanotubes contribute to both the Stokes and anti-Stokes spectra, (3) the regime (1.92 eV), where metallic nanotubes contribute to the Stokes spectra and not to the anti-Stokes spectra, and (4) the regime (1.49 eV), where the metallic nanotubes contribute only to the anti-Stokes spectra and not to the Stokes spectra. The plot in Fig. 19 is for a nanotube diameter distribution $d_t = 1.49 \pm 0.20$ nm assuming $\gamma_e = 0.04$ eV for the

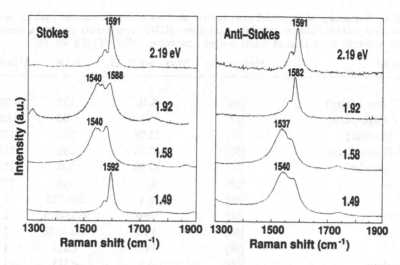

Figure 20. Resonant Raman spectra for the Stokes and anti-Stokes process for SWNTs with a diameter distribution $d_t = 1.49 \pm 0.20$ nm [55]

damping term [40, 55], and pertains to the nanotubes used in the spectra shown in Fig. 20. The arrow at 1.69 eV in Fig. 19 denotes the energy $E_{11}^M (d_t)$ in the relation $E_{11}^M (d_t) = 6a_{C-C}\gamma_0/d_t$ and thereby provides a sensitive determination of γ_0, the electronic overlap energy for carbon nanotubes of diameter d_t. Experiments on different carbon nanotube samples show γ_0 to be only slightly lower than in graphite (3.16 eV) and only weakly dependent on the nanotube diameter. The parameter γ_0 is equivalent to the transfer integral t in the tight binding approximation (see §3).

From Figs. 19 and 20 it is seen that by special selection of the laser excitation energy, the Raman spectra for a specific single wall nanotube sample with a specified diameter distribution can be taken from metallic nanotubes, based on diagrams such as Fig. 11, showing the energies for interband transitions for metallic nanotubes, and Fig. 19, showing the metallic windows for the Stokes and anti-Stokes processes. Likewise, by a different selection of laser excitation energy, the Raman spectra from semiconducting nanotubes can be observed without complication from the presence of metallic nanotubes. Through the identification of metallic windows in the Raman spectra for nanotube samples with different diameters, the map of the interband electronic transitions based on the van Hove singularities for the 1D electronic states (see Figs. 11 and 17) can be established experimentally.

TABLE 1. Summary of reported gravimetric storage of H_2 in various carbon materials. Note that certain tubular graphite nanofibers (GNF) apparently possess a structure distinctly different from that of multi-walled nanotubes (MWNT) [59, 60, 68].

Material	Reference	Max. wt.% H_2	T (K)	P(atm)
SWNT (low purity)	[56]	5-10	133	0.395
SWNT (high-purity)	[57]	~4	300	0.658
GNF[†] (tubular)	[59]	11.26	298	112
GNF[†] (herringbone)	[59]	67.55	298	112
GNF[†] (platelet)	[59]	53.68	298	112
Graphite	[59]	4.52	298	112
GNF[†]	[60]	0.4	298-773	1
Li-GNF[†]	[60]	20.0	473~673	1
Li-Graphite	[60]	14.0	473~673	1
K-GNF[†]	[60]	14.0	<313	1
K-Graphite	[60]	5.0	<313	1
SWNT (high purity)	[58]	8.25	80	70.9
SWNT (~50% pure)	[67]	4.2	300	100

[†]GNF = graphite nanofiber.

7. Hydrogen Storage in Carbon Nanotubes

The recent reports of very high reversible adsorption of molecular hydrogen in pure nanotubes [56, 57, 58], alkali-doped graphite, and pure and alkali-doped graphite nanofibers [59, 60] have aroused tremendous interest in the research community and also in industry. The US Department of Energy (DOE) Hydrogen Plan has set a standard for hydrogen adsorption by providing a commercially significant benchmark for the amount of reversible hydrogen adsorption that would be of interest for hydrogen based energy sources for transportation and other applications. This benchmark requires a system with a weight efficiency (the ratio of stored H_2 weight to system weight) of 6.5 wt.% hydrogen and a corresponding volumetric density of 63 kg H_2/m^3 [61]. If the encouraging experimental reports (summarized in Table 1) are reproducible, it may be possible to reach the goals of the DOE Hydrogen Plan in the near future. On the other hand, the community still awaits confirmation of these experimental results by researchers in other laboratories. Of additional concern is the fact that theoretical calculations [62, 63, 64, 65, 66] have been unable to identify adsorption mechanisms compatible with the requirements of the DOE Hydrogen Plan.

Some of the carbon materials in Table 1 involve single wall carbon nanotubes (SWNTs). Other entries involve graphite nanofibers, which for some of the host materials in Table 1, differ in a fundamental way from multi-wall carbon nanotubes. We confine our discussion here to these entries on carbon nanotubes.

Dillon *et al.* [56] were the first to investigate the hydrogen adsorption capacity of bundles of single-walled carbon nanotubes (SWNT). They reported their initial results on poorly characterized nanotubes [56] and in comparison with hydrogen adsorption in activated carbon, which has a high density of pores and therefore a high surface area for adsorption. Their work was initially stimulated by the theoretical description of capillarity in carbon nanotubes by Pederson and Broughton [69], suggesting that gaseous species such as hydrogen might also be drawn into the capillaries of nanotubes. Dillon, Heben and their coworkers subsequently developed [57, 70] a method to produce samples with a high concentration of short single wall nanotubes with open ends that are accessible to the entry of hydrogen molecules. With these specially prepared samples, they estimated the hydrogen adsorption to be ∼4 wt.% hydrogen at room temperature and 500 Torr hydrogen pressure. By opening and re-capping the nanotubes, Dillon *et al.* [70] concluded that most of the hydrogen is stored within the capillary, rather than in the interstitial spaces between the single wall nanotubes.

A similar hydrogen storage capacity of 4.2 wt.% was recently reported for single wall nanotubes (SWNTs) at a temperature of 300 K and pressure of 100 atm [67]. In this study, SWNT were synthesized by a semicontinuous hydrogen arc discharge process, using a sulfur promoter. The sample (containing ∼50% SWNT) was treated by soaking in HCl and subsequent heat treatment at 500 °C under vacuum. The relatively high hydrogen storage capacity was attributed to the large mean tube diameter of about 1.85 nm.

Ye *et al.* have undertaken experimental measurements of hydrogen adsorption on high-purity, "cut" SWNTs [58]. The highest gravimetric hydrogen storage capacity achieved in SWNT material treated in this manner was 8.25 wt.%, at a temperature of 80 K and pressure of ∼70 atm (see Table 1). At a pressure of ∼40 atm, a sudden increase in the adsoption capacities of the SWNT samples is reported; Ye and coworkers suspect that a structural phase transition is responsible for this effect. In their model, the ropes are split into individual tubes, thereby increasing the surface area available for physisorption.

Chen *et al.* [60] reported high hydrogen adsorption in alkali-metal-doped graphite nanofibers (GNF) at ambient pressure and slightly elevated temperatures. The GNF used in this work were synthesized by catalytic decomposition of methane and reportedly exhibited a structure akin to a stack of cones (for this reason, we refer to their host material as nanofibers rather than nanotubes). Although very large hydrogen uptake was reported (see Table 1), these results await experimental confirmation.

7.1. PHYSICAL ARGUMENTS FOR HYDROGEN ADSORPTION ON CARBON

To gain insight into the hydrogen adsorption problem, it is first necessary to review a few basic facts about hydrogen molecules and the carbon surfaces to which they might bind. In the ground state, the hydrogen molecule is nearly spherical with a kinetic diameter of 2.89 Å, and the intermolecular interactions between H_2 molecules are weak [66, 71]. Therefore, H_2 molecules at elevated pressures on a solid surface are expected to assume a close-packed configuration. We use the parameters of the experimentally observed structure to provide one simple estimate for the maximum packing density for hydrogen molecules on a surface. Each molecule in this close-packed structure has six nearest neighbors, 4 in-plane and 2 out-of-plane. Experimentally, solid hydrogen at 4.2 K forms a hexagonal close-packed structure, with lattice parameters a_H =3.76 Å and c_H =6.14 Å,[72] so that c_H/a_H equals 1.633, which is identical to $c/a=\sqrt{8/3}$=1.633 for the ideal hexagonal close-packing of spheres. This indicates that in the ground rotational state, the H_2 molecule is approximately spherical in shape.

Using purely geometric arguments we can thus gain a simple geometric estimate for the close-packing capacity of hydrogen molecules above a plane of graphite. Graphite has a honeycomb structure, with an in-plane lattice parameter a =2.46 Å and an interplanar separation of 3.35 Å. Since the value of the kinetic diameter for the hydrogen molecule is greater than a, the closest packing of hydrogen molecules on a graphite surface would have to be incommensurate with the graphene plane. Commensurate H_2 adsorption on a 2D $\sqrt{3} \times \sqrt{3}$ graphene-based superlattice [8] (see Fig. 21) would yield a lattice constant of a_H =4.26 Å, which is significantly larger than that for the low temperature close-packed structure of solid hydrogen [72]. The ratio of the number of hydrogen to carbon atoms (H:C) for the $\sqrt{3} \times \sqrt{3}$ commensurate stacking on a graphene surface would be H:C = 1:3. Given the 12:1 ratio between the masses of carbon to hydrogen, this commensurate stacking would yield a ~2.8 wt.% of hydrogen for one layer of H_2 adsorbed on a single graphene layer.

In a real system, there are intermolecular forces and molecule-surface interactions that come into play, and there are also many edge states that might provide additional adsorption sites for the H_2 molecules. Neutron scattering measurements [73] show that at sparse coverage, the commensurate $\sqrt{3} \times \sqrt{3}$ structure occurs (see Fig. 21, top). A denser triangular structure, which is incommensurate with the graphite, is observed for higher hydrogen pressures (see Fig. 21, bottom). Once a second layer begins to form, a lattice parameter of 3.51 Å is observed [74], smaller than the value of

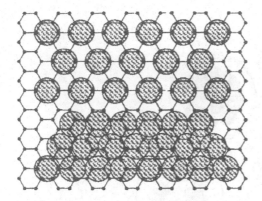

Figure 21. Relative density of a ($\sqrt{3} \times \sqrt{3}$) commensurate (top) and an incommensurate (bottom) monolayer of H_2 on a graphite surface [75].

3.76 Å measured for the bulk hexagonal close-packing of hydrogen molecules in solid hydrogen. For this more dense two-layer phase, we obtain an estimate of the H:C ratio = 0.49, thus yielding a 4.1 wt.% hydrogen adsorption on a flat sp^2 carbon surface.

Graphite intercalation compounds have also been considered as a method for the uptake of hydrogen [8]. Although alkali-metal GICs (e.g., K, Rb, Cs) are capable of the intercalation of H_2, the weight percent of hydrogen in the final compound for either a chemisorption route ($C_8KH_{2/3}$ [8, 76, 77, 78]) or a physisorption process ($C_{24}KH_2$ [8, 79]) is too low to be of interest in the context of the DOE Hydrogen Plan. Recent experimental efforts using electrochemical techniques to intercalate hydrogen directly between the graphene layers [80] have been unsuccessful. These experimental results are consistent with a molecular orbital calculation carried out by the same group showing that hydrogen ions, unlike lithium ions, do not form stable intercalation compounds because of the absence of 3-dimensional p-orbitals which are presumably needed to bond the hydrogen to the adjacent graphene layer. However, the use of low concentrations of alkali metals (in a metal/carbon ratio of ∼1:15) has been reported (see Table 1) to catalyze high hydrogen uptake [60], though these findings have not been confirmed by other groups.

Simple geometric arguments can be used to estimate the filling of a rope (crystalline lattice) of single wall carbon nanotubes. We present here two simple geometrical arguments. The first assumes the hydrogen to be a completely deformable fluid that fills the space not occupied by the carbon nanotube. Assume that we have a rope of (10,10) nanotubes each tube having a diameter 13.8 Å, and a unit cell dimension of 17.2 Å for the rope. This geometry yields an area of 147.4 Å^2 occupied by the nanotube and an area of 108.8 Å^2 of unoccupied area available for the hydrogen fluid. Assuming a hydrogen density of 0.071 g/cm^3 (corresponding to that of liquid hydro-

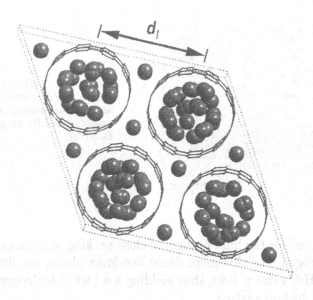

Figure 22.   A typical configuration of H_2 molecules adsorbed on a triangular array of carbon nanotubes with a lattice constant of d_t. This configuration resulted from a classical Monte Carlo calculation in which the simulated storage pressure was 10 MPa and the simulated temperature was 50 K. The computed gravimetric storage capacity in this "periodic infinite lattice" is ~3.1 wt.% H_2 [68].

gen) and a density of 2.26 g/cm³ for the carbon atoms within the nanotube volume yields a 2.3 wt.% hydrogen uptake.

A second geometrical model takes into account the accessible volume to hydrogen molecules by considering the packing of hydrogen molecules of kinetic diameter 2.9 Å on the inner walls and in the interstitial volume of the nanotubes as shown in Fig. 22. For a (10,10) nanotube array, 8 hydrogen molecules fit along the inside wall of the nanotube, which is estimated to have a diameter of (13.8 – 3.4) Å = 10.4 Å, taking into account the charge cloud of the π-electrons at the inner carbon surface. The diameter of a circle through the center of the adsorbed hydrogen molecules is (10.4 – 2.9) Å = 7.5 Å. One hydrogen molecule fits into the interstitial volume of tubes with diameters greater than or equal to that of the (9,9) armchair tube. Using this geometrical model with close packing of the H_2 molecules within the core of the (10,10) tube leads to a 3.3 wt.% hydrogen adsorption within the tube and a 0.7 wt.% adsorption within the interstitial space or a total of 4.0 wt.% hydrogen adsorption. As the tube diameter increases, the C–H van der Waals interaction at the wall interface decreases, but at the same time there is also an increase in the unoccupied central core volume, which for the (10,10) nanotube is described by the volume within a cylinder of diameter $d_c = (7.5 - 2.9)\,\text{Å} = 4.6\,\text{Å}$ [68].

Some have argued that the curved inner surface of the tube leads to a reduction in the accessible geometric volume for the hydrogen, and detailed calculations in support of this viewpoint have been presented [66]. On the other hand, one can argue that under high pressure conditions, the high compressibility of hydrogen and the attractive intermolecular interactions should lead to a closer packing of hydrogen molecules, especially at low temperature. Detailed calculations in support of this concept have also been carried out [63], yielding hydrogen densities higher than the simple geometric picture. This latter concept seems to be supported by the experimental observations of Dillon *et al.* [56, 70]

We can further argue, using a 2.9 Å kinetic diameter for H_2, that hydrogen molecules should be adsorbed effectively within the interstitial space between nanotubes (see Fig. 22). Since a hydrogen molecule in this interstitial space is in close proximity to three graphene surfaces, it should therefore experience a stronger surface attraction than on a single planar graphene surface and this site would be expected to exhibit a higher binding energy [75].

Molecular sieving calculations on the use of ropes of single wall nanotubes for the separation of tritium from hydrogen indicate that the interstitial spaces between the nanotubes do provide suitable pores for hydrogen isotope separation [81, 82], so that the interstitial spaces therefore should also be available for direct hydrogen molecular adsorption. Calculations of hydrogen uptake on carbon surfaces and in carbon nanotube bundles have become an active field of investigation [68]. It has, however, been difficult to identify an absorption mechanism that would allow hydrogen uptake and release near room temperature meeting the DOE Hydrogen Plan. Calculations in support of the experimental results [8, 66] are also being attempted.

Acknowledgements

The author gratefully acknowledges the helpful and informative discussions with Drs. G. Dresselhaus, J.-C. Charlier, A. Jorio, A. M. Rao, Professors M. Endo, R. Saito, H. Kataura, M.A. Pimenta, Mr. K.A. Williams, and Dr. Sandra Brown. She is also thankful to many other colleagues for their assistance with the preparation of this article. The research is supported by NSF grant DMR 98-04734.

References

1. Dresselhaus, M. S., Dresselhaus, G., and Saito, R. (1995) Physics of carbon nanotubes, *Carbon* **33**, 883–891.

46

2. Dresselhaus, M. S., Dresselhaus, G., and Saito, R. (1992) Carbon fibers based on C_{60} and their symmetry, *Phys. Rev. B* **45**, 6234–6242.

3. Saito, R., Dresselhaus, G., and Dresselhaus, M. S. (1998), *Physical properties of carbon nanotubes*, Imperial College Press, London.

4. Saito, R., Fujita, M., Dresselhaus, G., and Dresselhaus M. S. (1992) Electronic structure of chiral graphene tubules, *Appl. Phys. Lett.* **60**, 2204–2206.

5. Iijima, S. and Ichihashi T. (1993) Single shell carbon nanotubes of 1-nm diameter, *Nature (London)* **363**, 603–605.

6. Iijima, S. (1991) Helical microtubules of graphitic carbon, *Nature (London)* **354**, 56–58.

7. Wallace, P. R. (1947) The band theory of graphite, *Phys. Rev.* **71**, 622–634.

8. Dresselhaus, M. S. and Dresselhaus, G. (1981) Intercalation compounds of graphite, *Advances in Phys.* **30**, 139–326.

9. Saito, R., Fujita, M., Dresselhaus, G., and Dresselhaus,M. S. (1992) Electronic structures of carbon fibers based on C_{60}, *Phys. Rev. B* **46**, 1804–1811.

10. Painter, G. S. and Ellis, D. E. (1970) Electronic band structure and optical properties of graphite from a variational approach, *Phys. Rev. B* **1**, 4747–4752.

11. Saito, R., Dresselhaus, G., and Dresselhaus, M. S. (2000) Trigonal warping effect of carbon nanotubes, *Phys. Rev. B* **61**, 2981–2990.

12. McClure, J. W. (1956) Diamagnetism of graphite, *Phys. Rev.* **104**, 666–671.

13. Mintmire, J. W., Dunlap, B. I., and White, C. T. (1992) Are fullerene tubules metallic?, *Phys. Rev. Lett.* **68**, 631–634.

14. Hamada, T., Furuyama, M., Tomioka, T., and Endo M. (1992) Preferred orientation of pitch precursor fibers and carbon fibers prepared from isotropic pitch, *J. Mater. Res.* **7**, 1178–1188; *ibid.*, 2612-2620.

15. Venkataraman, L. (1993) *Calculation of the phonon modes in carbon nanotubes*, Bachelor of Science Thesis, Department of Physics, MIT, Cambridge, MA.

16. Jishi, R. A., Venkataraman, L., Dresselhaus, M. S., and Dresselhaus, G. (1993), Phonon modes in carbon nanotubules, *Chem. Phys. Lett.* **209**, 77–82.

17. Jishi, R. A., Inomata, D., Nakao, K., Dresselhaus, M. S., and Dresselhaus, G. (1994) Electronic and lattice properties of carbon nanotubes, *J. Phys. Soc. Jpn.* **63**, 2252–2260.

18. Hamada, N., Sawada, S., and Oshiyama, A. (1992) New one-dimensional conductors : graphitic microtubules, *Phys. Rev. Lett.* **68**, 1579–1581.

19. Dresselhaus, M. S., Jishi, R. A., Dresselhaus, G., Inomata, D., Nakao, K., and Saito, R. (1994) Group theoretical concepts for carbon nanotubes, *Molecular Materials* **4**, 27–40.

20. Dresselhaus, M. S., Dresselhaus, G., and Eklund, P. C. (1996) *Science of fullerenes and carbon nanotubes*, Academic Press, New York, N.Y.

21. Dresselhaus, M. S., Dresselhaus, G., Saito, R., and Eklund, P. C. (1992), C_{60}-*Related balls and fibers*, Elsevier Science Publishers, B.V., New York, Chapt. 18, pp. 387–417.

22. Saito, R., Dresselhaus, G., and Dresselhaus, M. S. (1993) Electronic structure of double-layered graphene tubules, *J. Appl. Phys.* **73**, 494–500.

23. Charlier, J.-C. (1998) Theory of electronic structure of carbon nanotubes, in Delhaès, P. and Ajayan, P. M. (eds.), *Fullerenes and Nanotubes*, Vol. 2, Gordon and Breach, Paris, France.

24. Wildöer, J. W. G., Venema, L. C., Rinzler, A. G., Smalley, R. E., and Dekker, C. (1998) Electronic structure of carbon nanotubes investigated by scanning tunneling spectroscopy, *Nature (London)* **391**, 59–62.

25. Odom, T. W., Huang, J. L., Kim, P., and Lieber, C. M. (1998) Atomic structure and electronic properties of single-walled carbon nanotubes, *Nature (London)* **391**, 62–64.

26. Odom, T. W., Huang, J. L., Kim, P., Ouyang, M., and Lieber, C. M. (1998) STM and spectroscopy studies of single-walled carbon nanotubes, *J. Mater. Res.* **13**,

2380–2388.

27. Odom, T. W., private communication.

28. Kim, P., Odom, T., Huang, J.-L., and Lieber, C. M. (1999) Electronic density of states of atomically resolved single-walled carbon nanotubes: Van Hove singularity and end states, *Phys. Rev. Lett.* **82**, 1225–1228.

29. Kataura, H., Kumazawa, Y., Kojima, N., Maniwa, Y., Umezu, I., Masubuchi, S., Kazama, S. , Zhao, X., Ando, Y., Ohtsuka, Y., Suzuki, S., and Achiba, Y. (1999) Optical absorption and resonant raman scattering of carbon nanotubes, in Kuzmany, H., Mehring, M., and Fink, J. (eds.), *Electronic Properties of Novel Materials – Science and Technology of Molecular Nanostructures*, AIP Conf. Proc. 486, American Institute of Physics, Woodbury, N.Y., pp. 328-332.

30. Ajiki, H. and Ando, T. (1994) Aharonov-Bohm effect in carbon nanotubes, *Physica B, Condensed Matter* **201**, 349–352.

31. Kazaoui, S., Minami, N., Jacquemin, R., Kataura, H., and Achiba, Y. (1999). *Phys. Rev. B* **60**, 13339–13342.

32. Rao, A. M., Richter, E., Bandow, S., Chase, B., Eklund, P. C., Williams, K. W., Menon, M., Subbaswamy, K. R., Thess, A., Smalley, R. E., Dresselhaus, G., and Dresselhaus, M. S. (1997) Infrared and Raman spectroscopic studies of single-wall carbon nanotubes, *Science* **275**, 187–191.

33. Kasuya, A., Sasaki, Y., Saito, Y., Tohji, K., and Nishina, Y. (1997) Evidence for size-dependent discrete dispersions in single-wall nanotubes, *Phys. Rev. Lett.* **78**, 4434–4437.

34. Pimenta, M. A., Marucci, A., Brown, S. D. M., Matthews, M. J., Rao, A. M., Eklund, P. C., Smalley, R. E., Dresselhaus, G., and Dresselhaus, M. S. (1998) Resonant Raman effect in single-wall carbon nanotubes, *J. Mater. Research* **13**, 2396–2404.

35. Pimenta, M. A., Marucci, A., Empedocles, S., Bawendi, M., Hanlon, E. B., Rao, A. M., Eklund, P. C., Smalley, R. E., Dresselhaus, G., and Dresselhaus, M. S. (1998) Raman modes of metallic carbon nanotubes, *Phys. Rev. B Rapid* **58**, R16016–R16019.

36. Alvarez, L., Righi, A., Guillard, T., Rols, S., Anglaret, E., Laplaze, D., and Sauvajol, J.-L. (2000) Resonant Raman study of the structure and electronic properties of SWNTs, *Chem. Phys. Lett.* **316**, 186–190.

37. Mintmire, J. W. and White, C. T. (1998) Universal DOS in carbon nanotubes, *Phys. Rev. Lett.* **81**, 2506–2509.

38. White, C. T. and Todorov, T. N. (1998) Carbon nanotubes as long ballistic conductors, *Nature (London)* **393**, 240–242.

39. Kataura, H., Kumazawa, Y., Maniwa, Y., Umezu, I., Suzuki, S., Ohtsuka, Y., and Achiba, Y. (1999) Optical properties of single-wall carbon nanotubes, *Synthetic Metals* **103**, 2555–2558.

40. Dresselhaus, G., Pimenta, M. A., Saito, R., Charlier, J.-C., Brown, S. D. M., Corio, P. , Marucci, A., and Dresselhaus, M. S. (2000) On the $\pi - \pi$ overlap energy in carbon nanotubes, in Tománek, D. and Enbody, R. J. (eds.) *Science and Applications of Nanotubes*, Kluwer Academic, New York, pp. 275–295.

41. Saito, R., private communication.

42. Wang, S. and Zhou, D. (1994) Microscopy of single layer carbon nanotubes, *Chem. Phys. Lett.* **225**, 165–169.

43. Ge, M. and Sattler, K. (1993) STM and properties of fullerenes and carbon nanotubes, *Science* **260**, 515–518.

44. Dravid, V. P., Lin, X., Wang, Y., Wang, X. K., Yee, A., Ketterson, J. B., and Chang, R. P. H. (1993) Buckytubes and derivatives: Their growth and implications for buckyball formation, *Science* **259**, 1601–1604.

45. Amelinckx, S., Bernaerts, D., Zhang, X. B., Van Tendeloo, G., and Van Landuyt, J. (1995) A structure model and a growth mechanism for multishell carbon nanotubes prepared by the arc discharge method, *Science* **267**, 1334–1338.

46. Zhang, Z. and Lieber, C. M. (1993) Nanotube structure and electronic properties

probed by STM, *Appl. Phys. Lett.* **62**, 2792–2794.

47. Sattler, K. (1995) STM analysis of carbon nanotubes and nanocones, *Carbon* **33**, 915–920.

48. Kataura, H. (unpublished).

49. Kataura, H., Y. Kumazawa, N. Kojima, Y. Maniwa, I. Umezu, S. Masubuchi, S. Kazama, X. Zhao, Y. Ando, Y. Ohtsuka, S. Suzuki, and Y. Achiba: 1999a, 'Optical Absorption and Resonant Raman Scattering of Carbon Nanotubes'. In: H. Kuzmany, M. Mehring, and J. Fink (eds.): *Proc. of the Int. Winter School on Electronic Properties of Novel Materials (IWEPNM'99)*. Woodbury, N.Y., American Institute of Physics. AIP conference proceedings (in press).

50. Saito, R. and Kataura, H. (2000), in Dresselhaus, M. S., Dresselhaus, G., and Avouris, P. (eds.) *Carbon Nanotubes*, Springer-Verlag, Berlin, to be published.

51. Dresselhaus, M. S. and Eklund, P. C. (2000) Phonons in carbon nanotubes, *Advances in Physics*, in press.

52. Charlier, J.-C., private communication.

53. Rao, A. M., Jorio, A., Pimenta, M. A., Dantas, M. S. S., Saito, R., Dresselhaus, G., and Dresselhaus, M. S. (2000) Polarized Raman study of aligned multiwalled carbon nanotubes, *Phys. Rev. Lett.* **84**, 1820–1823.

54. Jorio, A., Dresselhaus, G., Dresselhaus, M. S., Souza, M., Dantas, M. S. S., Pimenta, M. A. , Rao, A. M., Saito, R., Liu, C., and Cheng, H. M. (2000) Polarized Raman study of single wall semiconducting carbon nanotubes, *Phys. Rev. Lett.* **85**, in press.

55. Brown, S. D. M., Corio, P., Marucci, A., Dresselhaus, M. S., Pimenta, M. A., and Kneipp, K. (2000) Anti-Stokes Raman spectra of single-walled carbon nanotubes, *Phys. Rev. B Rapid* **61**, R5137–R5140.

56. Dillon, A. C., Jones, K. M., Bekkedahl, T. A., Kiang, C. H., Bethune, D. S., and Heben, M. J. (1997) Storage of hydrogen in single wall carbon nanotubes, *Nature (London)* **386**, 377–379.

57. Dillon, A. C., Gennett, T., Alleman, J. L., Jones, K. M., and Heben, M. J. (1999), private communication (in preparation).

58. Ye, Y., Ahn, C. C., Witham, C., Fultz, B., Liu, J., Rinzler, A. G., Colbert, D., Smith, K. A., and Smalley, R. E. (1999) Hydrogen adsorption and cohesive energy of single-walled carbon nanotubes, *Appl. Phys. Lett.* **74**, 2307–2309.

59. Chambers, A., Park, C., Baker, R. T. K., and Rodriguez, N. M. (1998) Hydrogen storage in graphite nanofibers, *Physical Chemistry B* **102**, 4253–4256.

60. Chen, P., Wu, X., Lin, J., and Tan, K. (1999) High H_2 uptake by alkali-doped carbon nanotubes under ambient pressure and moderate temperatures, *Science* **285**, 91–93.

61. Hynek, S. J., Fuller, W., and Bentley, J. (1997), *Int. J. Hydrogen Energy* **22**, 601.

62. Stan, G. and Cole, M. W. (1998) Hydrogen adsorption in nanotubes, *J. Low Temp. Phys.* **110**, 539–544.

63. Stan, G. and Cole, M. W. (1998) Low coverage adsorption in cylindrical pores, *Surface Science* **395**, 280–291.

64. Wang, Q. and Johnson, J. K. (1999) Computer simulations of hydrogen adsorption on graphite nanofibers, *J. Phys. Chem. B* **103**, 277–281.

65. Wang, Q. and Johnson, J. K. (1999) Molecular simulations of hydrogen adsorption on single-walled carbon nannotubes and idealized carbon slit pores, *J. Chem. Phys.* **110**, 577–586.

66. Rzepka, M., Lamp, P., and de la Casa-Lillo, M. A. (1998) Physisorption of hydrogen on microporous carbon and carbon nanotubes, *J. Phys. Chem. B* **102**, 10894–10898.

67. Liu, C., Fan, Y. Y., Liu, M., Cong, H. T., Cheng, H. M., and Dresselhaus, M. S. (1999) Hydrogen storage in single-walled carbon nanotubes at room temperature, *Science* **286**, 1127–1129.

68. Dresselhaus, M. S., Williams, K. A., and Eklund, P. C. (1999) Hydrogen adsorption in carbon materials, *MRS Bulletin* **24**, 45–50.

69. Pederson, M. R., Jackson, K. A., and Boyer, L. L. (1992) Enhanced stabilization of C_{60} crystals through doping, *Phys. Rev. B* **45**, 6919–6922.

70. Heben, M. J. (1999), private communication.
71. Van Kranendonk, J. (1982), in "Solid hydrogen: Theory of the properties of solid H_2, HD, and D_2", Plenum Press, New York, N.Y.
72. Hellwege, K.-H. and Hellwege, A. M. (eds.) (1988) *Landolt-Börnstein Numerical Data and Functional Relationships in Science and Technology, New Series III/14a*, Springer-Verlag, Berlin, p. 18.
73. Nielsen, M., McTague, J. P., and Passell, L. (1979), in Dash, J. G. and Ruvalds, J. (eds.) *Phase Transitions in Surface Films*, Plenum press, New york, N.Y., p. 127.
74. Nielsen, M., McTague, J. P., and Ellenson, W. (1977) Adsorbed layers of molecular deuterium, hydrogen, oxygen, and helium-3 on graphite studied by neutron scattering, *J. Phys.* **38**, C4-10.
75. Brown, S. D. M., Dresselhaus, G., and Dresselhaus, M. S. (1998) Reversible hydrogen uptake in carbon-based materials, in Rodriguez, N. M., Soled, S. L., and Hrbek, J. (eds.) *Recent Advances in Catalytic Materials: MRS Symposium Proceedings, Boston, Vol. 497*, Materials Research Society Press, Pittsburgh, PA, pp. 157–163,
76. Colin, M. and Hérold, A. (1972), *Bull. Soc. Chem. Fr.* p. 1982.
77. Furdin, G., Lagrange, P., Hérold, A., and Zeller, C. (1976) Magnetic susceptibility of the ternary phases potassium graphite hydride KC_8H_x $(0 < x < 2/3)$, *C.R. Acad. Sci. Paris* **282**, C563–566.
78. Enoki, T., Sano, M., and Inokuchi, H. (1983) Hydrogen in aromatics. III. Chemisorption of hydrogen in graphite-alkali metal intercalation compounds, *J. Chem. Phys.* **78**, 2017–2029.
79. Inokuchi, H., Wakayama, N., Kondo, T., and Mori, Y. (1967) Activated adsorption of hydrogen on aromatic-alkali-metal charge-transfer complexes, *J. Chem. Phys.* **46**, 837–842.
80. Watanabe, M., Tachikawa, M., and Osaka, T. (1997) On the possibility of hydrogen intercalation of graphite-like carbon materials-electrochemical and molecular orbital studies, *Electrochimica Acta* **42**, 2707–2717.
81. Wang, Q., Challa, S. R., Sholl, D. S., and Johnson, J. K. (1999) Quantum sieving in carbon nanotubes and zeolites, *Phys. Rev. Lett.* **82**, 956–959.
82. Beenakker, J. J. M., Borman, V. D., and Krylov, S. Y. (1995), Molecular transport in subnanometer pores: zero-point energy, reduced dimensionality and quantum sieving, *Chem. Phys. Lett.* **232**, 379–382.

FROM VAPOR-GROWN CARBON FIBERS (VGCFs) TO CARBON NANOTUBES

M. Endo, Y.A. Kim, T. Matusita, T. Hayashi

Faculty of Engineering, Shinshu University
Wakasato4-17-1, Nagano 380-8553, Japan
e-mail: endo@endomoribu.shinshu-u.ac.jp

Abstract. Careful growth control during chemical vapor deposition makes it possible to obtain various morphologies of carbon fibers: from normal vapor-grown carbon fibers (VGCFs) through submicron VGCFs to nanofibers or carbon nanotubes. The combined effects of excellent physical properties and low production costs have spurred applications research for these fibers in various fields.

1. Introduction

Carbon fibers may be defined as a one-dimensional filament morphology consisting of sp^2-bonded graphitic carbon oriented along an axis parallel to the basal plane, and having a length-to-diameter ratio, "aspect ratio", of more than 100. Fibers are frequently classified in terms of their precursor materials, since their structure, microstructure, and properties devolve from their precursor materials and subsequent processing steps, including heat treatment at elevated temperatures. Besides vapor-grown carbon fibers [1,2], the main categories of carbon fibers are polymer precursor-based fibers fabricated from either PAN (polyacrylonitrile) or pitch.

VGCFs obtained by the pyrolysis of hydrocarbons such as benzene or methane at temperatures around 1100°C using ultra-fine metal particles catalysts are a new class of carbon fibers that are distinctively different in method of production, unique physical properties, and prospect for low-cost fabrication. The most interesting feature of these VGCFs is that the diameters of the fibers are controllable over a wide range by changing the manufacturing conditions (Figure 1). The distribution of diameters strongly affects the physical and chemical properties of carbon fibers. Moreover, these fibers typically are discontinuous, which enables one to prepare interesting composites with plastic, metal, or ceramics. Therefore, it is very important to understand the manufacturing parameters to obtain carbon fibers with optimum properties for each specific application. Among these parameters, the carrier gas, flow rate, and injection conditions all have major effects on the resultant carbon fibers obtained.

As prepared VGCFs have circular cross-sections of micrometer size with a central hollow tube of diameter less than several tens of nanometers. The graphene planes are arranged like the annual rings of a tree, and are accordingly parallel to the fiber axis [4]. Surprisingly, thick VGCFs with a diameter of ~15μm have a core structure very similar to that of arc-grown carbon nanotubes (Figure 2) [3]. A VGCF is identical to a carbon nanotube in the nanoscale domain of initial formation, with a highly graphitic structure arising from a high degree of preferred orientation of the basal planes along the fiber axis. VGCFs are dissimilar to arc-grown carbon nanotubes in that a

51

L.P. Biró et al. (eds.), Carbon Filaments and Nanotubes: Common Origins, Differing Applications, 51–61.
© 2001 *Kluwer Academic Publishers. Printed in the Netherlands.*

52

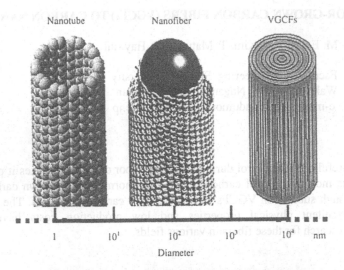

Figure 1. Schematic morphologies from VGCFs to Carbon Nanotubes

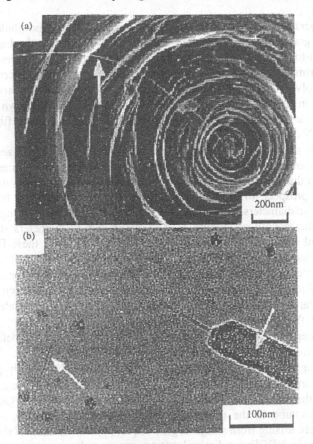

Figure 2. FE-SEM (a) and TEM (b) photographs showing the core tube in
the cross section of VGCFs

catalyst is routinely used to initially form the fibers, and, typically, the catalyst particle remains at the tip of the fiber during growth.

Although carbon nanotubes can be grown in a variety of ways, one synthesis method based on chemical vapor deposition is similar to that which has been used for the growth of VGCFs. In this chapter, we compare carbon nanotubes with VGCFs in terms of the synthesis conditions, microstructure, range of diameters, etc.

2. Synthesis of VGCFs and Carbon nanotubes (PCNT) by the pyrolysis of hydrocarbons

VGCFs may be obtained through the decomposition of benzene at 1150-1300°C in an electric furnace in the presence of H_2 as the carrier gas (see Figure 3) [1]. Ultra-fine particles of Fe (ca. 10nm diameter) or its compounds, specifically $Fe(NO_3)_3$ or ferrocene, may be introduced into the chamber as a catalyst. Hydrocarbon decomposition will take place at the catalytic particle, leading to a continuous carbon uptake by the catalytic particle and a continuous output by the particle of well-organized tubular filaments of the carbon honeycomb lattice (see page ?). The rapid growth rate of several tens of μm/min or more, 10^6 times faster than for the growth of common metal whiskers, allows commercially available systems to produce useful quantities of vapor grown carbon fibers. Evidence in support of this growth model is the presence of catalytic particles in the resulting thin vapor grown carbon fibers, as shown in Figure 4. The primary fiber is firstly formed by the catalytic process (with a diameter of several nanometers), and the fiber is then thickened by a CVD (chemical vapor deposition) process in which carbon layers are deposited on the primary core filament.

It is known that the temperature of the chamber and the partial pressure of the benzene vapor need to be controlled for effective VGCF growth. The preparation of carbon nanotubes from benzene or hydrocarbons is essentially similar to that used for VGCFs described above, but without a significant wall-thickening stage. Thus nanotubes can be regarded as the precursor for VGCFs, but with a diameter less than 1nm. An apparatus, basically similar to that shown in Figure 3 can be employed for nanotube preparation but it is most crucial to exercise meticulous control over parameters such as temperature, gas residence time, benzene partial pressure, hydrogen gas flow rate and so on. It has been reported that pyrolysis of hydrocarbon vapor in the presence of a Fe catalyst at 1040°C yields carbon nanotubes with 2-3 nm diameters [3]. Thermal decomposition of a gaseous mixture of benzene and H_2 similar to that used in VGCF preparation yields multi-walled nanotubes; heating them to 2800°C (Figure 5) reveals at least two coaxial tubes. Such uniform growth of small diameter nanotubes requires the partial pressure of benzene to be as low as possible.

These as-grown and heat-treated samples were observed by high-resolution transmission electron microscopy (HR-TEM), field emission scanning electron microscopy (FE-SEM), scanning tunneling microscopy (STM) and Raman spectroscopy to distinguish their structures.

54

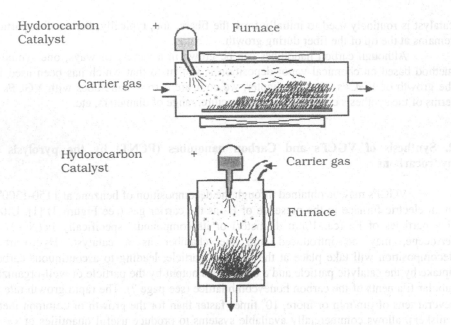

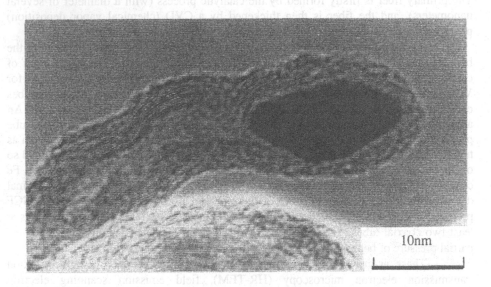

Figure 3. Two floating catalytic particle systems for volume growth of vapor-grown carbon fibers in a reaction chamber

10nm

Figure 4. Very early stage of fiber growth in which the catalytic particle is still actively promoting elongation. The primary fiber thus formed acts as a core for vapor grown carbon fibers

Figure 5. A HR-TEM image of a heat-treated carbon nanotube (and an enlarged insert), showing a nanotube without any deposited carbon.

3. Comparison of VGCFs and pyrolytic carbon nanotubes

Because of the structural similarities between carbon nanotubes and VGCFs, similar properties might be expected, and this expectation is partially substantiated by experimental observations. Some notable exceptions are the electronic structure and transport properties of carbon nanotubes, which are dominated by a quantum size effect, and the mechanical properties, which show a remarkable flexibility on bending (Figure 6). In contrast to this, carbon fibers, which are more brittle, tend to crack when subjected to bending forces

Though much smaller in diameter, multi-walled carbon nanotubes bear a close resemblance to VGCFs in microstructure and morphology, and often are grown concurrently with them, as shown in Figure 7, which depicts a carbon nanotube of ~3 nm diameter along with a VGCF. Such VGCFs tends to have rather large hollow cores and thick tube walls comprised of well-organized carbon layers, whereas a carbon nanotube tends to have very thin walls consisting of several perfect graphene cages. Carbon nanotubes frequently have sections of outer surface which are bare as well as other section which are covered with deposited carbon layers (Figure 8). At the tips of carbon nanotubes there is no evidence of metal particles, even immediately after growth, in contrast to the tips of VGCFs. The larger size of the cores and the presence of opaque particles at the tips of VGCFs suggest a different growth mechanism in which the catalyst is indispensable [4].

For a range of materials obtained by carefully varying growth conditions from those appropriate for VGCFs through nanotubes, the inter-shell spacings (d_{002}) vary from 0.34 nm to 0.36 nm, while the ideal 100 spacing is 0.213 nm, as for crystalline graphite. It is found that the median inter-shell spacing decreases as the tube diameter increases, as shown in Figure 9. This result suggests that for tubes with larger diameters the energy gained by the local graphene stacking is higher than the energy introduced by the defects associated with the stacking.

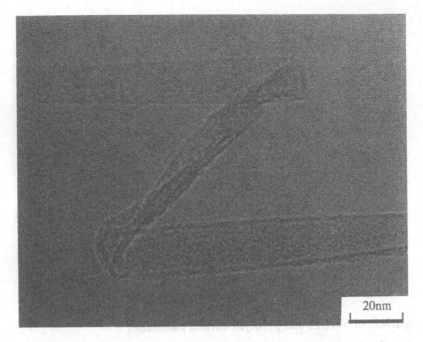

Figure 6. A HR-TEM image of a bent and twisted carbon nanotube , showing its flexibility.

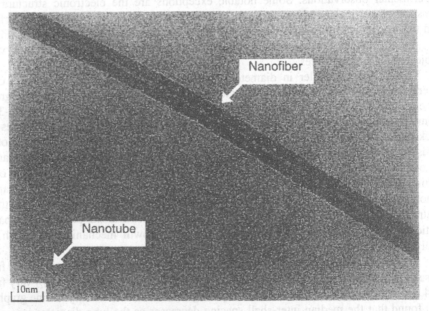

Figure 7.A TEM photograph showing a straight carbon nanofiber and a carbon nanotube grown simultaneously

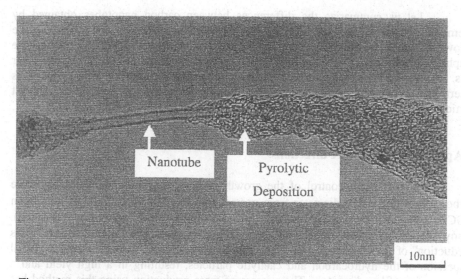

Figure 8.A TEM photograph of a vapor-grown carbon nanotube at an early stage of the thickening deposition

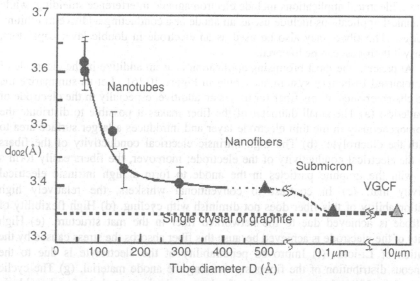

Figure 9. Interlayer spacing (d_{002}) as a function of tube diameter D.

Let us summarize the differences between carbon nanotubes obtained by chemical vapor deposition and VGCFs: (1) The diameter of carbon nanotubes must be below 30 nm, thin enough so that no faceting phenomena may occur during the graphitization process. (2) Graphene layers in nanotubes are aligned along the fiber axis. (3) Nanotubes have no micro-domains of amorphous carbon on their surfaces, as determined by scanning tunneling microscopy. (4) Nanotube tips contain no catalyst particles.

4. Applications and future directions

Through exact control of the growth process, it is possible to manufacture carbon fibers ranging in diameter from normal VGCFs (~10μm), through submicron VGCFs (~0.1μm), through nanofibers (30~150nm) to carbon nanotubes (< 30nm). Many researchers have contributed their efforts to reach the final goal of "mass production" of these fibers at low cost. The floating reactant method allows a spatial dispersion of the hydrocarbon and catalytic particles, resulting in a high yield and a rather uniform fiber diameter. The success of mass production using this method has spurred application research in various fields. The combined effects of excellent physical properties and low production costs promise both military and commercial applications.

The high graphitizability and surface to volume ratio caused by their small diameters make these fibers potentially useful for a number of applications. For example, they may be used as reinforcements in injection molded plastics or cement composites. Electrical applications include electromagnetic interference shielding, while electrochemical applications include use as an anode or a conducting additive in lithium-ion batteries. The fibers may also be used as an electrode in double layer capacitors, particularly if their cost can be lowered.

At present, the most promising application is as an additive in the electrode of a commercialized Li-battery system, as shown in Figure 10 [6]. Let us summarize the desirable characteristics of this fiber for use as an additive, especially in the electrode of Li-ion batteries: (a) The small diameter of the fiber makes it possible to distribute the fibers homogeneously in the thin electrode layer and introduces a larger surface area to react with the electrolyte. (b) The high intrinsic electrical conductivity of the fibers improve the electrical conductivity of the electrode; moreover, the fibers easily form a network with the graphite particles in the anode to form a high intrinsic electrical conductivity mat. (c) In contrast to conventional whiskers, the relatively high intercalation ability of this fiber does not diminish with cycling. (d) High flexibility of the electrode is achieved due to the networked fiber in the mat structure. (e) High cyclability of the electrode is achieved because this fiber absorbs the stress caused by the intercalation of Li-ions. (f) Improved penetrability of the electrolyte is due to the homogeneous distribution of the fibers surrounding the anode material. (g) The cyclic efficiency of the Li-ion battery is superior for long cycle times to that of carbon black. The core of a vapor-grown carbon fiber, whose morphology resembles a carbon nanotube (see Figure 2), was employed as the tip of scanning tunneling microscope (STM). The VGCF, with a carbon nanotube protruding, was attached to a standard platinum iridium tip, using silver paste as the binder (Figure 11). The key element in this system is the extruded central nanocore (shown on the right of Figure 11), which interacts with a sample. When a carbon nanotube is used as the tip, it is difficult to

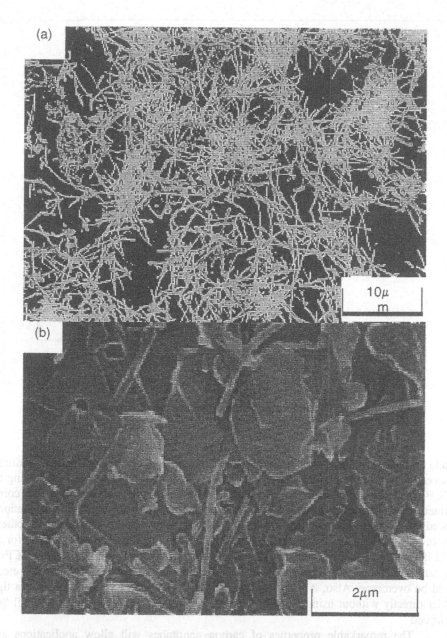

Figure 10. (a) SEM photographs of submicron VGCFs obtained by the floating catalysis process and (b) graphite flakes mixed with the fibers in a commercial cell

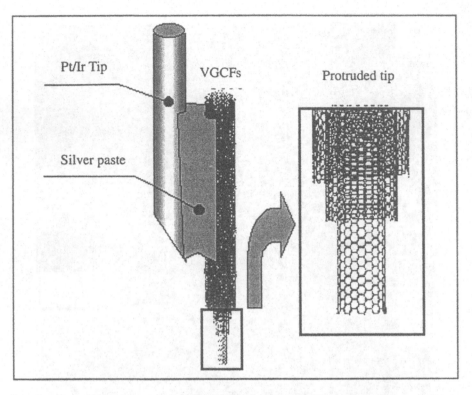

Figure 11. A schematic diagram of the VGCF-tip assembly

obtain an image having atomic resolution, as there are relatively large thermal vibrations observed at room temperature. It is expected that this problem can be solved by using a VGCF tip because the sheath has a high thermal conductivity and surrounds the core filament, absorbing and dissipating the thermal vibrations. Moreover, the preparation procedure used in this application was rather simple compared with those of other nanotube tips. Thus, this study shows that the core of a VGCF can be used as a tip for a STM if some technical problems such as the binding strength between Pt/Ir and VGCFs, the cutting method, optimization of the conditions for acquiring the STM images, etc., could be overcome. Also, it is expected that when VGCF-tips can be attached to a tip holder directly without using the Pt/Ir brace, the quality of the image obtained will be improved

The remarkable properties of carbon nanotubes will allow applications as lightweight structural fibers of high ultimate strength, conducting or semiconducting components in electronics, and as connectors between macroscopic and nanoscopic size regimes [7]. Therefore, it is critically important to invent new industrial methods for inexpensive production of high quality carbon nanotubes. In this sense, large-scale synthesis of carbon nanotubes through chemical vapor deposition is a very promising technique.

In order to exploit the enormous potential of nanofibular materials in the near future, these fibers must successfully compete with other types of carbon fibers in many future applications areas

Acknowledgements

This work was partially supported by the Research for the Future (RFTF) Program "Nano carbon for advanced energy devices" of the Japan Society for the Promotion of Science (JSPS),.

References

1. Endo, M., Koyama, T. and Hishiyama, Y. (1976) Structural Improvement of Carbon Fibers Prepared from Benzene, *Japanese Journal of Applied Physics*, 15, 2073-2076.
2. Endo, M. (1988) Grow Carbon Fibers in the Vapor Phase, *CHEMTECH*, 18, 568-576.
3. Endo, M., Takeuchi, K., Kobori, K., Takahashi, K., Kroto, H.W. and Sarkar, A. (1995) Pyrolytic carbon nanotubes from vapor-grown carbon fibers, *Carbon*, 33, 873-881.
4. Dresselhaus, M.S., Dresselhaus, G., Sugihara, K., Spain, I.L. and Goldberg, H.A. (1988) *Graphite Fibers and Filaments*, Springer-verlog, Berlin Heidelberg New York London Paris Tokyo.
5. Endo, M., Takeuchi, K., Hiraoka, T., Furta, T., Kasai, T., Sun, X., Kiang, C.H. and Dresselhaus, M.S. (1997) Stacking nature of graphene layers in carbon nanotubes and nanofibers, *Journal of Physical Chemistry Solids*, 58, 1702-1712.
6. Showa denko's catalog, (1997) Fine Carbon, V.G.C.F.
7. Dresselhaus, M.S., Dresselhaus, G. and Eklund, P.C. (1996) Science of Fullerenes and Carbon Nanotubes, Academic Press (1996).

NUCLEATION AND GROWTH OF CARBON FILAMENTS AND VAPOR-GROWN CARBON FIBERS

GARY G. TIBBETTS

General Motors Research & Development Center
30500 Mound Rd., Warren, MI, 48090-90555, USA

Abstract

Fifty years of research on the growth of carbon filaments has elucidated many aspects of filament production and worked out mechanisms that provide guidance for experiment and invention. The role of the catalytic particles has been investigated in many ways. It is known that the particle must dissolve carbon, and that any carbide phases formed should be of low stability, so as to minimize competition with the filament growth process. The particle must also become highly supersaturated in carbon in order that filament nucleation can occur. A useful growth procedure must optimize the chances of growing filaments of the proper length, diameter and morphology. The catalytic particle must become highly supersaturated in carbon in order that filament nucleation can occur. Since filaments are only one of a variety of structures that may be precipitated by the catalytic particle to relieve its supersaturation, the growing conditions must be carefully optimized. Encouraging filament growth requires that small diameter catalyst particles be protected from aggregation.

Cessation of filament growth may be due to the deposition of a passivating layer of carbon from the vapor phase or to a simple phase change in the catalytic particle.

1. Introduction

Despite the immense literature that has been published on the growth of carbon filaments [1], quantitative studies of the nucleation and growth mechanisms have been rare. It is easy to forgive the earliest observers for simply marveling at and cataloguing the beautiful and exotic objects they found, but it is disturbing that nearly 50 years after the original microscopic observation of carbon filaments [2], so little quantitative and conclusive work has been published. It is sadly true that carbon filament research papers at the taxonomic level are still submitted and published.

As a consequence of the lack of generality and rigor achieved in this field, it has been hard for the new practitioners of nanotube growth to fit their discoveries into perspective with the body of established knowledge. Fundamental questions such as "Are catalytic particles necessary for carbon filament growth?" and "How large

63

L.P. Biró et al. (eds.), Carbon Filaments and Nanotubes: Common Origins, Differing Applications, 63–73.

must a catalytic particle be to grow a carbon filament?" can still inspire heated argument.

In this chapter, we will begin by describing the growth mechanism for carbon filaments. We will then attempt to describe in quantitative terms the kinetics of the lengthening and the poisoning of growing filaments.

2. A growth mechanism

We will begin by considering a more or less classical interpretation of filament growth proposed by Baker *et al.* [3] and Lobo *et al.* [4], and further elaborated by Rostrup-Nielsen and Trimm [5]. Consider a metallic catalyst particle smaller then 1 µm in diameter exposed to a gas containing available carbon atoms, such as CO/CO_2 mixtures, or various hydrocarbon gases. Because of the slow kinetics of gas phase decomposition, it is quite easy to produce a gas mixture at near 1000°C and one atmosphere whose methane or CO composition is many times larger than the equilibrium concentration [6].

The high activity of carbon in the gas phase drives carbon to dissolve in the catalyst particle. The particle may thus easily become supersaturated with carbon atoms arising from the decomposition of vapor phase molecules containing carbon (Figure 1).

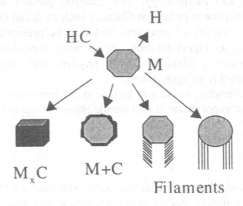

Figure 1. A metallic particle **M** saturated with carbon by the gaseous atmosphere can convert to the carbide phase, precipitate a layer of carbon, or grow filaments of several morphologies

For a particle with more carbon atoms dissolved in it than allowed by carbon's solubility, several avenues are open to lower its chemical potential. It may either form a carbide phase or precipitate carbon in some manner, preferably in the graphite phase.

Stable carbide phases are therefore the competitors of filament formation, and elements that form stable carbides are poor filament formers. For both Ni and Fe particles, the difficulty of forming carbides and their instability ensure that many particles remain supersaturated until they precipitate carbon.

So how does the carbon precipitation proceed? Graphite is the lowest energy carbon containing precipitate, but the precipitation process must proceed with considerable ordering if the carbon atoms are to crystallize in such a well-defined lattice. When catalyst particles are rapidly raised to a very high supersaturation, our experience is that they rapidly precipitate amorphous or poorly ordered carbon.

Suppose the catalyst particle is raised to an intermediate carbon supersaturation by a carefully mixed gas phase hydrocarbon. How can a long filament form? Figure 1 shows two possible morphologies. In the case where gas phase hydrogen is available to bond to the graphene plane edges, filaments whose graphite planes are canted with respect to the longitudinal axis of the filament [7] may be formed. If hydrogen atoms are not available to stabilize the outer edges of these canted planes, the filament creation process minimizes the exterior surface energy of the graphite precipitate by creating nested cylindrical graphite basal planes [8]. In the latter case, the core is hollow because too much energy is required to precipitate curved planes of very large curvature to occupy the center of the filament. Note that with either of these morphologies, the release of the original supersaturation does not terminate the growth, as the growing filament does not cover most of the particle area. The catalyst particle can continue to work as a tiny chemical plant, decomposing gas phase hydrocarbons, allowing the carbon atoms to diffuse through its bulk, and encouraging the carbon atoms to attach themselves to an ever lengthening graphite precipitate. It is clear that the latter process can survive only if the carbon atom flux to the active interface is between well-defined limits.

Filament growth will conclude when the particle is deactivated through a phase change, through uncontrolled precipitation that covers the active surface, or through gas phase carbon deposition which deactivates the particle surface.

In the sections that follow, we will examine the steps of lengthening and deactivation of the growing filaments in some detail.

3. Lengthening

We will now describe a series of experiments that optimized fiber growth on 12 nm magnetite particles sprayed on alumina substrates in a methane-hydrogen mixture flowing through a tubular reactor. Such experiment could be very useful if it could specify, for a given hydrocarbon/hydrogen mixture, the optimum temperature, hydrocarbon concentration, and gaseous flowrate.

After a lengthy series of experiments with different growth conditions [9], we found that catalyst particle sintering is a dominant effect, and that the exact conditions undergone by the particles before they reach the growth temperature are important to growth.

Under our conditions, we found that the most useful and reproducible procedures for fiber growth required the temperature to be ramped up at a fixed rate $R=dT/dt$ under a flow of inert gas. At a fixed starting temperature T_s, a lengthening gas mixture of methane and hydrogen was introduced at the gaseous flowrate F and the methane concentration $[CH_4]$. The four parameters, R, $[CH_4]$, T_s and F are the experiment's independent variables. The highest temperature reached during growth was 1120°C; in these experiments fiber growth proved to be

66

insensitive to this parameter. The experimental conditions are diagrammed in Figure 2.

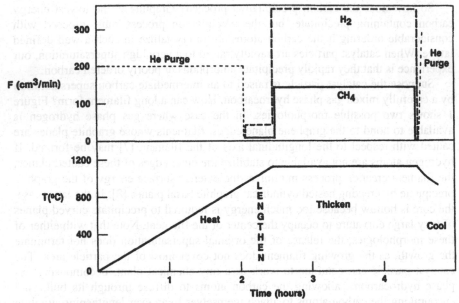

Figure 2. Gaseous flow rates F and temperature T as a function of time for the optimized fiber growth conditions. Duration of lengthening time is 24 min, the time required to heat the furnace from 950°C to 1120°C at 7.1°C/min. (Reprinted from reference 9 with permission from Elsevier Science Ltd.)

The computer controlled furnace was first heated as rapidly as possible (7.5°C/min) to 950°C under flowing He purge gas. To grow the optimum mass of fibers, lengthening begins at 950°C, when a mixture of 8 cc/min of methane and 42 cc/min hydrogen was introduced. During all the lengthening period the furnace warmed up at the slightly lower rate of 7.1°C/min. When the thickening temperature of 1120° was reached a mixture of 145 cc/min of methane and 355 cc/min hydrogen was introduced in the reactor. The purpose of this step was to thicken the filaments to a diameter thick enough (about 3.2 μm) so that their length and diameter could be readily evaluated under an optical microscope.

Average fiber length was approximated by microscopic observation, and the mass was determined by weighing the fibers after shaving them from the substrate, allowing the total number of fibers grown to be determined.

The first experiments were designed to measure the dependence of fiber mass and length on methane concentration. The results of two sets of experiments, both of which were performed at a total flow rate of 50 cc/min, are shown in Figure 3.

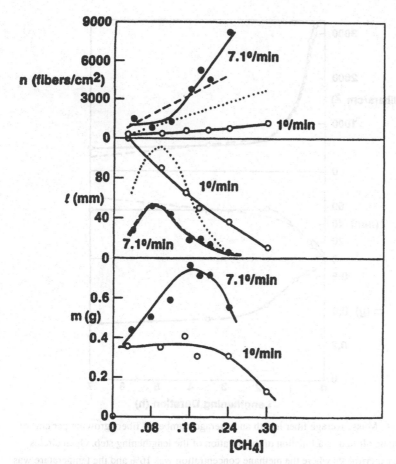

Figure 3. Mass, average fiber length, and average number of fibers growing per cm² of substrate plotted as a function of methane concentration. Solid circles describe experiments where the temperature was ramped form 950° to 1120°C at 7.1°C/ min. Open circles describe experiments where the temperature was ramped from 920°to 1120°C at 1°C/min. Modeling calculations are dotted and dashed lines. (Reprinted from reference 9 with permission from Elsevier Science Ltd.)

One set (dark circles) was obtained with *R*= 7.1°C/min, with the methane/hydrogen mixture first provided at 950°C. Another set (open circles) was obtained at R=1°C/min with the lengthening mixture first provided at 920°C.

These data show that low methane concentrations give long fibers, but poor nucleation, while high methane concentrations give short but densely nucleated fibers. The more rapid temperature sweep rate produces shorter fibers, but it grows more of them. The optimum mass for a temperature sweep rate of 7.1°C/min is obtained near 16% methane concentration.

Figure 4 shows experiments explicitly designed to investigate the effect of changing the rate of temperature increase.

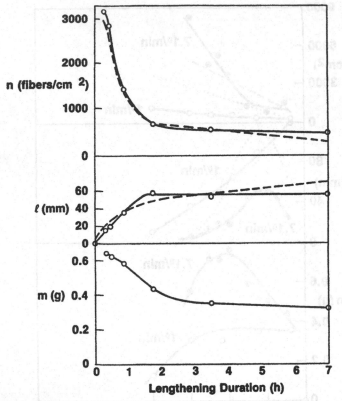

Figure 4. Mass, average fiber length and average number of fibers growing per cm² of substrate plotted as a function of the duration of the lengthening step. Open circles describe experiments where the methane concentration was 16% and the temperature was linearly ramped from 920°to 1120°C. Modeling calculations are the dashed lines. . (Reprinted from reference 9 with permission from Elsevier Science Ltd.)

All of the experiments were performed at 16% methane concentration, a total flow rate of 50 cc/min, and lengthening gas flow beginning at 920°C. The abscissa is the time elapsed in ramping between 920°C and 1120°. Despite the fact that 1.4 hours are required before the filaments reach their maximum length under these conditions, fiber mass is maximum for very rapid growth experiments. The enhancement in the number of fibers at very rapid temperature ramps more than compensates for the short fibers produced. The apparatus did not allow more rapid heating than 7.5°C/min.

Figure 5 shows the effect of varying the temperature at which the methane-hydrogen lengthening gas is first provided. These experiments were all performed at a total flow rate of 50 cc/min, a methane concentration of 16%, and a temperature ramp rate of 7.1°C/minute. The abscissa shows the temperature at which the methane mixture begins to flow into the growth tube (lengthening start temperature).

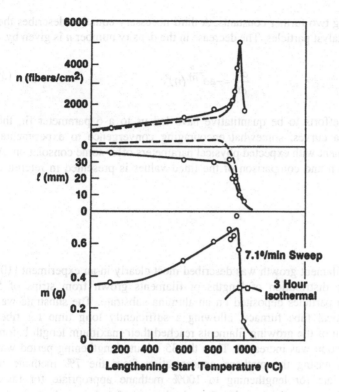

Figure 5. Mass, average fiber length, and average number of fibers growing per cm² of substrate plotted as a function of the temperature at which the lengthening gas mixture first flows into the reactor. Open circles describe experiments where the methane concentration was 16% and the temperature was linearly ramped from the start temperature to 1120°C at 7.1°C/min. Modeling calculations are the dashed lines. (Reprinted from reference 9 with permission from Elsevier Science Ltd.)

The center panel shows that rapid fiber lengthening begins at about 900°C. Starting the lengthening period below that temperature only increases the fiber length in a minor way. However, delaying lengthening beyond 900°C critically reduces the duration of lengthening and hence total length. The nucleation density shows a sharp peak at 965°C.

Modeling calculations incorporating the following key equations are superimposed on the data, showing various degrees of accuracy. The equation for filament lengthening rate,

$$\frac{dl}{dt} = C(CH_4)^2 e^{-\frac{B}{RT}},$$

(1)

uses constants C and B with R as the gas constant. The quadratic behavior arises from the desirability of forming a more reactive C-C hydrocarbon for rapid decomposition at the catalyst particle. The duration of lengthening τ is given by

$$\tau = \tau_0 e^{-\frac{E}{RT}} e^{-A(CH_4)},$$

(2)

with A and E being two further constants. A third necessary equation describes the sintering of the catalyst particles. The decrease in the density number n is given by

$$\frac{dn}{dt} = -ae^{-\frac{b}{RT}}(n)^2 .$$ (3)

Although our efforts to be quantitative have led us to a 6 parameter fit, the abundance of data curves, somewhat encouraging convergence to experimental values, and agreement with expected physical parameters offer some consolation. A complete exposition and comparison of the fitted values is presented in reference [9].

4. Deactivation

The cessation of filament growth was described most clearly in an experiment [10] that measured the distribution of lengths of filaments grown from strips of 5 different Fe based particles deposited on an alumina substrate. The substrate was inserted in a vertical tube furnace, allowing a sufficiently long time for fiber lengthening that all of the growing filaments reached their maximum length before methane concentration was increased. Then the 35 minute lengthening period was terminated by increasing the methane concentration from the 7% methane in hydrogen appropriate for lengthening to 100% methane appropriate for rapid thickening. The fibers were thickened to 7 μm in diameter for easy evaluation with an optical micrograph (Figure 6). The entire experiment was performed at 1054°C

A diligent tabulation of the length histogram of all the fibers longer than 0.3 mm showed that the fiber length histogram (Figure 7) may be fitted to an exponential function describing the number of fibers n having a given length l as proportional to e^{-cl}. This functional form is really a consequence of the growth and poisoning rates of the filaments. Previous work has showed that during periods of filament growth, the lengthening velocity is approximately constant. Thus, the increase in filament length dl in time dt is given by

$$dl / dt = \beta \quad \text{for } t < t_c.$$ (4)

After a time t_c, the filament ceases growth, so that

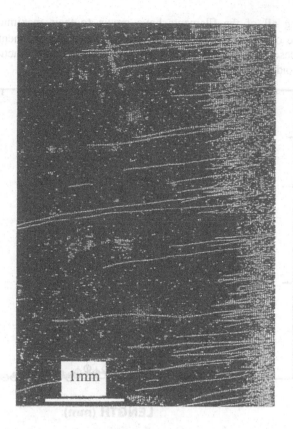

Figure 6. SEM of fibers growing from the metal particle painted strip visible along the right hand edge of the photo. (Reprinted from reference 10 with permission from Elsevier Science Ltd.)

$$dl / dt = 0 \ \text{ for } t \geq t_c. \tag{5}$$

Now consider a population of growing filaments. If the probability per unit time that any filament stops growing has a fixed value α then the number of filaments $-dn$ ceasing their growth in a period dt will be

$$- dn / dt = \alpha n. \tag{6}$$

Thus, if N filaments begin growing at t=0, then the population still growing at t will be

$$n = Ne^{-\alpha t}. \tag{7}$$

Expressed in terms of length, the number of filaments that grow to a length l (from equation 1) will be given by

$$n = Ne^{-(\alpha/\beta)l}. \tag{8}$$

Assuming all of the filaments have grown to their maximum lengths, this describes the observed histogram with $c = \alpha/\beta$. Thus, the exponential fiber length histogram describes a situation where only a negligible fraction of growing filaments avoids poisoning and grows to maximum length.

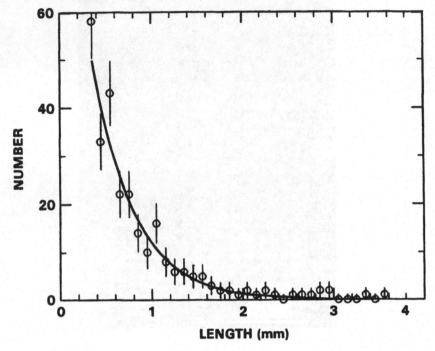

Figure 7. Fiber length distribution histogram for all measured fibers. Solid line is a least squares exponential fit to the data points

We have also shown [11] how the rate of poisoning of growing filaments is related to the rate of deposition of carbonaceous layers on a substrate.

5. Generalizations

We will conclude this chapter with a general guide to fiber growth that we hope will be useful to experimenters. With the general caveats which always must accompany such generalizations:

1. For longer fibers, lower hydrocarbon concentrations or decrease temperatures.

2. For more profuse fibers, increase hydrocarbon concentrations or temperatures.

3. Poisoning of catalyst particles by deposition from the vapor phase can result in an exponentially shaped fiber length histogram.

6. Acknowledgements

Thanks for experimental help to M. G. Devor, E. J. Rodda, and D. W. Gorkiewicz. Thanks also to T. W. Lorenzen for helpful discussions.

7. References

1. Baker, R. T. K., and Harris, P. S. (1978) The formation of filamentous carbon, *Chemistry and Physics of Carbon*, Vol. 14, p. 83-165, P. L. Walker and P. A. Thrower (eds.), Dekker, New York (1978).
2. Davis, W. R., Slawson, R. J., and Rigby, G. R. (1953) An unusual form of carbon, *Nature* **171**, 756.
3. Baker, R. T. K., Barber, M. A., Harris, P. S., Feates, F. S., and Waite, R. J. (1972) Nucleation and growth of carbon deposits from the nickel catalyzed decomposition of acetylene, *J. Catalysis* **26**, 51-62.
4. Lobo, L. S., Trimm, D. L., and Figueiredo, J. L. (1973) Kinetics and mechanisms of carbon formation from hydrocarbons on metals, in J. W. Hightower (ed.) *Proc. 5th Int. Congress on Catalysis*, North Holland, Amsterdam, Vol. 2, 1125-1135.
5. Rostrup-Nielsen, J., and Trimm, D. L. (1977) Mechanisms of carbon formation on nickel-containing catalysts, *J. Catalysis* **48**, 155-165.
6. Audier, M., Coulon, M., and Bonnetain, L. (1983) Disproportionation of CO on ion-cobalt alloys-III, *Carbon* **21**, 105-110.
7. Nolan, P. E., Lynch, D. C., and Cutler, A. H. (1998) Carbon deposition and hydrocarbon formation on group VIII metal catalysts, *J. Phys. Chem. B* **102**, 4165-4175.
8. Tibbetts, G. G. (1984) Why are carbon filaments tubular? *Jour. Cry. Growth* **66**, 632-637.
9. Tibbetts, G. G. (1992) Growing carbon fibers with a linearly increasing temperature sweep: experiments and modeling, *Carbon* **30**, 399-406.
10. Tibbetts, G. G. (1985) Lengths of carbon fibers grown from iron catalyst particles in natural gas, *Jour. Crystal Growth* **73**, 431-438.
11. Tibbetts, G. G., Devour, and M. G., Rodda, E. J. (1986) An adsorption-diffusion isotherm and its application to the growth of carbon filaments on iron catalyst particles, *Carbon* **25**, 367-375.

CARBON NANOTUBES FORMATION IN THE ARC DISCHARGE PROCESS

A. FONSECA AND J. B.NAGY

Laboratoire de Résonance Magnétique Nucléaire, Facultés Universitaires Notre-Dame de la Paix, 61 rue de Bruxelles, B-5000 Namur, Belgium

Abstract. The different aspects of the arc discharge production of carbon nanotubes are reviewed. Most of the attention is focussed on the production of single walled nanotubes. The purification techniques applied to nanotubes are also briefly described.

1. Introduction

The electric arc discharge was introduced in the early sixties by R. Bacon [1] to produce carbon fibers called whiskers. The same technique was adapted in 1990 by Krätsmer and Huffman [2] to produce fullerenes in good yields.

Multiwalled carbon nanotubes (MWNTs) were found by Iijima [3] in the cathode deposit obtained by DC arc discharge evaporation of pure graphite rods in He gas, during the observation of fullerene soots. Later on, Ebbesen and Ajayan [4] optimized the arc discharge technique to produce carbon nanotubes in high yields.

In 1993, single walled carbon nanotubes (SWNTs) were synthesized independently by Iijima [5] and Bethune [6]. They can be produced nowadays by three methods: laser evaporation of composite containing carbon and metallic catalyst (Ni, Co, Fe, Y and other), evaporation of an anode rod of the same composition in the DC arc discharge [7] and by catalytic chemical vapour deposition (CCVD) [8, 9]. More recently, SWNTs could be synthesized in large scale both by the arc discharge technique [10] and by the CCVD method [11].

SWNTs present a state of carbon as graphite sheets of one single atom thickness which form microporous structure. The diameter of SWNTs (size of pores) varies between approx. 0.7 - 1.8 nm, depending on the growth conditions [12, 13]. The specific surface area of SWNTs approaches to the upper limit of the value any carbon material may reach. Its estimation for SWNTs (regarding only one side of the rolled sheet) exceeds 1300 m^2/g. Due to its micropore structure and high specific surface area, SWNTs should find a wide application in techniques such as: storage systems for ecological fuels such as hydrogen and methane, effective adsorbents for chromatography including separation of organic isomers, adsorbent for concentrating of substances in analysis of environmental pollutions, electrodes in chemical batteries,...

Even before their discovery, SWNTs were predicted to be metallic or semiconducting depending on their diameter and chirality [14]. Electronic transport measurements on SWNTs revealed that they were a new class of quantum wires [15]. MWNTs have also remarkable electric, thermal and elastic properties. They can be used as bright long-life field emitters for cathode ray tubes [16].

L.P. Biró et al. (eds.), Carbon Filaments and Nanotubes: Common Origins, Differing Applications, 75–84.

76

In this paper, the different aspects of the DC arc discharge production of carbon nanotubes and their purification techniques are reviewed.

2. Principles

Two graphite rods are generally used as electrodes, the anode being of lower diameter (ca. 6 mm) than the cathode (ca. 10 mm), in most cases (Figure 1).

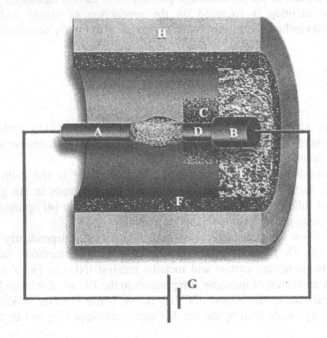

A: Anode
B: Cathode
C: *Collarette*
D: *Deposit*
E: *Web-like soot*
F: *Soot*
G: DC power supply
H: Water-cooled double wall reactor

Figure 1. Schematic representation of the DC arc discharge experimental set-up and possible products (*soot, web-like soot, deposit* and *collarette*).

The anode is either pure graphite or contains metals. In the latter case, the metals are mixed to graphite powder and introduced in a hole made in the anode center. The synthesis is carried out at low pressure (ca. 30-130 Torr [16] or 500 Torr [7]) in controlled atmosphere composed of inert and/or reactant gas. The distance between the electrodes is reduced until the flowing of a current (ca. 50-150 A).

The temperature in the inter-electrode zone is so high that carbon sublimes from the positive electrode (anode) that is consumed. A plasma is formed between the electrodes. The latter plasma can be stabilized for a long reaction time by controlling the distance between the electrodes by means of the voltage (ca. 25-40 V) control. The reaction time varies from 30-60 seconds [16] to 2-10 minutes [7].

The C_n^{m+} ions present in the plasma are attracted to the cathode where they condense to form carbon nanotubes and graphite carbon nanoparticles (GNP). When the arc discharge is carried out in the presence of metals, M_x^{y+} ions are also present in the plasma, so that metal nanoparticles (MNP), metal filled MWNTs (FMWNTs) and metal filled GNP (FGNP) are also formed at the cathode.

3. Production

In the arc discharge method, it has been observed that by changing parameters of the process not only MWNTs but also SWNTs can be produced. The physical and chemical factors impacting on the arc discharge process are the carbon vapour concentration, the carbon vapour dispersion in inert gas, the temperature in the reactor, the composition of catalyst, the addition of promoters and the presence of hydrogen. These factors affect the nucleation and the growth of the nanotubes, their inner and outer diameters and the type of nanotubes (SWNTs, MWNTs).

3.1. MWNTs

The graphite evaporation by a DC arc discharge in the presence of different gases [3, 16-19] produces MWNTs and carbon nanoparticles. The latter are undesired products and they are difficult to eliminate. However, some preliminary results have been obtained to elucidate the influence of hydrogen addition in the atmosphere on the process of nanotube's synthesis under arcing conditions [20]. The structure of MWNTs fabricated by carbon evaporation at high gas pressure has also been studied by the same authors [20].

It was found that the carbon nanoparticles content is reduced if the DC arc discharge is carried out in the presence of a hydrocarbon gas [21], the decomposition of which produces hydrogen. The lowest carbon nanoparticles content in the crude MWNTs production is achieved when the gas is pure hydrogen [16, 22-23].

The microstructure of the condensed products which are formed in the process of carbon vapour condensation has been thoroughly studied [7] :

- When no catalyst is used, only the *soot* and the *deposit* are formed (Figure 1). The soot contains fullerenes but no carbon nanotubes, while MWNTs are found in the *deposit*, together with GNP. Most of the MWNTs are found in aligned bundles, along the electric field, in the inner core of the *deposit*. The inner diameter of the MWNTs varies from 1 to 3 nm and the outer diameter varies from 2 to 25 nm, depending on the number of walls (ca. 10). Most of the MWNTs do not exceed 1 μm in length and have closed tips.

- When metal catalysts are co-evaporated with carbon in the DC arc discharge, the macroscopic and microscopic structure of the products are affected depending on the catalyst (Table 1). Hence, the *soot* and the *deposit* are formed together with a *collarette* and some *web-like soot* both around the *deposit* (Figure 1). The *deposit* and, mainly its core, contains MWNTs, and FMWNTs, together with more or less GNP, FGNP and MNP, depending on the catalyst (Table 1). The *soot* varies from powder-like to spongiest and may contain MWNTs, FMWNTs and SWNTs, depending on the catalyst. The latter SWNTs have closed tips, are free of catalyst and are isolated or in bundles by tens (ca.

10-20 nm thick). Most of the SWNTS have diameters of 1.1 - 1.4 nm and are several microns long. The *web-like soot* has the same composition as the spongiest *soot* but it is only formed in the presence of certain catalysts (Table 1). The *collarette* is mainly constituted of SWNTs (up to 80%), isolated or in bundles, but it is only formed in the presence of certain catalysts (Table 1). The rest of the *collarette* is essentially constituted of FGNP but contains also GNP and MNP. The latter nanoparticles have diameters of ca. 2-20 nm.

Table 1. Products obtained using arc discharge, depending on the catalyst, at different localizations in the DC arc discharge reactor.

Catalyst	Products	Localization[a]	Quality	Reference
none	MWNTs, GNP	*deposit*	+++	3
Co, Co/Ni, Co/Y, Co/Fe, Ni, Ni/Y, Ni/Lu, Fe, Mn, Li, B, Si, Cr, Zn, Pd, Ag, W, Pt, Y, Lu	MWNTs, FGNP, GNP, MNP	*deposit*	++	26-37
Sn, Te, Bi, Be, Sb, Pb, Al, In, S, Se, Cd, Gd, Hf	FMWNTs FGNP, GNP, MNP	*deposit*	+	37-39
Mn	FMWNTs FGNP, GNP, MNP	*soot*	+	28-29, 34
Co, Co/Ni, Co/Y, Co/Fe, Co/Pt, Co/Cu, Co/Bi, Co/Pb, Ni, Ni/Y, Ni/Lu, Ni/Mg, Ni/Fe, Ni/Cu, Ni/Ti, Fe	SWNTs FGNP, (GNP, MNP)[b]	*soot*	+	5, 10, 35, 40
Co, Co/Ni, Co/Y, Co/Fe, Co/Pt, Ni/Y, Ni/Lu, Ni/Fe	SWNTs FGNP, (GNP, MNP)[b]	*web-like soot*	++	10, 25-26
La, Ce, Pr, Nd, Tb, Dy, Ho, Er, Y, Lu, Gd	SWNTs FGNP, (GNP, MNP)[b]	*soot*	+	26, 36, 41
Co, Co/Ni, Co/Y, Co/Fe, Ni/Y, Ni/Lu, Ni/Fe	SWNTs, FGNP, (GNP, MNP)[b]	*collarette*	+++	10, 25, 26

[a]: some *soot* and a *deposit* containing MWNTs are also formed in each case.
[b]: minor products.

3.2. SWNTs

The arc discharge method, in comparison with the laser evaporation method, is cheaper and the yield is higher. Unfortunately, even in the most advanced version of its implementation [10, 24] condensed products have a large fraction of non-tubes nanoparticles. Most of the SWNTs are present in the *collarette* (if any) followed by the *web-like soot* (if any) and the *soot*. The collarette can measure up to 5 cm in diameter and weight up to 500 mg.

The effects of all control parameters of the arc process and of composition of catalyst on the yield of SWNTs have been studied (Table 1) and the results clearly showed the inability to succeed with empirical optimization of the process to produce pure SWNTs or pure MWNTs.

Recently, Zhang et al. [42] observed chains of C_{60} molecules present within SWNTs formed by Ni/Y catalysed carbon arc evaporation. Latter-on, Sloan et al. [43] carried out capillary filling of SWNTs by heat treatment in the presence of fullerenes. The latter fullerenes were found to be present as close-packed continuous chains inside of ca. 5-10% of the observed SWNTs [44].

3.3. QUALITY OF THE NANOTUBES

The MWNTs produced by carbon arc are generally of very "good turbostratic quality", meaning that the concentric tubes of each MWNT are continuous in structure and are separated by the graphite interplanar distance (3.4 Å). Their abundance in the *soot* or *deposit* is characterized by the "Quality", improving from + to +++, in Table 1. Very small amount of amorphous carbon is generally found on the external wall of the MWNTs. In most of the cases, GNP, FGNP and MNP are present in the pristine MWNTs samples (Table 1). These nanoparticles are very difficult to eliminate.

The SWNTs are generally found as bundles of tens of nanotubes, arranged in hexagonal lattice, but some isolated nanotubes are also present. The tubes have round shaped closed tips (supposed to be half fullerenes) and are continuous in structure. When the "quality" of the SWNTs is poor (ca. + or ++ in Table 1), soot and amorphous carbon are present along the isolated and bundles of SWNTs. Nanoparticles do also contaminate the pristine SWNTs (Table 1) and their elimination is even more difficult than for MWNTs.

3.4. CARBONIZATION AND GRAPHITIZATION OF NANOTUBES

When carbon materials are heated in inert atmosphere at temperatures of 1600-2000 °C, secondary carbonisation takes place [45]. Applied to carbon nanotubes, the latter carbonisation repairs the structural defects of the nanotubes, whatever their number of walls. Further heating of carbon materials at temperatures of 2000-3000 °C causes a rearrangement of the material called graphitisation [45]. Applied to MWNTs, it suppresses the distortions of the walls, making them more regular. Nevertheless, the MWNTs become of better turbostratic structure but not graphitic because within a MWNT, it is impossible to recover the AB-AB (hexagonal) or ABC-ABC (rhombohedral) characteristic orientation of graphite.

Hence, MWNTs formed by the carbon arc method must be, and are, of very good turbostratic quality because of the high temperature of the *deposit* region where they are formed and graphitised in situ.

3.5. CONVERSION OF SWNTs INTO MWNTs

In 1997, Nicolaev et al. observed the increase of the SWNTs diameter caused by their coalescence under heating at 1400-1500 °C in the presence of hydrogen [46]. Later on, Metenier did the heat treatment of *collarette* samples at temperatures of 1600-2800 °C, in inert atmosphere, converting the SWNTs first into double wall nanotubes and further into MWNTs as the temperature is increased [47]. The main observations when increasing the temperature are :

- 1600-1800 °C: The bundles get separated and the average diameter of the SWNTs increases from 1.4 to 1.6 nm.

- 2000 °C: The average diameter of the SWNTs increases to 2.0-3.7 nm and some MWNTs are observed.

- 2200 °C: The average diameter of the remaining SWNTs increases to 4.0 nm and most of the tubes became MWNTs.

- 2400 °C: Some large SWNTs (ca. 4 nm) are rarely observed but most of the tubes are MWNTs with average inner/outer diameters of 3.5/7.0 nm.

- 2800 °C: Only MWNTs of very good turbostratic quality are observed.

This evidence can explain the absence of SWNTs in the *deposit* that is formed at the heart of the arc discharge, where the temperature exceeds 3000 °C. Moreover, if there in no cooling system close to the arc discharge (Figure 1) very few or no *collarette* and *web-like soot* will be formed. Hence, the cooling system of the arc discharge experimental device is responsible of the low temperature region close to the arc discharge where the vaporised carbon can condense as SWNTs to form the *collarette* and the *web-like soot*.

4. Purification

4.1. MWNTs

The carbon nanoparticles can be removed by oxidation in air at 500 °C [48, 49]. The air oxidation at 500 °C can also be carried out under infrared irradiation for 30 minutes [16, 50]. Purifications by controlled oxidation in solution can also be applied with high efficiency [51]. Note that all the oxidation methods generate MWNTs with open tips.

MWNTs can also be purified and fractionated according to length using a multistep microfiltration process on polycarbonate membranes of various pore sizes [52, 53] or size exclusion chromatography [24, 54]. The components of the crude sample which include MWCNTs, GNP, FGNP and MNP, are suspended in aqueous solution of 1 wt% sodium dodecyl sulfate surfactant and are separated from each other during the filtration.

4.2. SWNTs

The laser evaporation method is marked by low productivity, while its condensed products in optimised conditions contain up to 60-90 weight % of SWNTs [12]. Simple and effective procedures have been developed in order to remove non-nanotubes components out of condensed products [55].

The different oxidation techniques have also been applied with limited success (ca. 1% yield) to the purification of SWNTs. Microfiltration is the most appropriated technique for the purification of SWNTs [12]. It is even more efficient if applied after a mild oxidation of the crude sample by refluxing in nitric acid for 24 hours [7]. Unfortunately, the SWNTs yield of the purification process is lower than 10%.

Purification of SWNTs is a much more complex procedure. SWNT's purification yields small quantities of nanotubes keeping their price very high (at the level of $1000 per gram of pure SWNTs).

5. Conclusions

The arc discharge process can be applied successfully to produce selectively SWNTs or MWNTs. Nevertheless, the production is limited by the size of the electrodes and the scaling-up of the technique remains a problem.

Due to the lack of experimentally supported ideas on the mechanism and kinetics of SWNT's formation it is actually impossible to tell why the diameter of nanotubes depends on the catalyst and why some conditions [20] do favour formation of SWNTs or MWNTs. Hence, as a general rule, temperatures higher than 2000 °C favour the formation of MWNTs, while SWNTs are formed at lower temperatures.

The purification of the crude SWNTs or MWNTs can be achieved using oxidation or microfiltration but the efficiency of the purification techniques is still limited.

As far as applications is concerned, employing materials based on SWNTs in scientific and technical applications is restricted because the diameter and chirality of nanotubes cannot be controlled yet and because there is no cheap procedure of fabrication of such materials. For bulk applications such as polymer reinforcement, thermal and electric conductivity, and field emission, MWNTs that are readily accessible can already be used.

Acknowledgements

This work has been supported by the Belgian Ministry of Research in the framework of the PAI-IUAP program P4/10. The authors are grateful to F. Vallette for the drawings.

References

1   Bacon, R. (1960) Growth, structure and properties of graphite whiskers, *J. Appl. Phys.* **31**, 284-290.

2   Krätsmer, W., Lamb, L. D., Fortiropoulos, K., Huffman, D. R. (1990) Solid C_{60} : a new form of carbon, *Nature* **347**, 354-358.

3   Iijima, S. (1991) Helical microtubules of graphitic carbon, *Nature* **354**, 56-58.
4   Ebbesen, T. W., Ajayan, P. M. (1992) Large-scale synthesis of carbon nanotubes, *Nature* **358**, 220-220.
5   Iijima, S., Ichihashi, T. (1993) Single-shell carbon nanotubes of 1 nm diameter, *Nature* **363**, 603-605.
6   Bethune, D. S., Kiang, C. H., de Vries, M. S., Gorman, G., Savoy, R., Vazquez, J., Beyers, R. (1993) Cobalt-catalyzed growth of carbon nanotubes with single-atomic-layer walls, *Nature* **363**, 605-606.
7   Journet, C. (1998) La production de nanotubes de carbone, *Ph D thesis*, Université de Montpellier II Sciences et Techniques du Languedoc, Montpellier, France.
8   Cheng, H. M., Li, F., Su, G., Pan, H. Y., He, L. L., Sun, X., Dresselhaus, M. S., (1998) Large-scale and low-cost synthesis of single-walled carbon nanotubes by catalytic pyrolysis of hydrocarbons, *Appl. Phys. Letters* **72**, 3282-3284.
9   Rice group (1999) *Chem. and Eng. News* **77**, 31.
10   Journet, C., Maser, W. K., Bernier, P., Loiseau, A., Lamy de la Chapelle, M., Lefrant, S., Deniard, P., Lee, R., Fisher, J. E. (1997) Large-scale production of single-walled carbon nanotubes by the electric arc technique, *Nature* **388**, 756-758.
11 Colomer, J.-F., Stephan, C., Fefrant, S., Van Tendeloo, G., Willems, I., Konya, Z., Fonseca, A., Laurent, Ch., B.Nagy, J. (2000) Large-scale synthesis of single-wall carbon nanotubes by catalytic chemical vapor deposition (CCVD) method, *Chem. Phys. Lett.* **317**, 83-89.
12   Rinzler, A. G., Liu, J., Dai, H., Nikolaev, P., Huffman, C. B., Rodrigues-Macias, F. J., Boul, P. J., Lu, A. H., Heymann, D., Colbert, D. T., Lee, R. S., Fischer, J. E., Rao, A. M., Eklund, P. C., Smalley, R. E. (1998) Large-scale purification of single-wall carbon nanotubes: Process, product, and characterization, *Appl. Phys. A* **67**, 29-37.
13   Liu, J., Rinzler, A. G., Dai, H., Hafner, J. H., Brandley, R. K., Boul, P. J., Lu, A., Iverson, T., Shelimov, K., Huffman, C. B., Rodriguez-Macias, F., Shon, Y.-S., Lee, T. R., Colbert, D. T., Smalley, R. E. (1998) Fullerene pipes, *Science* **280**, 1253-1256.
14   Saito, R., Fujita, M., Dresselhaus, G., Dresselhaus, M. S. (1992) Electronic structure of graphene tubules based on C_{60}, *Phys. Rev. B* **46**, 1804-1811.
15   Tans, S. J., Devoret, M. H., Dai, H., Thess, A., Smalley, R. E., Geerligs, L. J., Dekker, C. (1997) Individual single-wall carbon nanotubes as quantum wires, *Nature* **386**, 474-477.
16   Ando, Y., Zhao, X., Kataura, H., Achiba, Y., Kaneto, K., Tsuruta, M., Uemura, S., Iijima, S. (2000) Multiwalled carbon nanotubes prepared by hydrogen arc, *Diamond and Related Materials* **9**, 847-851.
17   Ando, Y., Iijima, S. (1993) Preparation of carbon nanotubes by arc-discharge evaporation, *Jpn. J. Appl. Phys.* **32**, L107-109.
18   Ando, Y. (1994) Preparation of carbon nanotubes, *Full. Sci. Technol.* **2**, 173-180.
19   Zhao, X., Wang, M., Ohkohchi, M., Ando, Y. (1996) Morphology of carbon nanotubes prepared by carbon arc, *Jpn. J. Appl. Phys.* **35**, 4451- 4456.
20   Blank, V. D., Gorlova, I. G., Hutchison, J. L., Kiselev, N. A., Ormont, A. B., Polyakov, E. V., Sloan, J., Zakharov, D. N., Zybtsev, S. G. (2000) The structure of nanotubes fabricated by carbon evaporation at high gas pressure, *Carbon* **38**, 1217-1240.

21   Wang, M., Zhao, X., Ohkohchi, M., Ando, Y. (1996) Carbon nanotubes grown on the surface of cathode deposit by arc discharge, *Full. Sci. Technol.* **4**, 1027-1039.

22   Zhao, X., Ohkohchi, M., Whang, M., Iijima, S., Ichihashi, T., Ando, Y. (1997) Preparation of high-grade carbon nanotubes by hydrogen arc discharge, *Carbon* **35**, 775-781.

23   Ando, Y., Zhao, X., Ohkohchi, M. (1997) Production of petal-like graphite sheets by hydrogen arc discharge, *Carbon* **35**, 153-158.

24   Duesberg, G. S., Muster, J., Krstic, V., Burghard, M., Roth, S. (1998) Chromatographic size separation of single-wall carbon nanotubes, *Appl. Phys. A* **67**, 117-119.

25   Ebbesen, T. W. (1994) Carbon nanotubes, *Annu. Rev. Mater. Sci.* **24**, 235-264.

26   Maser, W. K., Bernier, P., Lambert, J. M., Stephan, O., Ajayan, P. M., Colliex, C., Brotons, V., Planeix, J. M., Coq, B., Molinie, P., Lefrant, S. (1996) Elaboration and characterization of various carbon nanostructures, *Synth. Metals* **81**, 243-250.

27   Kiang, C. H., Goddard III, W. A., Beyers, R., Bethune, D. S. (1995) Carbon nanotubes with single-layer walls, *Carbon* **33**, 903-914.

28   Ajayan, P. M., Lambert, J. M., Bernier, P., Barbetette, L., Colliex, C., Planeix, J. M. (1993) Growth morphologies during cobalt catalysed single-shell carbon nanotubes, *Chem. Phys. Lett.* **215**, 509-517.

29   Lambert, J. M., Ajayan, P. M., Bernier, P. (1995) Synthesis of single and multi-shell carbon nanotubes, *Synth. Metals* **70**, 1475-1476.

30   Kiang, C. H., Goddard III, W. A., Beyers, R., Salem, J. R., Bethune, D. S. (1994) Catalytic synthesis of single-layer carbon nanotubes with a wide range of diameters, *J. Phys. Chem.* **98**, 6612- 6618.

31   Seraphin, S., Zhou, D. (1994) Single-walled carbon nanotubes produced at high yield by mixed catalysts, *Appl. Phys. Lett.* **64**, 2087-2089.

32   Lambert, J. M., Ajayan, P. M., Bernier, P., Planeix, J. M., Brotons, V., Coq, B., Castaing, J. (1994) Improving conditions towards isolating single-shell carbon nanotubes, *Chem. Phys. Lett.* **226**, 364-371.

33   Lin, X., Wang, K., Dravid, V. P., Chang, R. P. H., Ketterson, J. B. (1994) Large scale synthesis of single-shell carbon nanotubes, *Appl. Phys. Lett.* **64**, 181-183.

34   Ajayan, P. M., Colliex, C., Lambert, J. M., Bernier, P., Barbedette, L., Tencé, M., Stephan, O. (1994) Growth of manganese filled carbon nanofibers in the vapor phase, *Phys. Rev. Lett.* **72**, 1722-1725.

35   Seraphin, S., Zhou, D., Jiao, J., Minke, M. A., Wang, S., Yadav, T., Withers, J. C. (1994) Catalytic role of nickel, palladium, and platinum in the formation of carbon nanoclusters *Chem. Phys. Lett.* **217**, 191-198.

36   Saito, Y., Kawabatta, K., Okuda, M. (1995) Single-layered carbon nanotubes synthesized by catalytic assistance of rare-earths in a carbon arc, *J. Phys. Chem.* **99**, 16076-16079.

37   Loiseau, A., Pascard, H. (1996) Synthesis of long carbon nanotubes filled with Se, S, Sb and Ge by the arc method, *Chem. Phys. Lett.* **256**, 246-252.

38   Guerret-Plécourt, C., Le Bouar, Y., Loiseau, A., Pascard, H. (1994) Relation between metal electronic structure and morphology of metal compounds inside carbon nanotubes, *Nature* **372**, 761-765.

39   Ata, M., Hudson, A. J., Yamaura, K., Kurihara, K. (1995) Carbon nanotubes filled with gadolinium and hafnium carbides, *J. Appl. Phys.* **34**, 4207-4212.

40　Kiang, C. H., Goddard III, W. A., Beyers, R., Salem, J. R., Bethune, D. S. (1996) Catalytic effects of heavy metals on the growth of carbon nanotubes and nanoparticles, *J. Phys. Chem. Solids* **57**, 35-39.

41　Ruoff, R. S., Lorents, D. C., Chan, B, Malhotra, R., Subramoney, S (1993) Single crystal metals encapsulated in carbon nanoparticles, *Science* **259**, 346 – 348.

42　Zhang, Y., Iijima, S., Shi, Z., Gu, Z. (1999) Defects in arc-discharge-produced single-walled carbon nanotubes, *Phil. Mag. Lett.* **79**, 473-479.

43　Sloan, J., Wright, D. M., Woo, H. G., Bailey, S., Brown, G., York, A. P. E., Coleman, K. S., Hutchison, J. L., Green, M. L. H. (1999) Capillarity and silver nanowire formation observed in single walled carbon nanotubes, *Chem. Commun.* No 8, 699-700.

44　Sloan, J., Dunin-Borkowski, R. E., Hutchison, J. L., Coleman, K. S., Williams, V. C., Claridge, J. B., York, A. P. E., Xu, C., Bailey, S. R., Brown, Friedrichs, G. S., Green, M. L. H. (2000) The size distribution, imaging and obstructing properties of C$_{60}$ and higher fullerenes formed within arc-grown single walled carbon nanotubes, *Chem. Phys. Lett.* **316**, 191-198.

45　Oberlin, A. (1984) Carbonization and graphitization *Carbon* **22**, 521-541.

46　Nikolaev, P., Thess, A., Rinzler, A. G., Colbert, D. T., Smalley, R. E. (1997) Diameter doubling of single-wall carbon nanotubes *Chem Phys. Lett.* **266**, 422-426.

47　Metenier, K. (1999) Intercalation et stockage electrochimique dans les nanotubes de carbone. Evolution thermique des nanotubes monoparoi, *Ph D thesis*, Université d'Orléans, Orléans, France.

48　Ebbesen, T. W., Ajayan, P. M., Hiura, H., Tanigaki, K. (1994) Purification of nanotubes, *Nature* **367**, 519.

49　Ajayan, P. M., Ebbesen, T. W., Ichihashi, T., Iijima, S., Tanigaki, K., Hiura, H. (1993) Opening carbon nanotubes with oxygen and implications for filling, *Nature* **362**, 522-525.

50　Ando, Y., Zhao, X., Ohkohchi, M. (1998) Sponge of purified carbon nanotubes *Jpn. J. Appl. Phys.* **37**, L61.

51　Satishkumar, B. C., Govindaraj, A., Mofokeng, J., Subbanna, G. N., Rao, C. N. R. (1996) Novel experiments with carbon nanotubes: opening, filling, closing and functionalising nanotubes, *J. Phys. B: At. Mol. Opt. Phys.* **29**, 4925-4934.

52　Abatemarco, T., Stickel, J., Belfort, J., Franck, B. P., Ajayan, P. M., Belfort, G., (1999) Fractionation of multiwalled carbon nanotubes by cascade membrane microfiltration, *J. Phys. Chem. B* **103**, 3534-3538.

53　Bandow, S., Rao, A. M., Williams, K. A., Thess, A., Smalley, R. E., Eklund, P. C., (1997) Purification of single-wall carbon nanotubes by microfiltration, *J. Phys. Chem. B* **101**, 8839-1842.

54　Duesberg, G. S., Burghard, M., Muster, J., Philip, G., Roth, S. (1998) Separation of carbon nanotubes by size exclusion chromatography, *Chem. Commun.* 435-436.

55　Dillon, A. C., Gennett, T., Jones, K. M., Alleman, J. L., Parilla, P. A., Heben, M. J. (1999) Simple and complete purification of single-walled carbon nanotube materials, *Adv. Mater.* **11**, 1354-1358.

CATALYTIC PRODUCTION, PURIFICATION, CHARACTERIZATION AND APPLICATION OF SINGLE- AND MULTIWALL CARBON NANOTUBES

Z. KÓNYA

University of Szeged, Applied and Environmental Chemistry Department
Rerrich Béla tér 1, H-6720, Szeged, Hungary

Introduction

From the beginning of the catalysis, coke formation on catalysts was an important part of the process, and hence, it is rather well documented. With the industrial recovery, this field became much more important since in most cases coke formation is undesirable. The prevention of carbon deposit accumulation is a high priority objective in many processes involving transformation of hydrocarbons. In the latter industrial reactions coke formation can cause not only the deactivation of the catalyst but since it represents a large amount of solid material, it can bring on the blockage of the reactors and deteriorating heat transfer properties. However, because of the numerous different and interesting carbon structures (fibers, cones, tubular structures, etc.) generated as byproduct during the catalytic reactions, the new family of these materials started to be investigated[1,2,3].

The story of the carbon family is grandiose: subtlety in simplicity. The mystery is only coming from the different hybridization that carbon atoms can assume. Carbon has four valence electrons. When they are shared equally (sp^3), diamond is formed. When three electrons are shared in a plane and one is delocalized between the planes (sp^2), the carbon forms graphite. All the catalytically generated carbon forms belong to the structure of sp^2 bonded carbon.

Since Iijima's discovery[4], the field of carbon nanotubes (Single- and Multiwall Carbon Nanotubes – SWNTs and MWNTs) has become a separate subject in material science. These materials attracted more and more attention from physicists and chemists. Their surprisingly new and unique physical and chemical properties generated tremendous interest. Let us note that carbon nanotubes are also considered as members of the recently discovered carbon allotropes, the fullerenes, for which the Nobel Prize was given in 1996 to Curl, Kroto and Smalley[5].

At the beginning, the quantity of carbon nanotubes available for experiments was very low, since the early arc-discharge method produced only tiny amounts. The increasing demand on having more and more nanotubes brought the fast development in the ways of production, including not only improvements of the arc-discharge technique but also new methods like the laser evaporation and the catalytic vapor decomposition (CVD) of gaseous hydrocarbons.

85

L.P. Biró et al. (eds.), Carbon Filaments and Nanotubes: Common Origins, Differing Applications, 85–109.
© 2001 *Kluwer Academic Publishers. Printed in the Netherlands.*

Synthesis of MWNTs

The first successful experiments on the preparation of MWNTs by Yacaman[6] and Ivanov[7] were followed by hundreds of modified or new procedures. On the basis of these experiments, several fundamental principles were concluded. In most cases the syntheses are carried out at high temperature applying a stream of carbon source diluted with inert compounds such as nitrogen, argon, etc. The experimental set-up consists of a high temperature oven in which the catalysts are placed onto a highly resistant ceramic or metal plate (Figure 1.).

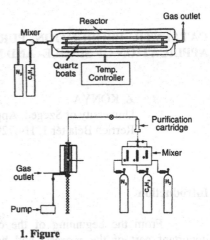

1. Figure

Scheme of horizontal and vertical furnaces used for nanotube production

The nature and yield of the deposit obtained in the reaction are controlled by varying different parameters such as the nature of the metals and the supports, the hydrocarbon sources, the gas flows, the reaction temperature, the reaction time, etc. By selecting the proper conditions, both the physical (e.g. length, shape, diameter) and chemical properties (e.g. number of defects, graphitization) of MWNTs can be designed in advance. For the catalyst preparation, various methods are applied, such as impregnation, ion-exchange and mechanical grinding of the support and the metal components. Activity of the catalyst samples prepared in different ways shows dependence on the nature of the catalyst support (silica of different pore diameter, zeolites of various structure) and on the pH of the solution during the synthesis of catalyst samples. The supported Co[7], Ni[8] and Fe[9] catalysts were found to be the most active in the CVD. When other elements (e.g. Cu, Cr, Mn) are used, only a negligible amount of carbon nanotubes (NTs) is formed. As far as the catalyst support is concerned, its quality is of considerable importance. While Si-, Mg- and Al-containing materials (silica, alumina, zeolites, magnesia) prove to be applicable as supports, the different forms of carbon (graphite, activated carbon) are not suitable in NTs production. These results suggest that, contrary to received wisdom, the support has a peculiar role in the reaction. The hydrocarbons most frequently used are acetylene, ethylene and benzene because of their high carbon content. It was also shown that generally, increasing reaction time results in the formation of larger amounts of amorphous carbon due to the deactivation of the catalytic sites. Determining the optimal reaction temperature is a very complicated task since several requirements come into conflict: while a high temperature favors proper graphitization, the homogeneous pyrolysis of the hydrocarbon becomes excessive above a certain temperature depending on the carbon source used. Moreover, the metals and supports demand different temperature ranges (Figure 2.).

Nanotubes of helical shape have been produced using a particular catalyst preparation[10]. Silica supported Co catalyst proved to give a large amount of helical

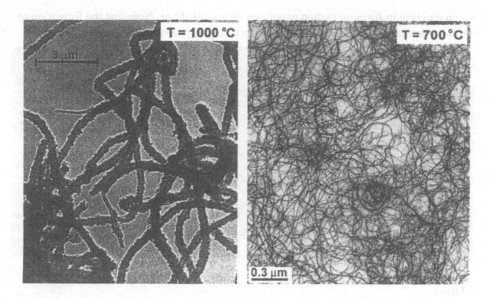

2. Figure

TEM and HREM pictures of helical nanotubes

tubes in the decomposition of acetylene at 700 °C (Figure 3.). According to results obtained over catalysts prepared by impregnation, quality of the tubes strongly depends on the pH of the initial solution and on the pretreatment during catalyst preparation. Those samples which contain acidic centers on the surface show low activity. Formation of carbon nanotubes is observed over catalysts produced from solutions of pH≥7 (Table I.). Although catalyst activity is increasing in parallel to the basicity of the solution, selectivity towards helical carbon nanotubes goes up suddenly when the catalyst surface has neutral character. The presence of acidic OH groups on silica support probably hinders carbon nanotubes formation, and side-reactions which do not involve carbon deposition become dominant.

Table I.

Carbon yield in the decomposition of acetylene over different Co/silica samples

	pH	4	5	6	7	8	9
Carbon yield	Calcined in air (450 °C)	0.07	0.09	0.51	0.71	0.74	0.47
	Prereduced	0.15	0.23	0.45	0.76	0.83	0.65

With further pH increase, the amount of spiral nanotubes becomes significantly higher. A possible explanation for this phenomenon is that the solubility of Co-acetate is good. Particles forming on the silica surface are symmetric. Consequently, the catalytic centers are symmetric, too. When increasing pH, precipitation of $Co(OH)_2$ begins, some particles get onto the surface of catalyst support in solid, probably asymmetric form. Presumably, these asymmetric particles contain regions of different catalytic activity, which can help in forming carbon nanotubes of helical shape. Note that the formation of spiral carbon nanotubes was observed only in catalytic processes.

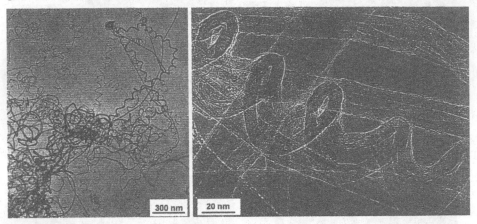

3. Figure

TEM and HREM pictures of helical nanotubes

Arrays of aligned multiwall nanotubes were synthesized in order to use them for flat panel display applications. Thess et al.[11] succeeded in producing more than 70 percent of single-wall nanotubes by condensation of a laser-vaporized carbon-nickel-cobalt mixture at 1200 °C. In parallel, iron particles embedded in mesoporous silica were used for the large scale synthesis of aligned MWNTs[12]. One of the most promising methods was reported by Ren et al.[13], where aligned carbon nanotubes were grown over areas up to square centimeters on nickel-coated glass below 700 °C using a special plasma-enhanced hot filament chemical vapor decomposition. Dai's group[14] presented the synthesis of massive arrays of length-controlled carbon nanotubes that are self-oriented on patterned porous and plain silicon substrate. The main points in this method are the use of CVD for nanotube production, the catalytic particle size control by special substrate design and the positioning of the array by patterning. Recently, Forró's group[15] produced patterned nanotube films (Figure 4.) using microcontact printing of pure transition metal (Fe, Ni, Co) catalysts and their mixtures.

Besides the demonstrated methods, there are peculiar procedures applied for synthesis of MWNTs such as growth from NiPc deposited on a quartz plate by vacuum chemical vapor decomposition[16], production of carbon tubes on Ni particles using 2-methyl-1,2'-naphtyl ketone[17] at temperatures between 600 and 1000 °C, pyrolytic decomposition of propylene using uniform and straight channels of anodic alumina film as a template[18] and many others. Although these techniques are able to produce MWNTs, it can be stated that their importance is rather theoretical than practical.

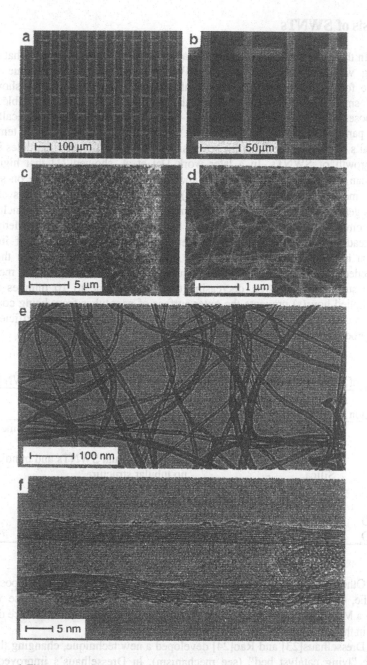

4. Figure

SEM images (a-d) of a surface with patterned carbon nanotubes at different scales and HREM images (e,f) of the nanotubes (ref [15])

Synthesis of SWNTs

In the early stage of research on nanotubes production it seemed that however promising was the CVD method for MWNTs syntheses, the technique was not applicable for SWNTs. In 1996, Dai et al.[19] and Fonseca et al.[20] showed that although small quantities were obtained, the modified CVD was practicable plan for this purpose as well. In these studies Fe, Mo, Co, Ni were the catalytically active particles particularly on Si- and Al-based supports in the 800-1200 °C temperature range. Dai's work suggested that the small size of the catalytic metal particles is crucial for the growth of SWNTs. In a later work Dai et al. showed[21] that high quality SWNTs can be produced by chemical vapor decomposition of methane on supported transition metal oxide catalysts (Table II.). The experimental set-up was similar to apparatus generally producing multiwall nanotubes, however some factors including the catalyst composition, the support and the hydrocarbon were different. Methane was used instead of the generally applied acetylene or ethylene because of its kinetic stability at high temperatures. Since there is no pyrolytic decomposition the carbon atoms needed for nanotube growth are produced by catalytic reaction from methane on the metal surfaces. Moreover, decreasing the reaction time to 10 minutes from the usually applied hour(s) prevents the outer surface of nanotubes from being coated with amorphous carbon. Finally, Fe_2O_3 was found to be significantly more efficient in the SWNTs production than CoO or NiO.

Table II.
Summary of Dai's results using supported catalysts for the production of SWNTs [21]

Catalyst composition	Support	Presence of SWNTs	Description
Fe_2O_3	Alumina	++	Abundant individual SWNTs; some bundles
Fe_2O_3	Silica	++	Abundant individual SWNTs
CoO	Alumina	+	Some individual SWNTs and bundles
CoO	Silica	-	no tubular structure
NiO	Alumina	-	Weakly graphitized MWNTs
NiO	Silica	-	no tubular structure
NiO/CoO	Alumina	-	no tubular structure
NiO/CoO	Silica	+	Some SWNT bundles

Other works by Rousset's group[22] accounted for using composite metal oxides (Fe, Co, Fe-Co) to produce SWNTs. In this case the active phase was most probably a Mg and Al containing spinel-type thin layer formed on the surface during the reaction in the reducing H_2/CH_4 atmosphere.

Dresselhaus[23] and Rao[24] developed a new technique, changing the idea of the basic "lying catalyst bed" (see mechanism). In Dresselhaus's improved floating catalyst method, two furnaces are used (Figure 5.). The organometallic compound (ferrocene) is vaporized in the first oven at low temperature (200-400 °C) in a hydrogen stream. The decomposition of the hydrocarbon (benzene) and the catalyst takes place in the second furnace (900-1200 °C). The addition of 0.5 to 5% thiophene to benzene results in purer SWNTs, but in weaker yield. One very important feature of this technique is its capability to produce both SWNTs and MWNTs depending on the sulfur content of benzene.

Rao's method[24] was slightly different from that proposed by Dresselhaus's. In this case the synthesis of SWNTs is achieved by the pyrolysis of mixtures of acetylene with metallocenes (Fe, Co, Ni) or

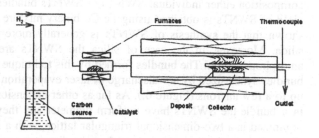

5. Figure

Scheme of a typical two-furnace system used for both Single- and Multiwall nanotube formation

Fe(CO)$_5$ at 1100 °C. When using metallocenes, a two-oven system is also used, while the Fe(CO)$_5$ is pyrolyzed along with acetylene in the same furnace.

In the latter two methods the organometallic precursor acts as the source of small metal particles, which are most probably necessary for the formation of SWNTs. It is noteworthy that the pyrolysis of benzene admixed with metallocenes under similar conditions primarily gives MWNTs. It seems that acetylene is a better source for SWNTs than benzene probably due to its smaller size.

Recently, Colomer et al.[25] published an improved method for SWNTs production. In this work the synthesis of SWNTs is carried out by ethylene decomposition on supported metal catalysts prepared by impregnation of various supports. Co, Ni and Fe and their binary and ternary mixtures are used as the metallic components, while the supports are silica or alumina. The role of the support is to disperse metal particles. Catalysts are prepared by combining sonication and impregnation. The reaction is carried out in a fixed bed quartz reactor at 1080 °C for one hour. It was shown by Transmission Electron Microscopy that depending on the metal

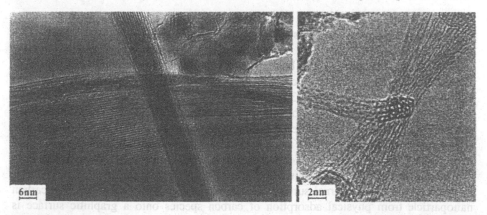

6nm   2nm

6. Figure

HREM images of large SWNT bundle and higher magnification of SWNT bundle cross section

composition either individual SWNTs or SWNTs bundles are formed. In fact, the best yield of SWNTs is obtained using Fe-Co binary mixture supported on alumina. It was shown that the synthesis of SWNTs is generally more efficient on alumina than on silica. Moreover, in the case of silica the SWNTs are coated by a thick layer of amorphous carbon. The bundles formed by this technique seem to be very similar to the bundles synthesized by arc-discharge or laser evaporation techniques and their length is up to a few microns (Figure 6.). As far as other dimensions of the tubes are concerned, in a bundle the SWNTs have uniform diameter and they form crystallite-like entities organized in a two-dimensional triangular lattice, with a lattice constant a of ca. 10 Å. This corresponds to nanotubes of 7 Å in diameter separated by the van der Waals intertube distances of 3.4 Å. In comparison with bundles produced by arc-discharge or laser evaporation techniques, this diameter was found to be the smallest.

On the basis of the observations Colomer et al. proposed that even if there is a difference in the abundance of bundles formed on silica or alumina, the support has no strong influence on the growth mechanism contrary to what has previously been suggested[21]. Most probably, the support plays an important role in the dispersion of the metal on the surface and can influence the catalytic cracking of the hydrocarbon, but the presence or the absence of SWNTs bundle is not governed by this. On the other hand, the preparation of catalysts with mixtures of metals can induce better dispersion as a result of alloy formation followed by segregation.

Applying a new catalyst and modified experimental set-up Colomer et al[26] very recently proposed an efficient method in which very high amounts of SWNTs can be produced in a short reaction time. The catalyst developed was a mixture of Co and Fe on magnesium oxide and the reaction was carried out for 10 minutes at a temperature of 1080 °C using methane as carbon source.

As it can be concluded from the literature about the production of nanotubes, there are various experimental techniques for production of both SWNTs and MWNTs. However, no method is currently available for scaling up the production of nanotubes for industrial purposes. We are convinced that the CVD is the most promising procedure for this.

Growth mechanisms of MWNTs

At the beginning, three basic mechanisms were proposed for the growth of multishell nanotubes: shell-by-shell growth[27,28], curling of graphite sheets[29] and simultaneous growth of all shells[30].

In the shell-by-shell model each concentric shell results from the carbon adsorption onto the surface of the preceding outer shell. Nevertheless, as discussed in several papers[31], carbon atoms and micro-clusters can be adsorbed on the surface of nanoparticles only at temperature less than 200 K. This temperature is not sufficient to cleave and reform the bonds between the carbon atoms and hence to form a newer graphitic shell on the external surface. As a result, the growth mechanism of a carbon nanoparticle from physical adsorption of carbon species onto a graphitic surface is impossible. Since the shell-by-shell growing carbon nanotubes is analogous to the case of nanoparticles, this mechanism is very doubtful.

Another possibility for the shell-by-shell mechanism is that each shell starts growing from the nanotube base attached to the substrate[32]. Assuming this possibility, the outer shell could stop growing before reaching the nanotube tip. However, the HREM (High Resolution Transmission Electron Microscopy) results showed that the outer wall of the nanotubes is closed in almost every case.

Another proposed mechanism was the rolling up of large graphite sheets along the lines of sp^3 defects forming nanotubes. However, such a graphite sheet is a parallelogram[33] in shape with an acute angle of 30°. It is almost impossible to roll up this sheet into a nanotube with equal number of shells along the length of the nanotube. Scanning Tunneling Microscopy did not reveal the sheet edges[34], which should be found on the surface with the roll structure. Moreover, different walls of MWNTs were found to have different helicity proving that the nanotubes cannot be formed from a single graphite sheet.

Lastly, the third model, i.e. the simultaneous growth of all walls remained as the only possibility for the formation of MWNTs. However, it has to be taken into consideration that up to now there is no proper explanation of the step-by-step growth mechanism of multiwall nanotubes formation. There are general conceptions but to this day no definitive proofs are provided.

It is postulated that the formation and growth of nanotubes are an extension of other known processes in which graphitic structures are formed over metal surfaces at temperature below 1000 °C from carbon. It is also very obvious that the shape of graphitic carbon produced depends on the chemical properties and the physical dimensions of the metal particles. The most efficient metals were found to be Fe, Co, Ni. The peculiar ability of the mentioned metals to form graphitic carbon layers is thought to be related to the combination of different factors. These include their catalytic activity towards the decomposition of carbon compounds, the possibility to form unstable carbides and finally, the speed of carbon diffusion in the metal particle, which must be extremely rapid. Derbyshire et al. showed that over thin metal foils carbon dissolves to form a solid solution[35]. After cooling, the carbon precipitates onto the surface, forming a continuous thin film of highly crystalline graphite parallel to the metal surface. The high crystallinity obtained in a very short time shows that carbon atoms are extremely mobile and able to move easily over and through the metal.

When the size of the metal is in the range of microns, instead of

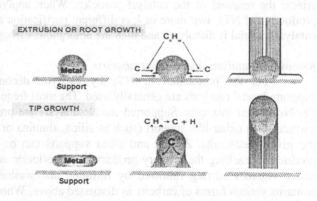

7. Figure

Schemes of tip-growth and root-growth mechanisms for carbon filament growth (ref [36])

graphite the carbon is produced as filaments of similar diameter. Baker[36] proposed a general scheme for the formation of such thin graphitic carbon structures on metal surfaces (Figure 7.). In this proposal, the metal particles are supported on a substrate and either strong or weak metal-support interaction is assumed. In the first case the carbon fiber grows upwards from the metal particles that remain attached to the substrate. This is the so called "extrusion" mechanism. In the case of weak metal-support interaction the particle is detached from the substrate and it moves to the head of the fiber, labeled "tip-growth" mechanism.

In a very recent paper[37], a detailed investigation has been reported on the influence of support and metals on the quality and quantity of carbon nanotubes produced by catalytic decomposition of acetylene. The authors attributed exceptionally important role to the size and nature of metal particles and additionally, to the alloy phase generation during the reaction. In opposition to the case of single metal particles, the carbon dissolves easily in this alloy phase, which results in an increase of one order of magnitude in the nanotubes yield.

PURIFICATION, CHARACTERIZATION AND APPLICATIONS OF MWNTs

The primary product of carbon nanotube fabrication is a mixture of various carbonaceous materials. Catalytic CVD (CCVD) processes result in around 80% conversion of substrate compound to carbon nanotubes as best values. From this follows that special purification procedure had to be developed for removal of amorphous carbon from the primary product.

Purification

When solid catalysts are used for the synthesis of NTs an additional problem arises: the removal of the catalyst particles. When applying CCVD process for the production of NTs, two more or less different purification steps are required. First, the catalyst material is dissolved, and then the amorphous carbon is removed.

Removal of catalysts and catalyst supports

For the production of NTs by catalytic decomposition of hydrocarbons supported metal catalysts are generally used. The most frequently applied metals are Co, Fe, Ni or their mixtures. Supported metal catalysts are prepared by impregnating the surface of an oxide-like support (such as silica, alumina or zeolites) with a solution of the given metal salts. Zeolite and silica supports can be separated from the carbon product by leaching the primary product in hydrofluoric acid (38-40 %) for 48 hours under vigorous stirring followed by filtration and washing. The resulting wet cake contains various forms of carbons as discussed above. When alumina is the support, its extraction differs from those previously described. A sodium hydroxide solution is used instead of HF solution since alumina can be easily dissolved in concentrated hydroxide solution. Upon treatment of the catalyst-carbon mixture, the metal component of the catalysts is generally also transferred into the solution. If not, further treatment in diluted mineral acid solution is recommended. It should be mentioned that metal particles encapsulated into the carbon nanotubes remain intact after this treatment. For their removal, we should first open the nanotubes, which are generally produced with closed ends (a procedure leading to opening carbon nanotubes will be discussed later).

Removal of amorphous carbon

There are chemical and physical methods to purify samples of carbon nanotubes containing by-product carbon material. The chemical methods generally result in products of higher purity.

In order to remove amorphous carbon, the catalyst-free material can be either oxidized or reduced[38]. Reduction is generally performed in hydrogen atmosphere at elevated temperature (above 1000 K). Oxidation of carbon of irregular structures can be performed either in gas or in liquid phase. Gas phase oxidation occurs at 500 °C in air. In 210 min such a treatment results in carbon nanotubes of purity higher than 95 %. Instead of air, oxygen or ozone can also be used. When purification is carried out in solution, the use of an acidic (0.5 M sulfuric acid) $KMnO_4$ solution (0.3 M) is recommended. After 1 hour reaction at 80 °C carbon nanotubes of 95 % purity are obtained.

In some instances these treatments lead to uncapping of nanotubes which are therefore ready to be filled with gases or metals. Some small amount of chemical functionalization, such as implantation by hydroxyl, carboxyl etc, groups can also take place.

Centrifugal separation is generally applied for further purification of SWNTs. First, a dilute aqueous NT solution is prepared and appropriate amount (0.1 %) ionic surfactant like benzalkonium chloride is sonicated in an ultrasonic bath for 4-6 hours. Then, the colloidal suspension is centrifuged first with 2000 g in order to separate large carbon particles (d=50-80 nm) followed by the second step with 20000 g in order to precipitate nano-spheres with d<50 nm. The result is a product with NTs of 70 % purity.

Microfiltration separation is appropriate for soot containing a higher yield of NT since the nanoparticles easily contaminate the membrane filter. Therefore, centrifugal purification is usually applied before microfiltration. First the soot is soaked in CS_2 to remove

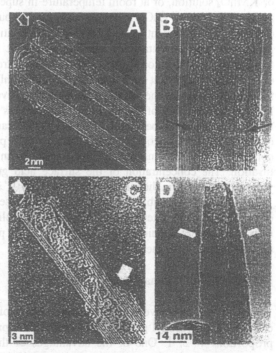

8. Figure

HREM images of different type of oxidation of MWNTs

fullerenes and other soluble materials from the sample. The next step is to disperse the material in water containing 0.1 % surfactant followed by agitation in ultrasonic bath. The suspension is then filtered on a 0.2 micrometer pore size membrane filter. The NTs are caught by the filter.

Opening up of carbon nanotubes
    The simplest method to open the nanotubes is their oxidative treatment. It is well known that graphite oxidizes primarily at defects of the hexagonal lattice to create etch pits. When such defect sites are present in the wall of the nanotubes, they become the center of preferential etching. However, nanotubes have additional structural features such as high curvature, helicity, and contain five and seven membered rings which modify the initiation and also the propagation of oxidation. Particularly for MWNTs the oxidation tends to start near the tips, providing a mechanism for opening the tubes. Oxidation proceeds layer by layer, resulting in thinner tubes. Once the tip is removed, the strain induced by the distribution of pentagons is no longer there. The concentric layers of MWNTs do not react at the same rate, since each shell has its own tip. It follows that the inner shells might persist longer than the outer. Therefore different oxidation rates are assumed for oxidation of an open MWNT.
    Upon treatment, carbon nanotubes are refluxed in concentrated HNO_3, H_2SO_4 or $KMnO_4$ solution, or at room temperature in superacidic HF/BF_3 solution, or even in aqueous solution of OsO_4 or OsO_4-NaIO_4 for 24 hours until their originally capped ends open up. The treatment in acidified $KMnO_4$ solution seems to be a somewhat better procedure. The advantage of HF/BF_3 treatment is that it is carried out at room temperature. As mentioned above, the oxidative treatment of nanotubes results in not only open nanotubes at their tips, but nanotubes thinner in diameter[39]. The extent of thinning depends on the duration of treatment. Several examples are seen in Figure 8. Further consequence of the oxidative treatment is the partial functionalization of the tubes, i.e. the nanotubes become covered with carboxyl or hydroxyl groups at their ending. These functional groups make NTs partially soluble. The number or concentration of the inserted carboxyl groups can be estimated by simple acid-base titration. The concentration of the surface acid groups in the nanotubes opened by various oxidants is in the range of $2*10^{20} - 10*10^{20}$ site/g of nanotube[40].
    Another possibility that is assured by these carboxyl groups is to do some sort of chemistry on these terminal groups. Several applications, for instance the use of NTs as polymer fillers, require or take advantage of the presence of chemical groups situated on their surface or at their tip.

Functionalization of nanotubes

    After oxidative treatment, the surface of the NT is covered by carboxyl, carbonyl and hydroxyl groups. These groups are present in the following approximate proportions 4: 2: 1. One of the advantages of these functional groups is that they increase the chemical reactivity of the material[41]. As Figure 9 shows when carboxyl groups are reacted with alkyl-aniline, a material ending with long alkyl chains can be obtained. The efficiency of this reaction has been proved by IR spectroscopy. Spectra of

A)

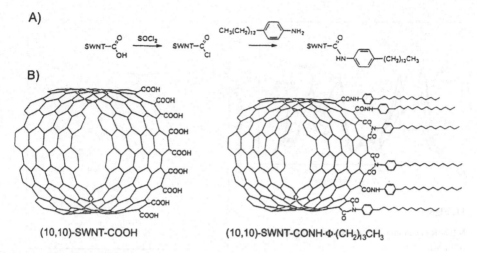

B)

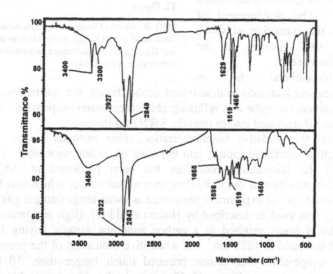

(10,10)-SWNT-COOH   (10,10)-SWNT-CONH-Φ-(CH$_2$)$_{13}$CH$_3$

9. Figure

Formation (A) and structure (B) of (10,10)-SWNT-amide (ref [42])

10. Figure

MID-IR spectra of 4-Tetradecylaniline and SWNT functionalized
by 4-Tetradecylaniline (ref [42])

the 4-tetradecylaniline and the product of reaction between SWNT and the alkylaniline
is shown in Figure 10.

The grafting of a long alkyl chain is utilized for producing atom-sharp tips for
Chemical Force Microscopy from end functionalized NT (see later). This year an
elegant oxidative treatment has been suggested by Yates et al[42]. They used ozone for

98

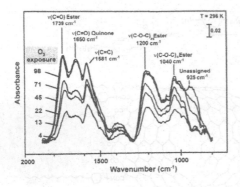

11. Figure

MID-IR spectrum of ozonized carbon nanotube
(ref [43])

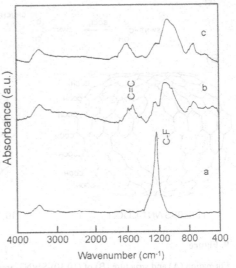

oxidative attack of SWNTs at 298 K.
The spectra taken during the reaction
using an in situ IR technique are shown
in Figure 11. The development of
bands due to the formation of C=O,
C=C and C-O-C groups on the
nanotube is clearly seen.

12. Figure

MID-IR spectra of crude carbon nanotubes
fluorinated at 500 °C (a), at room temperature (b)
and fluorination of graphitized nanotubes at room
temperature (c) (ref [44])

Dichlorocarbene is an
electrophilic reagent that adds to deactivated double bonds, but not to benzene. It was
reacted with carbon nanotube in a refluxing chloroform-water suspension. Around 5 %
of chlorine was incorporated into or onto the SWNTs[43].

Besides the oxidative functionalization, other methods are also known to
introduce reactive functional groups into the shell of carbon nanotubes. The detailed
description of the fluorination procedure has been published in 1997[44]. The
functionalization was carried out in F_2 flow, in a nickel reactor which was heated by a
conventional oven. In the experiment performed at room temperature a gas mixture of
F_2, HF and IF_5 was used as described by Hamwi et al.[45]. High temperature (500 °C)
fluorination for 4 hours resulted in a carbon nanotube sample having F/C=1. The
appearance of the band at 1223 cm^{-1} was a sensitive indication of the presence of C-F
bond. Room temperature fluorination required much longer time, 10 hours. The
respective spectra are seen in Figure 12. Fluorinated tubes became soluble in alcoholic
solvents. This enhanced solubility is due to the interaction of alcohol and fluorine atoms
on the surface of carbon nanotubes.

Several unsuccessful function-alization experiments were carried out in which
no clear-cut evidence was given, since the outer surface of nanotubes was covered by
amorphous carbon, the latter being also reactive. However, the importance of
functionalization makes no doubt and will be discussed again later from the polymer
filler point of view.

Characterization methods

Electron microscopy

Each technique of electron microscopy is applied for investigation of carbon nanotubes. This is the most demanded experimental method to prove the formation, shape, structure and purity of NTs. The use of SEM and TEM is also advantageous for investigation of the changes that occurred upon various treatments. High resolution TEM measurements allow the determination of the number of shells for MWNTs and the diameter for both SWNTs and MWNTs.

As it has been shown before, these techniques also provide basic information on the structural changes (opening, breaking and filling) of nanotubes.

X-ray diffraction (XRD)

X-ray diffraction patterns give structural information on the nanotubes. The following example shows what sort of changes can be detected by this technique. Ming et al.[46] investigated the deformation of catalytically grown MWNTs under high pressure. The microstructural changes of the MWNTs into a quasi-spherical onion occurred at 770 C° under 5.5 GPa. After treatment at 950 C° under the same pressure, a nanographitic ribbon-like structure could be observed by SEM and TEM. The collapse of the MWNT structure is evidenced in the XRD patterns (Figure 13).

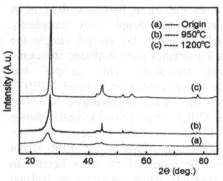

13. Figure

XRD patterns of the carbon NTs annealed at different temperatures under 5.5 Gpa (ref [46])

^{13}C-NMR spectroscopy

Being one of the most sophisticated instrumental analytical techniques, the ^{13}C-NMR might provide information on the structure of carbon nanotubes, particularly on the bonding of carbon atoms. Unfortunately, high quality NMR spectra can be obtained only for those samples that contain no paramagnetic nuclei. As the CCVD technology of carbon nanotubes production uses transition metals of paramagnetic properties and as some tiny amount of these metals (atoms or even small clusters) is encapsulated into the tubes or between the shells, the applicability of NMR spectroscopy is limited. However, there are successful measurements with this technique as well. When ^{13}C-labelled hydrocarbon is used for preparation of NTs, there are some chances to measure relaxation time and other parameters.

100

EPR spectroscopy

EPR spectroscopy may be applied for the investigation of magnetic and electronic properties of nanotubes. Here again difficulties may arise due to the presence of paramagnetic atoms or particles generally found in MWNTs and SWNTs prepared by CCDV methods. To illustrate this characterization opportunity, an example is taken from the work of Bandow et al[47]. The authors compared the EPR spectra of MWNT prepared by two different methods. After cleaning the soot obtained by arc burning carbon rod in hydrogen, the sample was completely dried. To obtain the second sample the dried nanotubes were dispersed in hexane using sonication. This sample was regarded as containing isolated MWNTs while the former agglomerated MWNTs. The EPR spectra of these samples, shown in Figure 14, are different. For the agglomerated MWNTs the EPR signal could be simulated by three Lorentzian components, while the signal of isolated MWNTs required only two Lorentzian peaks. These EPR spectra were interpreted in terms of the difference of the electron conduction in the two cases. For agglomerated tubes electrons can transfer between the tubes, while for isolated MWNTs electrons may only move around a nanometer-scale graphitic tube wall.

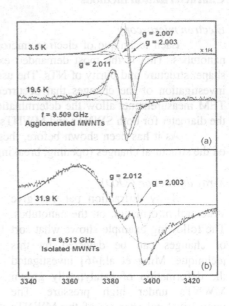

14. Figure

EPR spectra of isolated and aggregated MWNTs

Vibration and electron spectroscopy

Raman spectroscopy gives useful information on the geometry of SWNTs. SWNTs have strong signals in two regions of Raman spectral range[48]. The first is around 1600 cm^{-1} where the graphite-like modes are expected, and the second between 140 and 220 cm^{-1}, corresponding to the radial breathing mode (RBM). The vibration frequencies depend on the n,m indices of the tube. For RBM the frequency ν may be described as $\nu=A/D$ where D is the diameter of tube and A is a constant which is influenced by the macroscopic structure of tubes, i.e. whether they are armchair, zigzag or helical. Figure 15 shows Raman spectra of nanotubes prepared with various catalysts[49]. Data unambiguously prove that the diameter of SWNTs is influenced by the catalyst used for their synthesis.

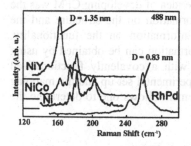

15. Figure

Raman spectra of breathing mode of SWNT synthesized on different catalysts

IR spectroscopy is generally used for characterizing of the modification of carbon nanotubes. It is very rarely applied for characterization of as made tubes.

Electronic spectra of nanotubes may serve to deduce information on the electronic conductivity of nanotubes[50,51].

STM, AFM (CFM) techniques

Among the methods used for characterization of nanotubes the STM and AFM (CFM) techniques proved to be the most useful ones. These techniques were developed in the past decades and became basic experimental tools in the nanotechnique. A review on the application of these methods have been published, recently[52].

For the STM measurement, carbon nanotubes after appropriate cleaning (removal of catalysts and the amorphous carbon) are suspended in toluene under sonication. The suspension thus prepared is placed on a freshly cleaved highly oriented pyrolytic graphite (HOPG) plate and the toluene is evaporated generally at room temperature. In most cases STM experiments are performed under ambient conditions, using mechanically prepared Pt tips. Typical tunneling voltages are in the range of 100-400 mV and tunneling currents usually are in the range of 0.2-1 nA. Horizontal and vertical calibration of STM is checked against the HOPG substrate. Figure 16 shows a typical STM image of carbon nanotubes. From this experimental data the diameter and even the orientation of nanotubes can be determined. For details of STM investigation of carbon nanotubes see ref [53].

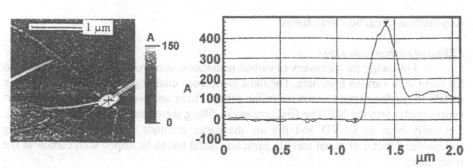

16. Figure

Constant current STM image of carbon nanotubes of a 10 nm diameter (plain view image and line cut along the black line) (ref [53])

Recently both MWNT and SWNT were applied as tips for AFM measurements since these sort of tips provide much better lateral resolution compared to the commercial Si and Si₃N₄ tips[54]. Two years ago Lieber's group developed a novel technique to get deeper insight into chemical and biological systems by Chemical Force

102

Microscopy that is a modified technique of AFM. The idea of developing CFM was the fact that AFM with the conventional tips gives information on the structure and the geometry of the substrate investigated, however, information on the functionality, binding and reactivity is lacking. Such chemical information can be obtained by using functionalized tips in CFM. They equipped CFM with a covalently functionalized SWNT as tip[55]. Figure 17 shows the scheme of experimental set-up and the mapping of a hydroxylated surface. This experimental arrangement allows one to determine the acid-base properties of tiny surfaces.

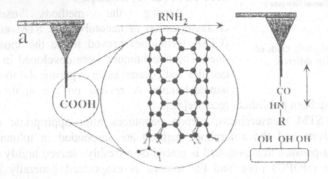

17. Figure

Schematic illustration of the modification of a SWNT tip by coupling an amine (R-NH₂) to a terminal –COOH group and the application of this probe. (ref [55])

Application of carbon nanotubes

Filling of carbon nanotubes

Following the discovery of carbon nanotubes, numerous attempts were made to fill them with various materials. The most prosperous challenge was to fill the NT with metals to produce metal nanowire in the tubes. There are two possible ways to insert nanoparticles into the NT. The first consists in filling the tubes during their preparation. Generally both in CCVD and the arc discharge methods result in NT containing encapsulated metal or metal carbide particles (metal has to be doped with carbon in the latter case).

Systematic studies showed that some metals are more suited than others to form continuous nanowires of constant diameter (Cr, Gd..), while some others induced a non-complete filling with encapsulated particles or cones (Pd, Fe, Co…). The ability to form wires depended on the electronic structure of the most stable ionic state of metal, since an increasing number of holes in the electronic shell led to long nanowires[56].

The filling by capillary forces involves in a first step the opening of the tubes. The capillary effect in carbon nanotubes was reported first by Ajayan and Iijima who succeeded in opening the tip of nanotubes and introducing low melting point metal (Pb, Bi) compounds into the tube interior[57].

Ugarte et al. reported that opening the nanotubes before filling can be avoided in some cases[58]. These authors used nitrate salt of Ag, Co and Cu for filling and assumed that oxygen liberated upon thermal treatment of the salt-tube system erodes the tip of tubes, giving access to the interior. Ni [59], Rh[60], Pd[61], Y, Sn[62] Au, Pt, Ag particles have been filled into the NT as well.

Successful filling of nanotubes is of great interest in the perspective of application. In the simplest case, the encapsulated materials are protected from causing environmental damage. For instance, encapsulated nickel nanoparticles were not damaged when stored for several months in aqueous solution[63].

Filled carbon nanotubes can be used in physics as potential nanoelectrodes or nanowires in nanoscale devices.

Metal or metal oxide encapsulated in the NT may be potential catalysts and/or catalyst supports. Ruthenium clusters incorporated into carbon NT were successfully applied as catalyst for liquid phase hydrogenation of cinnamaldehyde[64]. Nanotubes may serve as nanosized test tubes to carry out chemical reactions inside the tube[65].

Rh-phosphine supported on MWNTs exhibited excellent catalytic activity and enhanced regioselectivity in hydroformylation of propylene[66]. It was assumed that the nm size channels contribute dominantly to the appearance of regioselectivity towards formation of butyraldehyde.

Hydrogen storage in carbon nanotubes

It is well known that graphite and its modified derivatives adsorb hydrogen in quantities large enough to be worth studying as hydrogen storage reservoir. As the structure of carbon nanotube is similar to that of graphite, i.e. it can be regarded as a rolled up graphite sheet, a great research effort is focused at the investigation of hydrogen adsorption in SWNTs (and also in MWNTs). The question addressed is to determine which compound adsorbs the higher amount of hydrogen (SWNTs or slit pores graphite), and whether hydrogen adsorption takes place in the interstitial channels between adjacent nanotubes in a rope of SWNTs or in the interior of nanotubes. Calculations based on geometrical considerations showed that 2.3 w% hydrogen may be adsorbed between the tubes and around 3.3 w% in the tubes. Table III. gives some data about the experimental adsorption capacities of NTs (data taken from ref [67]).

Table III.
Adsorption capacities of different samples of SWNTs

Adsorbent	Max wt% H_2	T/K	P/MPa
SWNT low purity	**5-10**	**133**	**0.04**
SWNT high purity	4	300	0.04
SWNT high purity	8.25	80	7.18
SWNT 50% purity	4.2	300	10.1

Application as polymer fillers

Carbon nanotubes are extremely flexible. This characteristic property assures their application as polymer fillers. When Wagner et al. prepared a composite using epoxy resin and the product polymer was tested, very surprising behaviors were observed[68]. As it can be seen in Figure 18, carbon nanotubes in composites behave as elastic rods upon compression.

When a polymer-nanotube composite was uniaxially stretched at 373 K, the elongated polymer kept its length after removal of the load at room temperature[69]. The orientation and the degree of alignment were preserved as well. Comparison of TEM images taken on carbon nanofibers- and carbon nanotubes-containing composite materials show that the fibers pull out in an internal fracture surface of the composite after being sliced parallel to the stretching direction. The nanotubes were found to be aligned with their longitudinal axes parallel to the stretching direction. The stress transfer efficiency of the multiwall nanotubes-containing composite has been estimated to be at least one order of magnitude larger than that of conventional fiber-based composites[70].

Besides the mechanical reinforcement character of nanotubes, they show some peculiar electronic behavior in polymer composites. Less than 10 % nanotube loading can confer high conductivity to polycarbonates, nylon, polyesters and polyamides. This loading is much lower than the one generally required from carbon fibers, carbon black or stainless steel fibers[71] to obtain the same result. The future of such composites cannot be predicted yet.

Last but not least, a very intriguing application of nanotubes is discussed. Carbon nanotubes-reinforced aluminum composites have been produced[72]. It was found that the nanotubes in the composites are not damaged during the composite preparation and that no reaction products at the aluminum-nanotube interface are observable after treatment at 983 K for 24 hours. Authors concluded that due to the excellent mechanical properties and chemical stability of the nanotubes, the engineering application of nanotubes for metal matrix composites could be ideal.

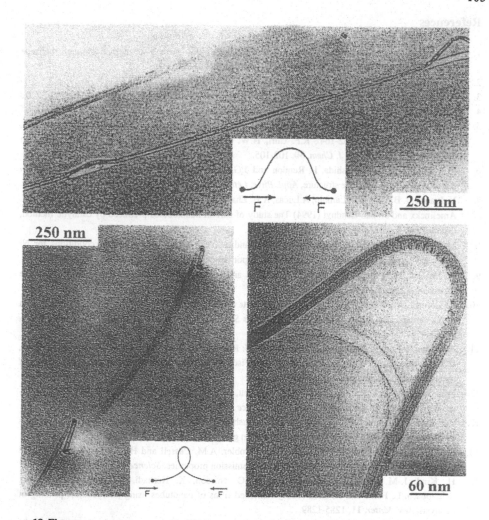

18. Figure

TEM micrographs of long and slender MWNTs under compression (ref [68])

106

References

1. A. Oberlin, M. Endo and T. Koyama (1976) Filamentous growth of carbon through benzene decomposition, *J. Cryst. Growth* **32**, 335-349.
2. R.T.K. Baker (1989) Catalytic growth of carbon filaments, *Carbon* **27**, 315-.323.
3. N.M. Rodriguez (1993) A review of catalytically grown carbon nanofibers, *J. Mater. Res.* **8**, 3233-3250.
4. S. Iijima (1991) Helical microtubules of graphitic carbon, *Nature* (London) **354**, 56-58.
5. H.W. Kroto, J.R. Heath, S.C. O'Brian, R.F. Curl, R.E. Smalley (1985) C-60 - Buckminsterfullerene, Nature (London) **318**, 162-164., R.F. Curl, H.W. Kroto and R.E. Smalley (1997) Nobel prize in chemistry for 1996, *S. Afr. J. Chem.* **50**, 102-105.
6. M.J. Yacaman, M.M. Yoshida, L. Rendon and J.G. Santiesteban (1993) Catalytic growth of carbon microtubules with fullerene structure, *Appl. Phys. Lett.* **62**, 202-204
7. V. Ivanov, J. B.Nagy, Ph. Lambin, A.Lucas, X.B. Zhang, X.F. Zhang, D. Bernaerts, G. Van Tendeloo, S. Amelinckx and J. Van Landuyt (1994) The study of carbon nanotubules produced by catalytic method, *Chem. Phys. Lett.* **223**, 329-335.
8. M. Yudasaka, R. Kituchi, T. Matsui, Y. Ohki and S. Yoshimura (1995) Specific conditions for Ni catalyzed carbon nanotube growth by chemical-vapor-deposition, *Appl. Phys. Lett.* **67**, 2477-2479.
9. K. Hernadi, A Fonseca, J. B.Nagy, D. Bernaerts and A. Lucas (1996) Fe-catalyzed carbon nanotube formation, *Carbon* **34**, 1249-1257.
10. K. Hernadi, A Fonseca, J. B.Nagy and D. Bernaerts (1998) Catalytic Synthesis of Carbon Nanotubes, in S. Yoshimura and R.P.H. Chang (eds.), *Materials Science, Supercarbon*, Springer-Verlag, Heidelberg, pp. 81-97.
11. A. Thess, R. Lee, P. Nikolaev, H.J. Dai, P. Petit, J. Robert, C.H. Xu, Y.H. Lee, S.G. Kim, A.G. Rinzler, D.T. Colbert, G.E. Scuseria, D. Tomanek, J.E. Fischer and R.E. Smalley (1996) Crystalline ropes of metallic carbon nanotubes, *Science* **273**, 483-487.
12. W.Z. Li, S.S. Xie, L.X. Qian, B.H. Chang, B.S. Zou, W.Y. Zhou, R.A. Zhao and G. Wang (1996) Large-scale synthesis of aligned carbon nanotubes, *Science* **274**, 1701-1703.
13. Z.F. Ren, Z.P. Huang, J.W. Xu, J.H. Wang, P. Bush, M.P. Siegal and P.N. Provencio (1998) Synthesis of large arrays of well-aligned carbon nanotubes on glass, *Science* **282**, 1105.
14. S. Fan, M.G. Chapline, N.R. Franklin, T.W. Tombler, A.M. Cassell and H. Dai (1999) Self-oriented regular arrays of carbon nanotubes and their field emission properties, *Science* **283**, 512-514.
15. H. Kind, J.-M. Bonard, C. Emmenegger, L.-O. Nilsson, K. Hernadi, E. Maillard-Schaller, L. Schlapbach, L. Forró and K. Kern (1999) Patterned films of nanotubes using microcontact printing of catalysts, *Adv. Mater.* **11**, 1285-1289.
16. M. Yudasaka, R. Kikuchi, Y. Ohki and S. Yoshimura (1997) Nitrogen-containing carbon nanotube growth from Ni phthalocyanine by chemical vapor deposition, *Carbon* **35**, 195-201.
17. M. Yudasaka, R. Kikuchi, T. Matsui, Y. Ohki, E. Ota and S. Yoshimura (1996) Graphite film formation by chemical vapor deposition on Ni coated sapphire, *Carbon* **34**, 763-767.
18. T. Kyotani, L. Tsai and A. Tomita (1996) Preparation of ultrafine carbon tubes in nanochannels of an anodic aluminum oxide film, *Chem. Mater.* **8**, 2109-2113.
19. H.J. Dai, A.G. Rinzler, P. Nikolaev, A. Thess, D.T. Colbert and R.E. Smalley (1996) Single-wall nanotubes produced by metal-catalyzed disproportionation of carbon monoxide, *Chem. Phys. Lett.* **260**, 471-475.
20. A. Fonseca, K. Hernadi, P. Piedigrosso, L.P. Biro, S.D. Lazarescu, P. Lambin, P.A. Thiry, D. Bernaerts and J. B.Nagy (1997) Synthesis of carbon nanotubes over supported catalysts, in K.M. Kadish and R.S. Ruoff (eds.), *Fullerens Volume IV.: Recent advances in the physics and chemistry of fullerenes and related materials*, The Electrochemical Society Inc., Proceedings Volume **97-14**, 884-906.

21. J. Kong, A.M. Cassell and H.J. Dai (1998) Chemical vapor deposition of methane for single-walled carbon nanotubes, *Chem. Phys. Lett.* **292**, 567-574.

22. A. Peigney, C. Laurent, F. Dobigeon and A. Rousset (1997) Carbon nanotubes grown in situ by a novel catalytic method, *J. Mater. Res.* **12**, 613-615., E. Flahaut, A. Govindaraj, A. Peigney, C. Laurent, A. Rousset and C.N.R. Rao (1999) Synthesis of single-walled carbon nanotubes using binary (Fe, Co, Ni) alloy nanoparticles prepared in situ by the reduction of oxide solid solutions, *Chem. Phys. Lett.* **300**, 236-242.

23. H.M. Cheng, F. Li, G. Su, H.Y. Pan, L.L. He, X. Sun and M.S. Dresselhaus (1998) Large-scale and low-cost synthesis of single-walled carbon nanotubes by the catalytic pyrolysis of hydrocarbons, *Appl. Phys. Lett.* **72**, 3282-3284.

24. B.C. Satishkumar, A. Govindaraj, R. Sen and C.N.R. Rao (1998) Single-walled nanotubes by the pyrolysis of acetylene-organometallic mixtures, *Chem. Phys. Lett.* **293**, 47-52.

25. J.-F. Colomer, G. Bister, I. Willems, Z. Konya, A. Fonseca, G. Van Tendeloo and J. B.Nagy (1999) Synthesis of single-wall carbon nanotubes by catalytic decomposition of hydrocarbons, *J. Chem. Soc., Chem. Commun.* 1343-1344.

26. J.-F. Colomer, C. Stephan, S. Lefrant, G. Van Tendeloo, I. Willems, Z. Kónya, A. Fonseca, Ch. Laurent and J. B.Nagy (2000) Large-scale synthesis of single-wall carbon nanotubes by catalytic chemical vapor deposition (CCVD) method, *Chem. Phys. Lett.* **317**, 83-89.

27. M. Endo and H. W. Kroto (1992) Formation of carbon nanofibers, *J. Phys. Chem.* **96**, 6941-6944.

28. D. Ugarte (1992) Morphology and structure of graphitic soot particles generated in arc-discharge C_{60} production, *Chem. Phys. Lett.* **198**, 596-602.

29. A.V. Eletskii and B.M. Smirnov (1995) Fullerines and the structure of carbon, *Usp. Fiz. Nauk.*, **165**, 977-1009.

30. Y. Saito, T. Yoshikawa, M. Inagaki, M. Tomita and T. Hayashi (1993) Growth and structure of graphitic tubules and polyhedral particles in arc-discharge, *Chem. Phys. Lett.* **204**, 277-282.

31. Y.E. Lozovik and A.M. Popov (1994) Carbon spheric nanoparticles – possible formation mechanism, *Phys. Lett. A.* **189**, 127-130., Y.E. Lozovik and A.M. Popov (1995) Mechanism of formation of carbon nanoparticles in an electric arc, *Teplofiz. Vys. Temp.* **33**, 534-539.

32. P.M. Ajayan, T. Ichihashi and S. Iijima (1993) Distribution of pentagons and shapes in carbon nanotubes and nanoparticles, *Chem. Phys. Lett.* **202**, 384-388.

33. T.W. Ebbesen, P.M. Ajayan, H. Hiura and K. Tanigaki (1994) Purification of nanotubes, *Nature* (London) **367**, 519-519.

34. M. Ge and K. Sattler (1993) Scanning-Tunneling-Microscopy of vapor-phase grown nanotubes of carbon, *J. Phys. Chem. Solids* **54**, 1871-1877.

35. F.J. Derbyshire, A.E.B. Presland and D.L. Trimm (1975) Graphite formation by the dissolution-precipitation of carbon in cobalt, nickel and iron, *Carbon* **13**, 111-113.

36. R.T.K. Baker and P.S. Harris (1978) Formation of Filamentous Carbon in Marcel Dekker (ed.) *Chemistry and Physics of Carbon 14*, New York, pp. 83.

37. Á. Kukovecz, Z. Kónya, N. Nagaraju, I. Willems, A. Tamási, A. Fonseca, J. B.Nagy and I. Kiricsi (2000) Catalytic Synthesis of Carbon Nanotubes Over Co, Fe and Ni Containing Conventional and Sol-gel Silica-Aluminas, *Phys. Chem. Chem. Phys.* **13**, 3071-3076.

38. T.W. Ebbesen (1994) Carbon nanotubes, *Annu. Rev. Mater. Sci.* **24**, 235-264.

39. N. Yao, V. Lordi, S.X.C. Ma, E. Dujardin, A. Krishnan, M.M.J. Treacy and T.W. Ebbesen (1998) Structure and oxidation patterns of carbon nanotubes, *J. Mater. Res.* **13**, 2432-2437.

40. B.C. Satishkumar, A. Govindaraj, J. Mofokeng, G.N. Subbanna and C.N.R. Rao (1996) Novel experiments with carbon nanotubes: Opening, filling, closing and functionalizing nanotubes, *J. Phys. B: At. Mol. Opt. Phys.* **29**, 4925-4934.

41. M.A. Hamon, J. Chen, H. Hu, Y.S. Chen, M.E. Itkis, A.M. Rao, P.C. Eklund and R.C. Haddon (1999) Dissolution of single-walled carbon nanotubes, *Adv. Mater.* **11**, 834-840.

42. D.B. Mawhinney, V. Naumenko, A. Kuznetsova and J.T. Yates Jr. (2000) Infrared spectral evidence for the etching of carbon nanotubes: Ozone oxidation at 298 K, *J. Am. Chem. Soc.* **122**, 2383-2384.

43. Y. Chen, R.C. Haddon, S. Fang, A.M. Rao, W.H. Lee, E.C. Dickey, E.A. Grulke, J.C. Pendergrass, A. Chavan, B.E. Haley, R.E. Smalley (1998) Chemical attachment of organic functional groups to single-walled carbon nanotube material, *J. Mater Res.* **13**, 2423-2431.

44. A. Hamwi, H. Alvergnat, S. Bonnamy, F. Beguin (1997) Fluorination of carbon nanotubes, *Carbon* **35**, 723-728.

45. A. Hamwi, M. Daoud, J.C. Cousseins (1988) Graphite fluorides prepared at room temperature 1. Synthesis and characterization, *Synthetic Metals* **26**, 89-98.

46. Z. Ming, C.L. Xu, L.M. Cao, D.H. Wu and W.K. Wang (2000) Deformation of catalytically grown carbon nanotubes induced by annealing under high pressure, *J. Mater. Res.* **15**, 253-261.

47. S. Bandow, S. Asaka, X. Zhao and Y. Ando (1998) Purification and magnetic properties of carbon nanotubes, *Appl. Phys. A..* **67**, 23-27.

48. J. Kurti, M. Milnera, M. Hulman, O. Zhou and H. Kuzmany (1999) About the composition of single wall carbon nanotube budles: An analysis from optical, Raman and IR experiments, AIP Conference Proceedings 486, XIII International Winterschool, Kirchberg, Austria, pp. 278-283.

49. H.Katura,Y. Kumazawa, N. Kojima, Y. Maniwa,, I. Umezu, S. Masubuchi, S. Kazama, X. Zhao, Y. Ando, Y. Ohtsuka, S. Suzuki and Y. Achiba (1999) Optical adsorption and resonance Raman scattering of carbon nanotubes, AIP Conference Proceedings 486, XIII International Winterschool, Kirchberg, Austria, pp. 328-332.

50. O. Jost, R. Friedlein, A.A. Gurbunov, T. Pichler, M. Reibold, H.-D. Bauer, M. Knupfer, M.S. Golden, L. Dunsch, J. Fink and W. Pompe (1999) The characterization of SWNTs containing soot by optical spectroscopy, AIP Conference Proceedings 486, XIII International Winterschool, Kirchberg, Austria, pp. 288-291.

51. H. Kataura, Y. Kumazawa, Y. Maniwa, I. Umezu, S. Suzuki, Y. Ohtsuka, Y. Achiba (1999) Optical properties of single-wall carbon nanotubes, *Synthetic Metals* **103**, 2555-2558.

52. M.S. Dresselhaus, G. Dresselhaus and P.C. Eklund (1996) *Science of Fullerenes and Carbon Nanotubes*, Academic Press, San Diego.

53. L.P. Biro, S. Lazarescu, P. Lambin, P.A. Thiry, A. Fonseca, J. B.Nagy and A.A. Lucas (1997) Scanning tunneling microscope investigation of carbon nanotubes produced by catalytic decomposition of acetylene, *Phys. Rev. B.* **56**, 12490-12498.

54. S.S. Wong, J.D. Harper, P.T. Lansbury and C.M. Lieber (1998) Carbon nanotube tips: High-resolution probes for imaging biological systems, *J. Am. Chem. Soc.* **120**, 603-604.

55. S.S. Wong, A.T. Wooley, E. Joselevich, C.L. Cheung and C.M. Lieber (1999) Functionalization of carbon nanotube AFM probes using tip-activated gases, *Chem. Phys. Lett.* **306**, 219-225.

56. C. Guerret-Piecourt, Y.LeBouar, A. Loiseau and H. Pascard (1994) Relation between metal electronic-structure and morphology of metal-compounds inside carbon nanotubes, *Nature* (London) **372**, 761-765.

57. P.M. Ajayan and S. Iijima (1993) Capillarity induced filling of carbon nanotubes, *Nature* (London) **361**, 333-334.

58. D. Ugarte, T. Stockli, J.M. Bonard, A. Chatelain and W.A. de Heer (1998) Filling carbon nanotubes, *Appl. Phys. A.* **67**, 101-105.

59. J. Li, M. Moskovits, T.L. Haslett (1998) Nanoscale electroless metal deposition in aligned carbon nanotubes, *Chem. Mater.* **10**, 1963-1967.

60. J. Cook, J. Sloan, R.J.R. Heesom, J. Hammer and M.L.H. Green (1996) Purification of rhodium-filled carbon nanotubes using reversed micelles, *J. Chem Soc., Chem Commun.*, 2673-2674.

61. R.M. Lago, S.C. Tsang, K.L.Lu, Y.K. Chen and M.L.H. Green (1995) Filling carbon nanotubes with small palladium metal crystallites – the effect of surface acid groups, *J. Chem Soc,. Chem Commun.*, 1355-1356.

62. T.W. Ebbesen, H. Hiura, M.E. Bisher, M.M.J. Treacy, J.L. ShreeveKeyer and R.C. Haushalter (1996) Decoration of carbon nanotubes, *Adv. Mater.* **8**, 155-157.

63. V.P. Dravid J.J. Host, M.H. Teng, B. Elliot, J.H. Hwang, D.L. Johnson, T.O. Mason and J.R. Weertman (1995) Controlled-size nanocapsules, *Nature* (London) **374**, 602-602.

64. J.M. Planeix, N. Coustel, B. Coq, V. Brotons, P.S. Kumbhar, R. Dutarte, P. Geneste, P. Pernier and P.M. Ajayan (1994) Application of carbon nanotubes as supports in heterogeneous catalysis, *J. Am. Chem. Soc.* **116**, 7935-7936.

65. D. Ugarte, A. Chatelain and W.A. de Heer (1996) Nanocapillarity and chemistry in carbon nanotubes, *Science* **274**, 1897-1899.

66. Y. Zhang, H.B. Zhang, G.D. Lin, P. Chen, Y.Z. Yuan and K.R. Tsai (1999) Preparation, characterization and catalytic hydroformylation properties of carbon nanotubes-supported Rh-phosphine catalyst Y, *Applied Catal.* **187**, 213-224.

67. M.S. Dresselhaus, K.A. Williams and P.C. Eklund (1999) Hydrogen adsorption in carbon materials, *MRS Bulletin* **24**, 45-50.

68. O. Lourie, D.M. Cox and H.D. Wagner (1998) Buckling and collapse of embedded carbon nanotubes, *Phys Rev. Lett.* **81**, 1638-1641.

69. L. Jin, C. Bower and O. Zhou (1998) Alignment of carbon nanotubes in a polymer matrix by mechanical stretching, *Appl. Phys Lett.* **73**, 1197-1199.

70. H.D. Wagner, O. Lourie, Y. Feldman and R. Tenne (1998) Stress-induced fragmentation of multiwall carbon nanotubes in a polymer matrix, *Appl. Phys. Lett.* **72**, 188-190.

71. R. Dagani (1999) Putting the 'nano' into composites, Chemical and Engineering News, 25-37.

72. T. Kuzumaki, K. Miyazawa, H. Ichinose and K. Ito (1998) Processing of carbon nanotube reinforced aluminum composite, *J. Mater Res.* **13**, 2445-2449.

OPTIMIZING GROWTH CONDITIONS FOR CARBON FILAMENTS AND VAPOR-GROWN CARBON FIBERS

J.L.FIGUEIREDO[1] and Ph. SERP[2]

1)Laboratório de Catálise e Materiais, Faculdade de Engenharia, Universidade do Porto, 4050-123 Porto, Portugal

2)Laboratoire de Catalyse, Chimie Fine et Polymères, Ecole Nationale Supérieure de Chimie de Toulouse, 118 Route de Narbonne, 31077 Toulouse France

1. Introduction

The production, properties and applications of vapor-grown carbon fibers (VGCF) and carbon filaments were reviewed in the course of a previous NATO ASI [1].

It is generally accepted that the growth of carbon fibers from gaseous hydrocarbons in the presence of a catalyst (VGCF) occurs as a result of the initial formation of carbon filaments or nanofibers, by a mechanism involving the dissolution and diffusion of carbon at the exposed surface of the catalyst particles and precipitation at the metal/support interfaces. In this way, the catalyst particles are carried out on top of the growing filaments [2,3]. This lengthening stage stops when the catalyst particles become covered with a carbon layer; thereafter, thickening occurs by chemical vapour deposition of pyrolytic carbon, leading to the production of VGCF exhibiting the well known "tree-trunk" structure of concentric carbon layers [4-6]. Originally, a Vapor-Solid process (V-S) was invoked, as the catalyst particles were considered to be in the solid state [7]. However, it was later recognised that, in order to account for the observed rates of fiber growth, a Vapour-Liquid-Solid (V-L-S) mechanism must be involved in the production of VGCF. Indeed, Benissad et al. [8] proposed that the catalyst particles can be molten, according to their sizes, in the temperature range where fiber lengthening occurs (1050-1100°C). More recently, Tibbetts and Balogh [9] showed that only molten particles are effective in catalyzing fiber growth.

Two different methods have been developed to produce VGCF: The substrate seeding method and the floating catalyst, or fluidization seeding method [10,11]. The latter has the advantage of being a continuous process, but it produces shorter fibers (some micrometers) with smaller diameters (hundreds of nanometers) when compared to the first method (which leads to fibers several centimeters long and some micrometers in diameter).

This chapter describes the influence of different experimental conditions on the

L.P. Biró et al. (eds.), Carbon Filaments and Nanotubes: Common Origins, Differing Applications, 111–120.

yields and properties of vapor-grown carbon fibers produced on a substrate seeded with different catalyst precursors. This in turn allows us to identify a number of guidelines which can be used to optimize the growth conditions of carbon filaments and VGCF. Among the parameters that have been studied, the following have been shown to be particularly important in determining the yields and morphology (fiber diameter and length, smooth or crenulated surface) of the VGCF produced: The nature of the catalyst precursor; the presence or absence of impurities such as sulfur; the presence of hydrocarbon during the catalyst reduction stage, the composition of the gas mixture, and the heating rate imposed during the fiber lengthening step. The effects of these parameters on other properties of the VGCF will also be addressed.

2. Production of VGCF on a Substrate

VGCF were grown from a methane-hydrogen mixture on a Grafoil® support seeded with an iron catalyst. The catalyst was deposited on the support by spraying a solution of the appropriate precursor. In the usual procedure, the volume of solution used was such that the resulting catalyst concentration on the support was $8.3 \ 10^{-5} \ g_{Fe}.cm^{-2}$. The iron catalyst precursors used in this work were: $Fe^{III}(NO_3)_3.9H_2O$, $Fe^{II}Cl_2.4H_2O$, $Fe^{III}Cl_3.6H_2O$, $Fe^{II}SO_4.7H_2O$, $Fe_2^{III}(SO_4)_3.5H_2O$, $Fe_3^{0}(CO)_{12}$. After drying at 110 °C, the substrate was inserted in a mullite tube externally heated by an electric furnace. Hydrogen (99.99995 %) and methane (99.995 %) were fed through mass flow controllers, and the linear flow velocity in the reactor was set at 0.4 cm/s [12].

Two different production cycles were used (Figure 1), based on the idea that the growth of VGCF consists of successive steps, namely the reduction and activation of the catalyst particles, followed by filament lengthening and by a final thickening stage. In cycle A, the events are clearly separated (decomposition of the precursor under nitrogen, reduction of the catalyst with hydrogen, and then fiber growth under the reaction mixture of methane and hydrogen); Cycle B corresponds to an optimized version of cycle A, after Benissad et al. [6]. In a more recent publication from the same group [13], it was shown that the yield of VGCF increases with the methane concentration in the gas phase up to 40% CH_4, decreasing thereafter. This behaviour is explained in terms of the competition between the formation of filaments (VGCF precursors) and of pyrolytic carbon (which is favoured at higher methane concentrations). On the other hand, 31% methane provided the best results in terms of fiber diameter (7-17 μm) and length (1-10 cm) with the $Fe_3(CO)_{12}$ precursor. Higher methane concentrations led to thicker and shorter fibers. These findings confirm that the composition of the reaction mixture (30% CH_4 - 70% H_2) selected for cycles A and B is an adequate choice.

The performance of each production cycle was evaluated in terms of the following parameters:

a) average fibre diameter and length (measured by SEM on 50 fibers);

b) number of crenulations per 100μm of fibre length (measured by SEM on 50 fibers);

c) mass of deposited carbon, measured by the difference in weight of the grafoil before and after reaction;

d) Mass of VGCF determined by shaving off the support, washing the fibers with ethanol in an ultra-sound bath for soot removal, and drying.

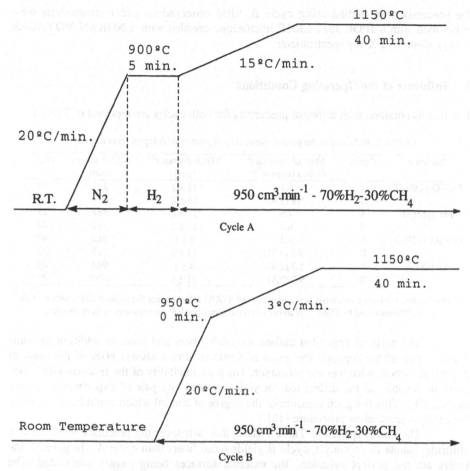

Figure 1 – Conditions for fiber growth in: Cycle A; Cycle B

During some experiments, the process was interrupted after the reduction stage in order to measure the particle size of the iron catalyst. Three different precursors: $Fe(NO_3)_3.9H_2O$, $Fe_2(SO_4)_3.5H_2O$ and $Fe_3(CO)_{12}$, were used in both cycles of VGCF growth (cycles A and B). In the case of cycle A, the run was stopped after the 5 minutes reduction step at 900°C; cycle B was interrupted at 950°C, this temperature corresponding to the end the calcination/reduction step. For each cycle, the three precursors were sprayed on three different zones of equal area, on the same substrate and with the same concentration of iron (8.3 10^{-5} $g_{Fe}.cm^{-2}$). Then, after each experiment, small portions of the support were cut and observed under SEM. The mean catalyst particle size was measured on 200 particles at least, although the precision of these measurements became poor in the case of the very small particles.

The influence of the production cycle (composition of the gas phase, heating rate) on the VGCF yield was studied using $Fe_3(CO)_{12}$ as precursor, and the influence of the precursor was studied using cycle B. SEM observations and microanalysis were performed with a JEOL JSM 6301F microscope coupled with a NORAN VOYAGER energy dispersive X-ray spectrometer.

3. Influence of the Operating Conditions

The results obtained with different precursors for both cycles are reported in Table 1.

TABLE 1. Influence of the production cycle and precursor [Adapted from ref. 12]

Precursor	Cycle	Mass of deposited carbon (mg/cm²)[a]	VGCF diameter[b] (μm)	VGCF length (mm)	NC[c]
$Fe(NO_3)_3.9H_2O$	A	4.1	16 ± 2	3±1	**57**
	B	5.9 (3.4)	15 ± 2	4±2	22
$FeCl_3.6H_2O$	A	2.9	21 ± 2	4±2	**25**
	B	6.1	21 ± 2	5±2	23
$Fe_2(SO_4)_3.5H_2O$	A	4.3	9 ± 1	5±2	40
	B	9.5 (5.3)	13 ± 2	7±3	32
$Fe_3(CO)_{12}$	A	5.2 (2.9)	9 ± 1	9±4	**48**
	B	9.7 (5.8)	11 ± 1	17±8	25

[a] The number between parenthesis is the real mass of VGCF, after shaving the support and removal of the soot. [b] Determined by SEM. [c] Number of crenulations per 100 micrometers of fiber (length).

The mass of deposited carbon includes fibers and soot, in addition to some densification of the support. The mass of fibers is almost always 60% of the mass of deposited carbon, whatever the precursor. The reproducibility of the reaction yields was quite reasonable, as the differences in yield between any pair of experiments did not exceed 10%. This is a good measure of the degree of control which could be achieved in the catalyst preparation procedure [12].

The results quoted in Table 1 show that, whatever the precursor used (nitrate, chloride, sulfate or carbonyl), cycle B yields more fibers than cycle A. In general, the fibers are not perfect cylinders, the external surfaces being mostly crenulated. The formation of crenulations has been explained as a stress relieving mechanism which occurs on cooling down [11]. The number of crenulations quoted in bold characters in Table 1 corresponds to pronounced crenulations, while the normal script refers to smooth crenulations. Analysis of Table 1 shows that the number of crenulations in the VGCF produced in cycle A is much larger for most precursors; moreover, pronounced crenulations were not observed in cycle B. Such crenulations have been shown to be responsible for a significant decrease in the tensile strength of the VGCF [14], and therefore the performance of cycle A in this respect is clearly lower when compared with cycle B.

3.1 INFLUENCE OF THE IRON PRECURSOR

We will compare the results obtained with the different precursors in cycle B, as it yields more fibers.

Table 1 shows that the precursor $Fe(NO_3)_3.9H_2O$ gives the worst yield of VGCF. An attempt to correlate this result with the size of the catalyst particles was tried first. Only the *active* particles, *i.e.* those with diameter less than 200nm, were considered. Indeed, from SEM observations it appears that the larger particles, which comprise only 10 to 15% of the total, are not active in the VGCF growth process. Although some uncertainty exists in the SEM measurements, it appears that there are no significant differences in the Fe particle sizes among the precursor used. Indeed, we found mean diameters of 25, 26, and 31nm for $Fe(NO_3)_3.9H_2O$, $Fe_2(SO_4)_3.5H_2O$ and $Fe_3(CO)_{12}$ respectively. Thus it appears that the particle size cannot explain by itself the differences in VGCF yield obtained using these three precursors.

We propose that the differences in yield that were observed with different precursors can be explained in terms of reactivity at the front face of the catalyst particle, and that this reactivity can be controlled by the absence or the presence of certain impurities. Thus, in the case of the sulfate precursor, some sulfur remains on the particles even after the reduction step at 950°C (Cf. Section 3.3 below). As sulfur is known to enhance filament nucleation [15], it is probable that it promotes also the yield of fibers. The fibers produced with this precursor are of medium average diameter (13 ± 2 μm) and medium length (≈7 mm). For the carbonyl precursor, it is well established that the decomposition-reduction process of this kind of complex leads to the production of small and pure aggregates of metal [16]. In this case the fibers are of medium average diameter (11 ± 1 μm) and relatively long (≈ 17 mm). In the case of the nitrate, as EDS does not allow the identification of nitrogen, it is quite difficult to identify a precursor. The VGCF produced with iron nitrate exhibit a large average diameter (15 ± 2 μm) and are relatively short (≈ 4 mm).

3.2 INFLUENCE OF THE PRODUCTION CYCLE

In this section we will compare the results obtained with the $Fe_3(CO)_{12}$ precursor in both cycles, because this precursor leads to pure iron particles and to higher VGCF yields. These cycles differ in both gas phase composition during the catalyst activation step and in heating rate.

The main role of the gas phase is to influence the particle sizes of the iron catalyst obtained on the support. SEM micrographs show that the diameters of the Fe crystallites obtained in cycle B are significantly lower, with filaments seen on the support even at 900°C. It is obvious that the presence of methane during catalyst activation (as in cycle B) is beneficial, as filament growth fixes the diameter of the Fe particles, preventing sintering [17]. However, it is difficult to correlate the yield with only the particle size; indeed, it was observed that it was not always the procedure with the smallest particle size which led to the best results [12].

After the catalyst activation stage, *i.e.* above 900°C, the composition of the gas phase is the same for both cycles; therefore, the effect of the different heating rates must be considered. In order to show that this is the most important factor as far as the yield of VGCF was concerned, a modified cycle B, with the same heating rate as cycle A (15°C/min) was tested. The result of this experiment was a mass of deposited carbon of 5.5 mg. cm^{-2} (3.6 mg. cm^{-2} of VGCF of 10 mm length), which is in the same range as cycle A. Therefore, the higher yields obtained with cycle B are a consequence of the lower heating rate used, which allows the fibers to grow longer (Cf. Table 1) before the catalyst particles become encapsulated with a layer of pyrolytic carbon, in agreement with the findings of other authors [18-19].

3.3 INFLUENCE OF SULFUR

Several authors report that the addition of sulfur to the gas phase (usually introduced as thiophene or hydrogen sulfide) promotes the formation of VGCF [9,15,20-23].

This section looks at the influence of sulfur on the VGCF yield and properties, using reaction conditions that allow the production of carbon fibers without any sulfur source. Hydrogen sulfide was added to the gas phase during the pretreatment of the catalyst by means of a permeation device (VICI Metronics ®), in such a way that the molar ratio S/Fe=0.5 [20]. A modifyed cycle B was used, with 60 or 120 minutes of thickening time at 1150°C.

Representative results obtained starting from different precursors are listed in Table 2. The diameters and lengths reported are representative of the longest fibers.

TABLE 2. Influence of sulfur [Adapted from ref. 20]

Precursor	H_2S	Mass of deposited carbon (mg/cm^2) [a]	VGCF diameter (μm) [b]	VGCF length (mm) [b]
Fe(NO$_3$)$_3$.9H$_2$O	no	3.6	15 ± 3	9
	yes	6.0	18 ± 4	20
FeCl$_3$.6H$_2$O	no	4.0	21 ± 7	12
	yes	5.4	20 ± 7	22
Fe(SO$_4$).7H$_2$O	no	6.5	16 ± 4	21
	yes	6.6	21 ± 7	20
Fe$_3$(CO)$_{12}$	no	5.0	11 ± 3	21
	yes	6.1	12 ± 4	26

[a] 60 min thickening, [b] 120 min thickening

When no sulfur is present in the gas phase, it may be observed that the masses of deposited carbon (mainly carbon fibers with some additional soot) are very similar when nitrate or chloride precursors are used. Fe$_3$(CO)$_{12}$ produces significantly longer fibers and better yield, which can be explained in terms of the smaller size and higher purity of the metal particles produced from this precursor. We believe that the results obtained with the sulfate precursor cannot be explained just by a particle size effect. They are also due to the presence of some sulfur in the iron particles even after the reduction stage. Thus, a temperature programmed reduction experiment performed under pure hydrogen on this precursor confirmed that the mass loss did not correspond to a complete reduction of the metal. Furthermore, semi-quantitative XPS analysis of the surface of the substrate, performed after the reduction step, showed the presence of sulfur when starting from sulfate (atomic ratio Fe/S: 30/1). No sulfur has been detected by XPS after the reduction step of the nitrate precursor. The promoting effect of sulfur on the yield and length of the fibers is also observed in the case of the other precursors when H$_2$S is added to the gas phase (Cf. results in Table 2). A more pronounced effect is noticed with the nitrate and chloride. No special effect of sulfur on the yield and length of the VGCF is found when the sulfate is used as starting material. Under the present experimental conditions, the promoting effect of sulfur was noticed provided the sulfur source was introduced during the pretreatment of the catalyst, i.e. before 950°C.

Indeed, no significant amount of fibers was formed when hydrogen sulfide was introduced along the entire duration of a run.

Table 2 shows that sulfur influences mainly the length of the fibers, and to a less extent the diameter. Thus, we believe that the promoting role of sulfur occurs before the thickening stage of the reaction (1150°C). Indeed, during an experiment stopped at 1150°C (0 min. of thickening) using the nitrate precursor on one part of the support and sulfate on the other part, fibers of some centimeters length were observed on the sulfate side, whereas the fibers on the nitrate side were only of some millimeters length. The diameters of these fibers were almost the same (1 μm). The thickening stage that followed did not affect the length of the fibers [20].

It has been argued that sulfur could promote the VLS mechanism by decreasing the melting point of the catalyst particles [15]; however, more recent work seems to indicate that sulfur is involved in fiber growth only after the initial melting of the catalyst particles composed of Fe+C. The suggestion is that sulfur encourages filament nucleation by raising the activity of dissolved carbon in the particles [9].

The mechanical properties of the VGCF produced from various precursors (according to cycle B) appear to be relatively independent of the precursor used, as shown in Table 3. The presence of sulfur affects markedly the properties of the VGCF obtained with the nitrate precursor, but has little influence in the case of the iron carbonyl [20]. It should be pointed out that the reported moduli were measured with only two gauge lengths, due to the short length of the fibers, and both the tensile strengths and moduli are somewhat lower than those reported by Tibbetts [11]. The BET surface areas reported in Table 3 (measured on all fibers present in each sample) are larger than the geometric area of the fibers ($\approx 1 m^2.g^{-1}$), implying a certain porosity.

TABLE 3. Properties of VGCF (120 min thickening) from different precursors [Adapted from ref.20]

Precursor	H$_2$S	BET Surface area (m^2/g)	Tensile strength (GPa) [a]	Modulus (GPa) [b]
Fe(NO$_3$)$_3$.9H$_2$O	no	5.7	0.87 ± 0.10	141
	yes	3.9	1.46 ± 0.20	155 (270)
Fe(SO$_4$).7H$_2$O	no	2.3	1.22 ± 0.22	99 (184)
	yes	6.2	1.32 ± 0.24	193 (274)
Fe$_3$(CO)$_{12}$	no	1.8	0.98 ± 0.16	139 (170)
	yes	2.6	0.88 ± 0.18	124 (162)

[a] given with a confidence interval of 95% [b] The value in parenthesis is the corrected modulus

4. Examination of the Substrate and Morphology of the Fibres

The morphology of the as-grown carbon fibers (on their own support) was determined by SEM. In addition, the Grafoil support was observed after removal of the longest fibers at several stages along cycle B [24].

The first observations, made after the reduction stage (950°C - 0 min.) show that carbon has already started to deposit on the metallic particles and fix their diameter, and very thin filaments (10 to 100 nm) are clearly visible. From numerous observations of the support performed at this stage of the reaction it may be concluded that at least two kinds of carbon filaments coexist: First, filaments of 50 to 100 nm in diameter having an iron particle at their tip (Figure 2a). Second, filaments presenting a very high aspect ratio, with an approximate diameter of ten nanometers and a length up to several micrometers (Figure 2b). These later filaments could correspond to the pyrolytic carbon nanotubes (PCNT) observed by Endo *et al.* [25]. Finally, filaments having the same size as the first category but which do not show any particles at their tip or inside their cores were also observed. These could be PCNT which were already thickened by carbon deposition. Indeed, the presence of very thin filaments (10 nm) was detected inside some broken thicker filaments (100nm) [24]. The experiments interrupted between 950 and 1150°C show the lengthening and thickening of the filaments, but the catalyst particles are not visible (Figure 3a). At the end of the thickening stage (at 1150°C), the support is depicted in Figure 3b (Magnification 10x lower than Fig.3a).

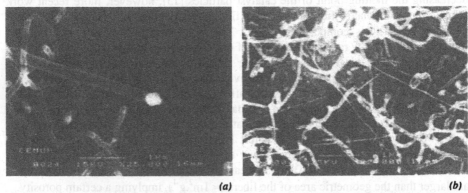

(a)   (b)

Figure 2. SEM micrographs: a) carbon filament showing the iron particle at the tip; b) carbon nanotubes (high aspect ratio) coexisting with carbon filaments [Reprinted from ref. 24 with permission from Elsevier Science].

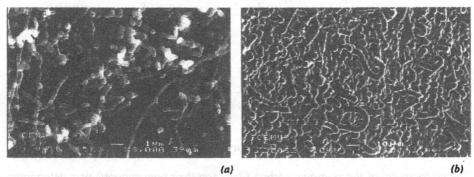

(a)   (b)

Figure 3. Aspect of the support: a) 1100°C (0 min); b) 1150°C (60 min) [Reprinted from ref. 20 with permission from World Scientific].

These observations confirm that during the early stages of fiber growth PCNT coexist with nanometer scale VGCF. Moreover, careful observation of the broken fibers present on the support allowed us to clearly observe a fine structure present in the core of the VGCF (Cf. Figure 4). In all cases, the diameter of this structure did not exceed 15 nanometers, which is comparable to the diameter range of the PCNT's .

Furthermore, from Figure 4 it is evident that the VGCF structure is more complex than previously thought, including a central nanometric filament, a micrometric inner core, and an outer pyrolytic carbon coating. These different types of carbon can be distinguished by their different reactivities towards oxidation [26].

So, these observations pose a question concerning the nature of the central structure in VGCF, upon which pyrolytic carbon is deposited in concentric layers: is it a "carbon nanotube" [25], or is it a "carbon filament" as originally proposed [4]? Although we have shown that in some cases the central structure looks like a PCNT, alternative mechanisms cannot be excluded. However, the recent work of Le et al. [13] seems to indicate that only the long/thin filaments (nanotubes) can be precursors of VGCF.

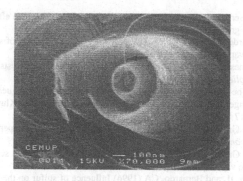

Figure 4. Carbon nanotube in the inner core of a VGCF [Reprinted from ref. 26 with permission from Elsevier Science].

Acknowledgments. Financial support by the EEC through the Human Capital and Mobility programme (contract number CHRX-CT94-0457) is gratefully acknowledged. The authors are indebted to Dr. Carlos Sá (CEMUP) for performing the XPS and SEM measurements, to Prof. C.Bernardo and Dr. C.Paiva for assistance with the mechanical characterization of VGCF, and to UCAR (USA) for providing the Grafoil ®.

References

1. Figueiredo, JL, Bernardo, CA, Baker, RTK and Hüttinger, KJ (1990) *Carbon fibers, filaments and composites*, Kluwer Academic Publishers, Dordrecht
2. Baker, RTK, Barber, MA, Harris, PS, Feates, FS and Waite, RJ (1972) Nucleation and growth of carbon deposits from the nickel catalyzed decomposition of acetylene, *J Catal* **26**,51-62.
3. Lobo, LS, Trimm, DL and Figueiredo, JL (1973) Kinetics and mechanism of carbon formation from hydrocarbons on metals, in: JW Hightower (ed.), *Proc. 5th Int. Congress on Catalysis, 1972*, North-Holland, Amsterdam, vol.2, pp.1125-1135
4. Oberlin, A, Endo, M and Koyama, T (1976) Filamentous growth of carbon through benzene decomposition, *J Crystal Growth* **32**,335-349
5. Tibbetts, GG (1985) Length of carbon fibers grown from iron catalyst particles in natural gas, *J Crystal Growth* **73**,431-438
6. Benissad, F, Gadelle, P, Coulon, M and Bonnetain, L (1988) Formation de fibres de carbone a partir du methane: I Croissance catalytique et epaississement pyrolytique, *Carbon* **26**,61-69

120

7. Tibbetts, GG, Devour, MG and Rodda, EJ (1987) An adsorption-diffusion isotherm and its application to the growth of carbon filaments on iron catalyst particles, *Carbon* 25,367-375
8. Benissad, F, Gadelle, P, Coulon, M and Bonnetain, L (1988) Formation de fibres de carbone a partir du methane: II Germination du carbone et fusion des particules catalytiques, *Carbon* 26,425-432
9. Tibbetts, GG and Balogh, MP (1999) Increase in yield of carbon fibres grown above the iron/carbon eutectic, *Carbon* 37, 241-247
10. Endo, M (1988) Grow carbon fibers in the vapor phase, *CHEMTECH*, September, 568-576
11. Tibbetts, GG (1990) Vapor-grown carbon fibers, in JL Figueiredo, CA Bernardo, RTK Baker and KJ Hüttinger (eds.), *Carbon fibers, filaments and composites*, Kluwer Academic Publishers, Dordrecht, pp. 73-94
12. Serp, Ph, Madroñero, A and Figueiredo, JL (1999) Production of vapour-grown carbon fibres: influence of the catalyst precursor and operating conditions, *Fuel* 78, 837-844
13. Le, QT, Schouler, MC, Garden, J and Gadelle, P (1999) Fe(NO₃)₃.9H₂O and Fe₃(CO)₁₂ as catalyst precursors for the elaboration of VGCF: SEM and TEM study – Improvement of the process, *Carbon* 37, 505-514
14. Van Hattum, FWJ, Serp, Ph, Figueiredo, JL and Bernardo, CA (1997) The effect of morphology on the properties of vapour-grown carbon fibres, *Carbon* 35, 860-863
15. Tibbetts, GG, Bernardo, CA, Gorkiewicz, DW and Alig, RL (1994) Role of sulfur in the production of carbon fibers in the vapor phase, *Carbon* 32, 569-576
16. Philipps, J and Dumesic, JA (1984) Production of supported metal catalysts by the decomposition of metal carbonyls, *Applied Catal.* 9, 1-30
17. Gadelle, P (1990) The growth of vapor-deposited carbon fibres. in JL Figueiredo, CA Bernardo, RTK Baker and KJ Hüttinger (eds.), *Carbon fibers, filaments and composites*, Kluwer Academic Publishers, Dordrecht, pp. 95-117
18. Tibbetts GG (1992) Growing carbon fibers with a linearly increasing temperature sweep: Experiments and modelling, *Carbon* 30, 399-406
19. Benissad-Aissani F, Gadelle P (1993) Influence des conditions operatoires sur la croissance catalytique des fibres de carbone vapodeposees a partir du methane, *Carbon* 31, 21-27
20. Serp, Ph, Figueiredo, JL and Bernardo, CA (1996) Influence of sulfur on the formation of vapor-grown carbon fibers produced on a substrate using different iron catalyst precursors, in KR Palmer, DT Marx and MA Wright (eds.) *Carbon and carbonaceous composite materials*, World Scientific, Singapore, pp. 134-147.
21. Jayasankar, M, Chand, R, Gupta, S and Kunzru, D (1995) Vapor-grown carbon fibers from benzene pyrolysis, *Carbon* 33, 253-258
22. Kato, T, Haruta, K, Kusakabe, K and Morooka, S (1992) Formation of vapor-grown carbon fibers on a substrate, *Carbon* 30, 989-994
23. Tibbetts, GG, Gorkiewicz, DW and Alig, RL (1993) A new reactor for growing carbon fibers from liquid- and vapor-phase hydrocarbons, *Carbon* 31, 809-814
24. Serp, Ph and Figueiredo, JL (1996) A microstructural investigation of vapor-grown carbon fibers, *Carbon* 34, 1452-1454
25. Endo, M., Takeuchi, K., Kobori, K., Takahashi, K., Kroto, H.W. and Sarkar, A (1995) Pyrolytic carbon nanotubes from vapor-grown carbon fibers, *Carbon* 33, 873-881
26. Serp Ph and Figueiredo JL. (1997) An investigation of vapor-grown carbon fibers behavior towards air oxidation, *Carbon*, 35, 675-683.

GASIFICATION AND SURFACE MODIFICATION OF VAPOR-GROWN CARBON FIBERS

J.L.FIGUEIREDO[1] and Ph. SERP[2]

1)Laboratório de Catálise e Materiais, Faculdade de Engenharia, Universidade do Porto, 4050-123 Porto, Portugal

2)Laboratoire de Catalyse, Chimie Fine et Polymères, Ecole Nationale Supérieure de Chimie de Toulouse, 118 Route de Narbonne, 31077 Toulouse France.

Abstract

Vapor-grown carbon fibers (VGCF) were produced from a methane-hydrogen mixture on a reconstituted graphite support using the $[Fe_3(CO)_{12}]$ complex as catalyst precursor. The fibers thus produced were submitted to different oxidative treatments: nitric acid, oxygen plasma and partial gasification with air or carbon dioxide. The original and the oxidised fibers were characterised by X-ray diffraction, SEM, AFM, nitrogen adsorption, XPS and ToF-SIMS. The use of nitric acid or plasma as oxidation agents does not affect significantly the surface morphology of the fibers, but greatly increases the number of surface oxygen functions. The air and carbon dioxide treatments do not lead to significant increase either of the surface area, or of the quantity of surface oxygen containing groups, despite the important weight loss attained (50%). This peculiar observation has been interpreted by considering the presence of traces of iron at the fibers surface, which catalyse the gasification of carbon. Removal of this iron by acid washing allows an improvement of the specific surface area. A detailed study of the gasification in air gave valuable informations on the intimate structure of the VGCF.

1. Introduction

Vapor-grown carbon fibers represent a variety of carbon fiber different from those produced from pitch and polyacrylonitrile (PAN) precursors. The production of VGCF involves the decomposition of a hydrocarbon (methane and benzene have been used with success), in the presence of hydrogen, on iron catalyst particles. Additionally, sulfur can be added to the gas phase as a promoter of the reaction [1-5]. The accepted mechanism for VGCF growth proceeds via the initial formation of a carbon filament, followed by a thickening stage by chemical vapor deposition of pyrolytic carbon. Recent observations by Scanning Electron Microscopy and Transmission Electron Microscopy have revealed that two kinds of filaments are formed in the initial stages of the growth process [6-7], and that only one of them can be precursors of VGCF. These filaments present a very high aspect ratio and consist of continuous straight concentric graphene layers with their

121

L.P. Biró et al. (eds.), Carbon Filaments and Nanotubes: Common Origins, Differing Applications, 121-132.

001 direction perpendicular to the tube axis, and therefore can be considered as carbon nanotubes.

First characterisations have revealed that these fibers present excellent thermal and electrical conductivities [8,9] (greatly exceeding those of PAN-based fibers) and good mechanical properties/cost ratio [2]. Thus, VGCF are extremely interesting as potential partial substitutes of PAN-based carbon fibers. Possible applications could be as reinforcement of reasonable strength level for composites, or as an adsorbent in the chemical industry. As a reinforcement, the standard technology to improve stress transfer within composites consists in surface oxidation of the fibers by various treatments (plasma, anodic oxidation, nitric acid oxidation,...). As an adsorbent, activation of conventional carbon fibers via carbon dioxide or steam treatment leads to the production of activated carbon fibers (ACF) which possess very large surface areas (>1000 $m^2.g^{-1}$) and many reactive carbon-oxygen and carbon-hydrogen surface groups [10]. In both cases, oxidative surface treatments of the fibers are essential for their application.

The present lecture deals with the activation of vapor-grown carbon fibers by plasma, nitric acid, carbon dioxide and air treatments in order to obtain modifications of their surface chemistry and/or morphology; and with their characterisation by TGA, nitrogen adsorption, XPS, SIMS, SEM and AFM.

2. Results and discussion

The fibers used in this study were grown on a reconstituted graphite support (Grafoil ®, 10x20 cm^2) impregnated by spraying 20 cm^3 of an acetone solution (0.025M) of [$Fe_3(CO)_{12}$]. After drying at 110°C in an oven, the substrate was placed in a horizontal, hot wall chemical vapor deposition reactor, having 70 mm i.d. and 1.5 m length. Reactants were hydrogen, and methane as the carbon source. In these experiments, gas flows were adjusted by mass flow controllers for H_2 an CH_4 and the support was heated from room temperature to 950°C under an atmosphere of 70%H_2 - 30%CH_4 (catalyst reduction), then the temperature was raised to 1150°C, and finally a thickening stage was realised at this temperature, in order to obtain fibers of significant diameter (3 - 40 μm). Typical properties of the fibers produced by this process are shown in Table 1.

TABLE 1. Typical properties of ex-[$Fe_3(CO)_{12}$] fibers (Reprinted from ref.[11] with permission from Elsevier Science)

Density (a) (g.cm⁻³)	d_{002} (nm)	ρ_{300} (μΩ.m)	BET ($m^2.g^{-1}$)	%C	%H	%N	%O
1.97	0.3476	9.1	1.8	99.5	0.25	0.15	0.1

(a) Measured with a helium pycnometer with 2 grams of material.

2.1. SURFACE TREATMENTS OF THE VGCF [11,12]

Four different treatments have been performed in order to activate the fibers. The activated VGCF were named ACF1 (nitric acid treatment), ACF2 (plasma treatment), ACF3 (air treatment) and ACF4 (carbon dioxide treatment). Oxidation with nitric acid (70 wt%) was performed on a Soxhlet at 80°C during five hours on two grams of material. Oxygen plasma treatments were realised at room temperature, at 1.0 mbar of oxygen with power treatments of 75 or 150 W for three minutes. Activation with air

(550°C - 60min.) or with pure CO_2 (900°C - 90min.) was carried out under atmospheric pressure on two grams of material, in order to reach a burn-off of about 50%.

X-ray diffraction data of the original and treated fibers showed that the carbon structure in the bulk of the VGCF remains untouched even when the morphology of the fiber changed with the oxidative treatment (as shown below).

2.1.1. *SEM and AFM observations of the fibers morphology and topography*

From the micrographs presented in Figure 1, we can see that the surface of the original fiber is quite smooth and regular (Fig. 1-a). By atomic force microscopy (Figure 2-a), (1000 x 1000 nm), the surface topography is characterised by a granulated texture, grains 4-8 nm high, which probably reflects the pyrolytic carbon deposition process corresponding to the thickening stage of fiber growth. Nitric acid (ACF1) treatments do not impair the fiber's microstructure. However, in the case of plasma treatment (150W), the surface of the fibers is seen to present a certain roughness due to the appearance of a granulated texture (Fig. 1-b). This is clearly seen from the AFM image (Figure 2-b), the size of the grains becoming much larger (20-60 nm high).

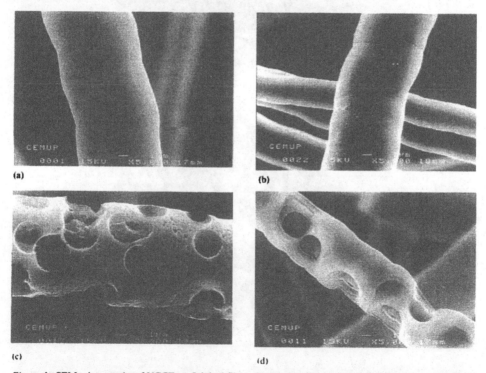

Figure 1. SEM micrographs of VGCF: a) Original fibers; b) oxygen plasma treated (150 W) fibers, ACF2; c) air treated fibers, ACF3; d) carbon dioxide treated fibers, ACF4. (Reprinted from ref.[11] with permission from Elsevier Science).

The effect of oxygen plasma treatments on the surface of pitch and PAN carbon fibers has been studied in a number of publications [13-17]. SEM observations suggest that, in general, the oxygen plasma peels off one complete outer layer of the carbon surface after another, in a very smooth way. However, these effects are expected to

124

depend upon the type of fiber (and the presence of surface heterogeneities) and the extent of the plasma treatment. Indeed, the development of some morphological changes like surface crenulation and slit-like porosity has also been reported [13]. Also, STM observations have shown the development of a certain roughness in pitch-based carbon fibers, the original ribbon-like surface features becoming less evident after an air plasma treatment [16]. With PAN-based carbon fibers, an initial surface roughening was observed, followed by smoothing as the duration of the plasma treatment increased [17].

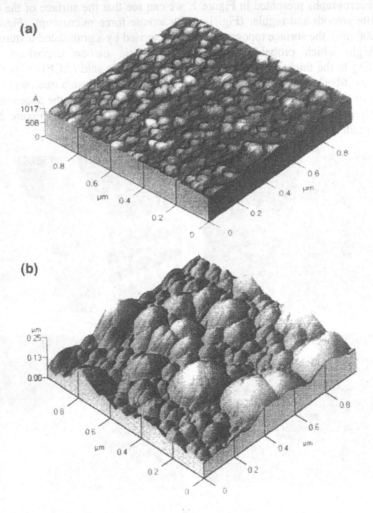

Figure 2. AFM 3-dimensional image profiles of a) Original fibers; b) oxygen plasma treated (150 W) fibers ACF2. (Reprinted from ref.[12] with permission from Elsevier Science).

The BET surface area of the fibers is not much affected by oxidation (Table 2). This result could be predicted in the case of the nitric acid and plasma treatments, as these methods are not supposed to create porosity, but for gaseous activation the result was unexpected. Indeed, in this latter case, analysis of the nitrogen adsorption isotherms revealed that no significant meso- or micro-porosity had been developed.

Activation with air or carbon dioxide led to the formation of macropores with diameters up to one micrometer (Fig. 1-c and 1-d). Very large pores are dominant in the case of CO_2 treatment. These macropores appear whatever the burn-off level and the temperature used for the reaction. Indeed, we have noticed that treatments in the range 400 to 900°C, in the case of air, and 800 to 1200°C, in the case of CO_2, give rise to the same fiber morphology. We have directly evidenced (see section 2.2 of this lecture) that this peculiar behaviour of VGCF towards oxidation can be explained by considering the presence of iron traces (resulting from the catalyst precursor) on or in the VGCF, which can catalyse the fiber gasification. Indeed, in the graphite-oxygen reaction the catalyst may operate by facilitating basal plane penetration in imperfect regions (pitting mode) [18]. Being unable to create some small porosity, the CO_2 or air treatments should not lead to any significant improvement of the adhesion of VGCF with polymer matrices, as confirmed in the recent work of Tibbetts et al. on the mechanical properties of VGCF composites [19].

In order to create some mesoporosity, the cleansing of the VGCF surface was undertaken prior to the activation step. Thus, an acid treatment (HCl conc., 5 hours reflux) was performed in order to remove any iron present on the fibers surface. ToF-SIMS analysis performed on the cleaned VGCF revealed that the treatment was efficient, the iron peaks disappearing. Upon activation of the acid-washed fibers with CO_2 (50% burn-off), some mesoporosity was seen to develop and a BET surface area of 32 $m^2.g^{-1}$ was measured. The large pores were not observed in this sample. The increase in the fiber surface area can improve the mechanical interlock with the matrix in a composite. However, it appears that the production of very high surface area VGCF of micrometric diameter, which could be used as an adsorbent, is a difficult task to be achieved.

2.1.2. XPS surface chemical analysis of the oxidised fibers.

Typical surface compositions of the VGCF obtained by XPS before and after the oxidative treatments are given in Table 2. The effect of oxidation is obvious as indicated by the increase of the surface oxygen concentration. Nitric acid and plasma appear to be the more efficient treatments to improve the surface oxygen content. This can favour the adhesion of VGCF to a matrix resin as far as the chemical bonding (functional groups) is concerned. The use of gaseous activators does not lead to a very significant improvement in the surface oxygen content. Similar results were obtained by Darmstadt et al. for VGCF oxidised in air (%O between 2.5 and 4.6) [20].

TABLE 2: Surface area and surface composition (by XPS) of the VGCF (Adapted from refs. [11,12])

	BET (m^2/g)	C (at.%)	O (at.%)	N (at.%)
Untreated Fibers	1.8	98.9	1.1	<0.1
ACF1	2.0	80.9	15.7	3.4
ACF2 - 75W	1.9	87.3	12.7	<0.1
ACF2 - 150W		80.1	19.9	<0.1
ACF3 (B.O. 50%)	4.2	92.9	7.1	<0.1
ACF4 (B.O. 50%)	7.2	97.5	2.5	<0.1
ACF4/HCl (B.O. 50%)	32	96.8	3.2	<0.1

* The H content, not detected by XPS, is neglected in these percentages.

In order to get further insight into the nature of the functional groups present on ACF1 and ACF2, a reconstruction of the O1s peak has been performed. For curve fitting of this peak, four different O functionalities were considered, as well as a small contribution of chemisorbed water [21].

Results of the fits of the O1s region are given in Table 3. The peak at 531.1eV corresponds to carbonyl oxygens, the peak at 532.2 eV to carbonyl oxygens in esters, amides, anhydrides and oxygens in hydroxyls or ethers, the peak at 533.0 eV to the ether oxygens in esters and anhydrides and the peak at 534.0 eV to the oxygens in the carboxyl groups. The contribution of water is located at 535.9 eV. The main difference between the two treatments comes from the intensity of the peaks at 533.0 eV and 534.0 eV, the plasma treatment giving rise to a higher amount of carboxylic groups to the detriment of the ether oxygen in esters or anhydrides. As carboxylic and hydroxyl groups are far more important for adhesion than esters and ethers (because the former can react with the matrix to form chemical bonds [22,23]), the plasma treatment appears to be more favorable in order to improve adhesion.

TABLE 3. Results of the fits of the O1s region, values given in % of total O1s intensity (Reprinted from ref.[11] with permission from Elsevier Science)

Fiber	Binding energy (eV)				
	531.1	532.2	533.0	534.0	535.9
ACF1	11	31	35	20	2
ACF2	6	38	15	39	2

2.2. STRUCTURE OF THE VGCF AND THEIR REACTIVITY IN AIR [12,24]

In the first part of this lecture, we have seen the effect of different oxidative treatments on the surface of the VGCF; in this second part we will focus our attention on the peculiar behaviour of these fibers when oxidised in air. Indeed, studies related to VGCF oxidation are relatively scarce [20,24-26]. Thermogravimetric analysis together with scanning electron microscopy observations were used to study this reaction and to get insight into the structure of these fibers.

Four types of vapor-grown carbon fibers were studied, namely FC3, FC6, FC12 and FC40 according to their mean diameters of 3, 6, 12 and 40 micrometers respectively. These different diameters have been obtained by adjusting the duration of the CVD carbon coating performed at 1150°C. The motivation we had to study in detail the rate of gasification of VGCF as a function of their diameter came from the fact that, in preliminary experiments, we had noticed some variations in the gasification rates of VGCF when we had not controlled the diameter. We have firstly attributed this result to heterogeneities induced in the production stage. But, taking into account the dependence of the fibers' mechanical properties on their diameter [27], we wanted to check if a similar phenomenon could occur during the oxidation reaction. Due to the nature of the growth mechanism (substrate seeding method), a distribution in the diameter of the fibers is inevitable. Thus, after shaving the support, if a careful selection of fibers is not realized, a wide range of diameters exists in a sample of VGCF. According to these observations, we have chosen to produce fibers with significantly different times of CVD coating in order to reach significantly different mean diameters. The mean diameters of FC3 (3 ± 1), FC6 (6 ± 2), FC12 (12 ± 4) and FC40 (40 ± 8) are given with a confidence interval of 70%.

Figure 3 displays the rates of gasification of these VGCF as a function of reaction time and of the diameters. These experiments were conducted in an isothermal mode at 700°C and the heating slope to reach the working temperature was realised under a dinitrogen atmosphere. First, it is worth noting that at this temperature the oxidation is complete in few minutes whatever the fiber's diameter. Furthermore, it is clear that different peaks predominate according to the fibers mean diameter.

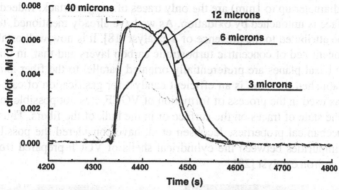

Figure 3. Rates of gasification of different mean diameter VGCF (T=700°C). (Reprinted from ref.[24] with permission from Elsevier Science)

For the large diameter (FC40) the majority of the total envelop consists of one peak centred at 4361s. When the average diameter is reduced to 12μm (FC12) an enlargement of the total envelop is observed. If the mean diameter reaches 6μm (FC6) one can see that the large envelop obtained for FC12 consists in fact of two large peaks. Finally, if the mean diameter is diminished to 3μm it becomes clear that several peaks are present in the total envelop. After careful deconvolution, four peaks appear in the total envelop of FC3 (see Fig. 4). The same peaks, with different intensities, are also present in the case of FC6. In a first hypothesis, this result could indicate that the VGCF are made up of different types of carbon presenting different reactivities.

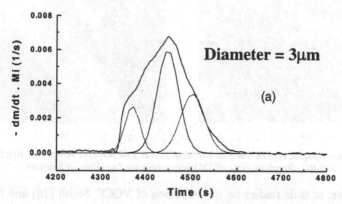

Figure 4. Deconvolution of the rate of gasification curve (T=700°C) for FC3. (Reprinted from ref.[24] with permission from Elsevier Science)

In order to gain further information on these different kinds of carbon, we have interrupted the oxidation experiments at different levels of burn-off in the case of FC6; then, we have observed the oxidised fibers by SEM. The experiments were stopped approximately at 25, 50 and 75% of burn-off, which correspond to the positions of the

128

first three peaks present in the gasification rate curve; for the last one no observation was realised, because it corresponds to an almost complete gasification (burn-off of 98.3 %).

In the first stage (Fig. 5-a) the attack of the fiber surface occurs by pitting, and some large pores (diameter up to 1mm) are the only traces of oxygen attack. Indeed, the rest of the fiber surface is untouched by oxidation. As we have already mentioned, this mode of attack may be attributed to the presence of a catalyst [18]. It is now well established that VGCF are constituted of concentric turbostratic carbon layers and that, in this particular structure, the basal planes are preferentially oriented parallel to the fiber axis [28]. It is also well established that iron is an efficient catalyst for gasification of carbon [18]. As this metal was used in the process of formation of VGCF, it is not possible to exclude its presence in the state of traces on the surface or in the bulk of the fibers. Thus, in a study on VGCF mechanical properties, Jacobsen et al. have considered the possibility of the presence of impurities between the cylindrical shells of VGCF prepared from methane pyrolysis on an iron catalyst [29].

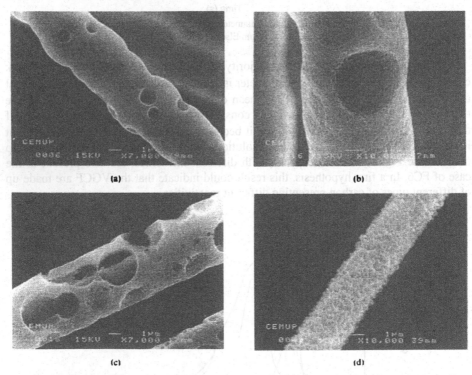

Figure 5. SEM micrographs: a) of FC6 at 25% BO; b) of FC6 at 50% BO; c) of FC6 at 75% BO; d) of FC3 at 50% BO. (Reprinted from ref.[24] with permission from Elsevier Science)

Furthermore, in their studies on the oxidation of VGCF, Smith [26] and Darmstadt [20] have reported iron concentrations of 0.05% and 0.02% by weight in their VGCF, respectively. In these works, the authors do not report any appreciable effect of iron on the VGCF behaviour towards oxidation; however, no observation of the fibers was performed. The analyses we have performed by XPS (surface analysis), by EDS or by Mössbauer spectroscopy (bulk analysis) were not convincing about the presence of iron in or on the fibers. But the fact to bear in mind is that the sensitivity of these methods is not high enough if we consider that iron can be present as traces. Direct evidence for the presence of iron on the VGCF was obtained by ToF-SIMS analysis. Indeed, the high sensitivity and the high molecular specificity of this technique are particularly suited for this study. The presence of iron was unambiguously evidenced by the presence of a peak at m/z = 55.92 and by the relative intensities of the ^{56}Fe (92 %), ^{54}Fe (5.8 %) and ^{57}Fe (2.2 %) peaks.

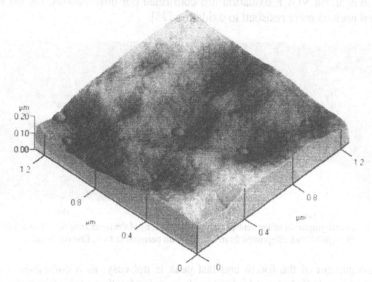

Figure 6. AFM observation of ACF4. (Reprinted from ref.[12] with permission from Elsevier Science)

We were also able to evidence by AFM (Figure 6) some nanoparticles at the bottom of the macropores created by CO_2 oxidation (ACF4). The size of these particles can be estimated from line profile measurements. Since the apex radius of the cantilever used was 50 nm, the heights of the particles are expected to be more reliably measured than their widths. From four of the particles shown if Figure 6, the average height of the particles was estimated as 10 nm.

The second stopping experiment was made at 50% of burn-off, a value which corresponds to the position of the second peak in the gasification rate curve. In this case, one can see in Fig. 5-b that, apart from the large holes, a second kind of attack of the surface occurs. We have attributed this second peak to the uncatalysed oxidative surface treatment which is a process whereby active carbon atoms located at the edges of planes or defects in a plane are oxidised selectively, generating new active sites, which are then oxidised successively. This kind of attack creates defects or holes of small depth and diameter on the fiber's surface.

The last stopping experiment (Fig. 5-c) performed at 75% of burn-off has allowed us to evidence that the VGCF structure is more complex than we previously thought. Indeed, it appears clearly from this micrograph that a harder inner core is present in the fiber, this part just starting to be attacked by oxygen as the outer shell is already seriously oxidized. In the case of FC6, the diameter of this inner core is almost one third of the total diameter. In a study on exfoliation of vapor-grown graphite fibers, Yoshida et al. [30] have already observed the presence of a central region, less easily exfoliated

130

than the outer part of the fiber, and the dimension of this central part was also one third of the total diameter. From the SEM observations we have performed, it appears that the core is generally attacked by a non catalytic process (general absence of large holes). It is worth noting that the catalytic attack is also rare in the case of very small diameter fibers, which may correspond to a pure inner core structure (see Fig. 5-d), as if the iron concentration was increasing with the fiber's diameter. So, we have ascribed the third peak to the non-catalytic attack of the VGCF's inner core. A recent work performed by Madroñero et al. on VGCF oxidation has confirmed our observations, i.e. the existence of a central nucleus more resistant to oxidation [25].

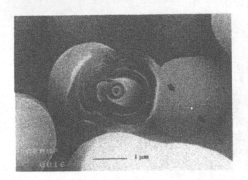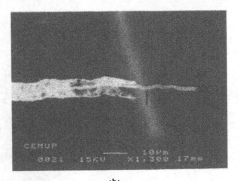

(a)   (b)

Figure 7. SEM micrographs: a) of the central carbon nanotube; b) of the inner core observed after oxidation of FC6 (BO 75%). (Reprinted from ref.[24] with permission from Elsevier Science)

The assignment of the fourth and last peak is not easy, as it corresponds to a very high level of burn-off, but not high enough in order for this peak to correspond to the attack of the central nanotube present in the VGCF's inner core (see Fig. 7-a). Nevertheless, it could correspond to a double contribution of, on one hand the attack of the carbon nanotube, and on the other hand the attack of inner cores of smaller diameter (Fig.7-b). The central nanotubes that we have observed in numerous broken VGCF present diameters around 10-50nm and have to be compared to those recently observed by Gadelle et al. [7].

From the previous SEM observations after oxidation, performed in an isothermal mode at 700°C, it appears that the core of the fibers is more resistant to air oxidation than the outer part. But the differences in reactivity between small and large diameter fibers are quite small as shown by data collected between 500 and 950°C at 20% of burn-off, from which the Arrhenius plots were constructed and the activation energies were calculated. Below 760°C, they were 40.0 ± 1.1, 38.3 ± 2.8 and 39.6 ± 1.7 kcal.mol^{-1} for FC3, FC6 and FC12 respectively. Thus, whatever the fiber diameter, below 760°C, the oxidation occurs exclusively in Zone I [31] (chemical reaction control and absence of diffusion limitations). As evidenced by nitrogen adsorption measurements performed on FC12 at 0% burn-off (SBET = 1.8 m^2.g^{-1}) and at 25 % burn-off (SBET = 3.5 m^2.g^{-1}), no significant microporosity develops at the beginning of the reaction, which could otherwise lead to diffusion limitations. At higher temperature, and whatever the fiber's mean diameter, the reaction rate reaches a constant value independent of the temperature, indicating that the reaction occurs in zone III (film diffusion regime) [31].

3. Conclusions

1. Nitric acid and oxygen plasma treatments of VGCF can be used to increase the concentration of surface oxygen groups without changing significantly the morphology of these fibers. An XPS investigation of the O1s region suggests that the plasma treatment might improve the adhesion of this type of fibers to polymeric matrices.

2. The oxidation treatments by air and by carbon dioxide are not able to improve either the concentration of surface oxygen or the surface area. About this last point, it has been shown that it is due to the presence of traces of iron left on the VGCF from the growth stages, which catalyse carbon gasification by a pitting mechanism. Evidence for the presence of iron on the fibers was obtained by TOF-SIMS. Cleansing of the VGCF with concentrated HCl effectively removes the iron contaminant, and eliminates the formation of large pits.

3. Vapor-grown carbon fibers present a triplex structure composed of a central carbon nanotube, an inner core, and an outer shell.

4. Acknowledgments

Financial support by the EEC through the Human Capital and Mobility programme (contract number CHRX-CT94-0457) is gratefully acknowledged. The authors are indebted to Dr. Carlos Sá (CEMUP) for XPS and SEM analyses, to Drs. J.M. Diez-Tascón and M.A. Martinez (INCAR, Oviedo, Spain) for the plasma treatments, to Prof. J.-P. Issi, Mrs. V.Wiertz, Dr. P.Bertrand and Dr. B.Nysten (Univ. Catholique de Louvain, Belgium) for assistance with TOF-SIMS measurements and AFM, and to UCAR (USA) for providing the Grafoil ®.

5. References

1. Tibbetts, GG, Bernardo, CA, Gorkiewicz, DW and Alig, RL (1994) Role of sulfur in the production of carbon fibers in the vapor phase, *Carbon*, 32, 569-576.
2. Tibbetts, GG (1990) Vapor-grown carbon fibers, in JL Figueiredo, CA Bernardo, RTK Baker and KJ Hüttinger (eds.), *Carbon Fibers Filaments and Composites*, Kluwer Academic Publishers, Dordrecht, pp. 73-94.
3. Serp, Ph, Figueiredo, JL and Bernardo, CA (1996) Influence of sulfur on the formation of vapor-grown carbon fibers produced on a substrate using different iron catalyst precursors, in KR Palmer, DT Marx and MA Wright (eds.), *Carbon and Carbonaceous Composite Materials*, World Scientific, Singapore, pp.134-147.
4. Tibbetts, GG Balogh, MP (1999) Increasing in yield of carbon fibres grown above the iron/carbon eutetic, *Carbon* 37, 241-247.
5. Fan, YY, Cheng, HM, Wei, YL, Su, G, Shen, ZH (2000) The influence of preparation parameters on the mass production of vapor-grown carbon nanofibers, *Carbon*, 38, 789-795.
6. Serp, Ph and Figueiredo, JL (1996) A microstructural investigation of vapor-grown carbon fibers, *Carbon* 34, 1452-1454.
7. Le, QT, Schouler, MC, Garden, J and Gadelle, P (1999) Fe(NO$_3$)$_3$.9H$_2$O and Fe$_3$(CO)$_{12}$ as catalyst precursors for the elaboration of VGCF. SEM and TEM study-improvement of the process, *Carbon*, 37, 505-514.
8. Piraux, L, Nysten, B, Haquenne, A, Issi, J-P, Dresselhaus, MS and Endo, M (1984) The temperature variation of the thermal conductivity of benzene-derived carbon fibers, *Solid State Communications*, 50, 697-700.
9. Issi, J-P and Nysten, B, (1998) Electrical and thermal transport properties in carbon fibers, in J-B Donnet, S Rebouillat, TK Wang and JCM Peng (eds.), *Carbon Fibers*, Marcel Dekker, New York, pp. 371-461.

132

10. Suzuki, M (1994) Activated carbon fiber: Fundamentals and applications, *Carbon*, **32**, 577-586.
11. Serp, Ph, Figueiredo, JL, Bertrand, P and Issi, J-P (1998) Surface treatments of vapor-grown carbon fibers produced on a substrate, *Carbon* **36**, 1791-1799.
12. Serp, Ph, Figueiredo, JL Nysten, B and Issi, J-P (1999) Surface treatments of vapor-grown carbon fibers produced on a substrate. Part II: Atomic force microscopy, *Carbon*, **37**, 1809-1816.
13. Kowbel, W and Shan, CH (1990) The mechanism of fiber-matrix interactions in carbon-carbon composites, *Carbon*, **28**, 287-299.
14. Ismail, IK and Vangsness, MD (1988) On the improvement of carbon fiber/matrix adhesion, *Carbon*, **26**, 749-751.
15. Jones, C and Sammann, E (1990) The effect of low power plasmas on carbon fibre surfaces, *Carbon*, **28**, 509-514.
16. Qin, RY and Donnet, J-B (1994) Study of carbon fiber surfaces by scanning tunneling microscopy, Part III. Carbon fibers after surface treatments, *Carbon*, **32**, 323-328.
17. Smiley, RJ and Delgass, WN (1993) AFM, SEM and XPS characterization of PAN-based carbon fibres etched in oxygen plasmas, *J.Mater.Sci.*, **28**, 3601-3611.
18. Baker, RTK (1986) Metal catalyzed gasification of graphite, in JL Figueiredo and JA Moulijn (eds.), *Carbon and Coal Gasification*, Martinus Nijhoff, Dordrecht, pp. 231-268.
19. Tibbetts, GG and McHugh, JJ (1999) Mechanical properties of vapor-grown carbon fiber composites with thermoplastic matrices, *J. Mater. Res.*, **14**, 2871-2880.
20. Darmstadt, H, Roy, C, Kaliaguina, S, Ting, JM and Alig, R.L (1996) Surface spectroscopic analysis of vapour-grown carbon fibres prepared under various conditions, *Carbon*, **36**, 1183-1190.
21. Zielke, U, Hüttinger, KJ and Hoffman, WP (1996) Surface-oxidized carbon fibers: I. Surface structure and chemistry, *Carbon*, **34**, 983-998.
22. Zielke, U, Hüttinger, KJ and Hoffman, WP (1996) Surface-oxidized carbon fibers: IV. Interaction with high-temperature thermoplastics, *Carbon*, **34**, 1015-1026.
23. Ehrburger, P (1990) Surface properties of carbon fibres, in JL Figueiredo, CA Bernardo, RTK Baker and KJ Hüttinger (eds.), *Carbon Fibers Filaments and Composites*, Kluwer Academic Publishers, Dordrecht, , pp. 147-161.
24. Serp Ph and Figueiredo JL (1997) An investigation of vapor-grown carbon fiber behavior towards air oxidation, *Carbon*, **35**, 675-683.
25. Madroñero, A, Merino, C and Hendry, A (1998) Characterisation of carbon fibres grown from carbonaceous gases by measurements of their density and oxidation resistance, *Eur. J. Solid State Inorg. Chem.*, **35**, 715-734.
26. Smith, GW (1984) Oxidation resistance of pyrolytically grown carbon fibers, *Carbon*, **22**, 477-479.
27. Tibbetts, GG and Beetz Jr, CP (1987) Mechanical properties of vapour-grown carbon fibres, *J. Phys. D: Appl. Phys.*, **20**, 292-297.
28. Oberlin, A, Endo, M and Koyoma, T (1972) *J. Cryst. Growth*, **32**, 335.
29. Jacobsen, RL, Tritt, TM, Guth, JR, Ehrlich, AC and Gillespie, DJ (1995) Mechanical properties of vapor grown carbon fiber, *Carbon*, **33**, 1217-1221.
30. Yoshida, A, Hishiyama, Y and Inagaki, M (1990) Exfoliation of vapor-grown graphite fibers as studied by scanning electron microscope, *Carbon*, **28**, 539-543.
31. Walker Jr, PL, Rusinko Jr, F and Austin, LG (1959) Gas reactions of carbon, in DD Eley, PW Selwood, PB Weisz (eds.), *Advances in Catalysis*, Vol. 11, Academic Press, New York, p. 133-221.

GROWTH OF NANOTUBES: THE COMBINED TEM AND PHASE-DIAGRAM APPROACH

A. LOISEAU[1], F. WILLAIME[2]

[1] *Laboratoire d'Etude des Microstructures, UMR n° 104 ONERA-CNRS, Onera, B.P. 72, 92322 Châtillon Cedex, France. e-mail: loiseau@onera.fr*
[2] *Section de Recherches de Métallurgie Physique, CEA/Saclay, F-91191 Gif-sur-Yvette, France. e-mail:@fwillaime@cea.fr*

1. Introduction

The main current challenges in the synthesis of nanotubes are on the one hand the optimization of the production of existing structures - in particular ropes of single wall carbon nanotubes (SWNT) [1] - and on the other hand the exploration of novel structures, such as multi-element nanotubes, including carbon nanotubes filled with foreign materials. Controlling their fabrication requires first to be able to determine their structural and chemical characteristics in a quantitative way. Transmission electron microscopy (TEM) provides a unique way for studying the morphologies, the structure and the chemistry of nanotubular materials and has therefore highly contributed to the development of the research on this new kind of structures. Understanding the growth mechanisms is expected to help controlling and optimizing the production of nanotubes with definite geometry and chemistry. No coherent scheme on the growth mechanism has indeed clearly emerged yet, in particular for SWNT ropes, in spite of intensive experimental [2 - 6] and theoretical researches [7]. In situ studies have just started for different reaction chambers. They are very promising for determining characteristics of the temperature gradient and the time evolution of the matter aggregation after the initial vaporization of the different chemical species [8 - 10], but for the moment one has to rely on studies on the soot after the synthesis. Many useful information about the formation of nanotubes can again be deduced from TEM observations.

In this paper we shall illustrate with two different examples how TEM can be used to study both the structure and the growth of nanotubular structures. These examples concern composite nanotubes: carbon multiwalled nanotubes filled with different metals on the one hand and BN - C nanotubes on the other hand. We shall show how

133

L.P. Biró et al. (eds.), Carbon Filaments and Nanotubes: Common Origins, Differing Applications, 133–148.
© 2001 *Kluwer Academic Publishers. Printed in the Netherlands.*

the knowledge on these multi element nanotubes provides clues for identifying the general physical processes which govern the growth of nanotubes. The TEM results obtained previously on these two types of tubes are summarized, and a metallurgical approach based on phase diagrams is proposed to deduce simple schemes accounting for the observed structures. Finally, the phase diagram arguments being very successful in the two cases considered here, a perspective is drawn to apply the same approach to the formation of SWNT in order to explain the role of catalyst.

2. Experimental

Various methods have been developed to synthesize carbon nanotubes (for a review see e.g. [10 - 13]) and among them the arc discharge method which consists, in its original form, in establishing an electric arc between two graphite electrodes in a He atmosphere. This method was adapted as follows to produce the multi-element nanotubes studied here.

The filled nanotubes [14] were obtained by drilling the graphite rod for the anode and filling it with a mixture of graphite and chosen element powders. Not less than 41 elements have been tested. We emphasize that in this type of method the tubes are produced and filled simultaneously, whereas other routes are based on the initial idea of using pre-existing tubes, opening them by oxidation and filling them by physical or chemical techniques (for a review see e.g. [15 - 17]).

The B-C-N nanotubes studied here were obtained by arcing a graphite cathode with an HfB_2 anode in a nitrogen atmosphere [18], following the idea used to produce successfully pure BN tubes from two HfB_2 electrodes [19]. In this configuration the three constituents (B, C and N) have different sources and this was the key for producing composite nanotubes with high B and N contents. When using other configurations involving doped carbon electrodes only low doping by B and N are obtained [20].

Both the structure and the chemical composition of the nanotubes were investigated in detail by Transmission Electron Microscopy (TEM): selected area electron diffraction (SAED), high resolution imaging (HRTEM) and Electron Energy Loss Spectroscopy (EELS). HRTEM and standard EELS characterizations were performed using a JEOL 4000FX working at 400 kV, equipped with a Gatan 666 parallel collection electron energy-loss spectrometer. The high spatial resolution EELS results which are recalled in this paper were obtained using a dedicated scanning transmission electron microscope (STEM) VG HB501, working at 100 kV, equipped with a field-emission source and a parallel collection electron energy-loss spectrometer [21].

3. Filled nanotubes

3.1 STRUCTURAL ANALYSIS

Long continuous fillings were obtained for twelve elements belonging to different groups [14]: transition metals (Cr, Ni, Re, Au), rare earth metals (Sm, Gd, Dy, Yb) and covalent elements (S, Ge, Se, Sb). Partial fillings were also obtained with elements like Mn, Co, Fe, Pd, Nb, Hf, Os, B, Te, Bi. In these cases, the filling is discontinuous and consists of a sequence of particles - not exceeding a few hundred nanometers - located at different places along the tube. These nanotubes are hollow at one end and often capped by a particle at the other end. These partially filled nanotubes closely resemble vapor grown carbon fibres [22]. The structural characteristics of both kinds of nanotubes, i.e. the number of graphitic layers, the degree of graphitization, the crystallinity of the filling material, depend on the chosen element as described in [23]. The variety of the microstructures is shown in detail in [15].

The encapsulated crystals were initially thought to be carbides since in most cases the corresponding SAED patterns were not consistent with the structure of the pure elements [23]. The chemical composition of the filling material has been then investigated in a rather systematic way using high spatial resolution. Surprisingly, these nanoanalyses revealed that the nanowires obtained with metals are not carbon-rich as initially assumed but contain sulfur. The source of sulfur was found to be the graphite electrodes (99.4 %): a refined analysis of this graphite revealed that the major impurities are Fe (0.3 %) and S ($\approx 0.25\%$). The analysis of different cases is presented in detail in [23] and shows that important amounts of sulfur were found in numerous filling materials along with the inserted element.

In some cases as Cr-based nanowires, nanowires are very long single crystals and a majority of them was found to contain sulfur in a ratio close to 1:1, excluding any other element. Figure 1 presents an example of concentration profiles deduced from a line-spectrum across a filled nanotube. The C profile is characteristic of a hollow carbon nanotube (i.e. it starts abruptly, reaches a maximum at the inner radius and decreases to the centre) and is perfectly anticorrelated to both the S and Cr profiles. The analysis of these profiles suggests that the tubular layers wrapping the nanowire are poor-to-free of sulfur and of metal and that the filling material does not contain a significant amount of carbon [23]. Furthermore, S and Cr profiles are very correlated, suggesting that the filling is an homogeneous chromium sulfide. The structure of these sulfides has been identified from SAED and HRTEM analyses to be the trigonal compounds Cr_5S_6 or Cr_2S_3.

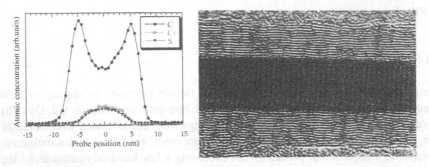

Figure 1 : Analysis of a filled nanotube obtained with a 99.4% graphite anode doped with Cr. Left: Concentration profiles of C, Cr and S deduced from the EELS line-spectra. Right: HRTEM image of the corresponding nanowire.

In some other cases as the case of Ni-based nanowires shown in Fig.2, the concentration in sulfur is not homogeneous along the tube axis and the filling consists of a succession of crystallites alternatively pure Ni and nickel sulfides with a S/Ni ratio ≈ 0.4. The compositional changes are associated with structural and morphological changes of the filling material as shown in Fig.2. In the areas where the S/Ni ratio is approximately 0.4, it is likely that a metastable sulfide was formed, since no stable nickel sulfide corresponding to such a S/Ni ratio is known. In more S-rich grains, the pseudo cubic compound Ni_3S_2 corresponding to a S/Ni ratio close to 0.7, was identified in certain cases from SAED analyses.

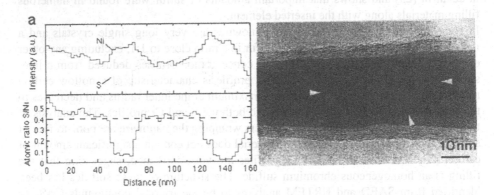

Figure 2 : Analysis of a filled nanotube obtained with a 99.4% graphite anode doped with Ni. Left: longitudinal concentration profiles along the tube axis of C, Ni and S deduced from EELS line-spectra recorded by scanning the probe along the wire axis. The filling material is alternatively pure Ni and a nickel sulfide. Right: HRTEM image of a nanowire showing a nickel crystallite inside between two nickel sulfide grains (or inside a nickel-sulfide matrix). Grain boundaries are arrowed.

The only exception to this spectacular concentration phenomenon of sulfur inside the nanotubes was observed for Ge: SAED and HRTEM analyses lead to the conclusion that the nanowires have the structure of pure Ge.

In order to understand the roles played by sulfur and the metal, we performed two kinds of experiments using in both cases high purity carbon rods (99.997%). We first successively doped the anode with Co (99.99%), Ni (99.9%), Cr (99.95%), Dy (99.9%) and S (sublimed). The other experimental conditions were unchanged [23]. It is striking that no filled nanotube was found in the cathode deposit. Co and Ni yielded the formation of single-walled nanotubes as usually observed [1] whereas only empty multi-walled nanotubes were produced with Cr, Dy and S.

In the second kind of experiments, we focused on the case of Cr which produced the longest nanowires and we added sulfur to Cr in a S/Cr=0.5% atomic ratio (i.e. S/C=0.1% in weight). This doping resulted in the abundant formation on the cathode of true nanowires encapsulated in carbon nanotubes similar to those obtained previously. EELS nanoanalysis revealed that some nanowires contained sulfur as those obtained with the 99.4% graphite rods but for a majority of them, the fraction of sulfur was below the limit of detection of the spectrometer [23].

All these results prove that sulfur is crucial for filling carbon nanotubes with Cr and that pure Cr nanowires are formed when S is added in catalytic quantity.

3.3 GROWTH MECHANISM AND THE ROLE OF SULFUR

There are obvious similarities between the present filled nanotubes and the carbon filaments grown by decomposition of gaseous hydrocarbons on catalytic particles [22]. First, it is likely that, in view of the typical filling length, filling does not occur after the growth of the nanotube is achieved: the nanotube is being filled - by a material which is most likely initially in a liquid state - while it is still growing. Therefore carbon and the metal are intimately related in the growth process, as in the case of the growth of carbon filaments. Second, the temperature in the very specific localized regions of the cathode where filled nanotubes are found is estimated to be rather low (between 1000°C and 2000°C) compared to that for the deposit facing the anode. This temperature range is comparable, although slightly higher, to that for the catalytic growth of carbon filaments. Finally, filled nanotubes, in particular those partially filled [14], display morphological similarities with carbon filaments [22]. For these reasons, the vapor-liquid-solid (VLS) scheme, generally adopted to describe the catalytic growth of carbon filaments [24], will be also used here to discuss the growth of filled nanotubes.

We will concentrate on the steady-state regime of the combined growth of the nanowire and the nanotube, the initial stage being a separate problem. We propose that a liquid-like metallic particle is attached to the growing end of the nanotube. The size of this particle may vary from the diameter of the tube to much larger dimensions. This particle contains metal, carbon and sulfur atoms which are incorporated either by direct condensation from the vapor phase, or by diffusion from the surface of the tube. The tube grows as carbon atoms come to complete the graphitic network at the particle/tube interface, but also as carbon is expelled from the metallic particle upon cooling. Indeed, as the temperature decreases, the solubility of carbon in the metal decreases resulting in a diffusion of carbon to the surface of the particle where it solidifies as concentric graphitic layers. The condensation of the vapor as a liquid and the solidification of the liquid are suggested to be both heat sources yielding a thermal gradient along the tube axis.

In contrast with the catalytic growth of carbon filaments, there is here a constant incoming flux of metal atoms. This is a necessary condition to allow for continuous fillings. The perfect fillings which are observed suggest that the filling material was in a liquid state as it filled the tube, while the tube was growing. It is striking to notice that the estimated growth temperature (1000°C to 2000°C) is close to the melting temperatures of some of the pure metals (Au, Ni) but it is much lower than that of others (T_m=2458°C for Nb, and T_m=3180°C for Re). This apparent contradiction can be explained by the role of sulfur as follows. The metal-sulfur phase diagrams are not known for all the elements which lead to a complete or partial filling, but it is striking to observe that in most cases an eutectic is formed such that the metal-sulfur particle remains liquid at a much lower temperatures than for the pure metal. The liquid-state is believed to be crucial in the tube filling process, through enhanced chemical reactivity or accelerated kinetics. In the case of nickel for instance, the melting temperature of the pure metal is 1453°C, whereas the eutectic temperature is 645°C. Assuming a growth temperature of 1000°C for the tube, we are precisely in a case where the pure metal is solid, whereas a particle at the eutectic composition (33% of sulfur) is liquid. We suggest that some indication about the growth temperature can be deduced from such criteria.

Another major difference between the properties of sulfides and metals which is likely to be important for tube filling, besides the above mentioned melting temperatures, is their surface tensions. Surface tension data for liquid sulfides are rather scarce to make a systematic comparison, but the examples of Ni and Fe and their sulfides are very instructive. The surface tensions of these two metals are rather high with values close to 2000 mN/m at the melting temperature in both the liquid and the solid state [25], whereas the surface tensions of Ni_xS_{1-x} and Fe_xS_{1-x} (for x≈0.7), measured in the liquid state at 1200°C, are respectively 443 mN/m and 311 mN/m [26]. These values have to be compared with the surface tension threshold values for the wetting of nanotubes [17]. It was indeed shown that pre-existing tubes can be filled by materials with a surface tension lower than typically 200 mN/m, with a dependence on the tube diameter [27, 28]. Once the material is inside the tube, it can be transformed (by chemical reaction assisted by annealing or electron irradiation) to another material with a higher surface tension, which has a non-wetting behavior, i.e. with a contact angle larger than 90° [28].

The importance of capillarity action in the case of tubes grown and filled simultaneously - as it is the case here - is less obvious and is still an open question [27]. TEM images of the ends of these nanowires clearly show that the meniscus of the filling material has always a convex shape (see Fig. 2 of Ref. [14a]),with contact angle close to 90°, attesting for a *weakly* non-wetting behavior. It should be emphasized that the wetting behavior under the synthesis conditions (i.e. at high temperature and with a filling material likely to be liquid) may differ from that under the observation conditions, i.e. at ambient temperature and in the solid state. Even if the tubes are not filled by capillarity action, it is clear that a large surface tension is a strong driving force against filling in particular in the case of liquid fillings. A liquid with a large surface tension, such as a transition metal liquid, emerging at an open end of a tube will indeed be sucked out. The values of the surface tension of liquid sulfides of iron and nickel reported above are much lower than that of pure transition metals and are close to the threshold value estimated for nanotube wetting. It is therefore suggested that, compared to a situation without sulfur, the addition of sulfur will cancel the strong driving force against filling, by drastically reducing the surface tension.

Sulfur may also favor the formation of the tubes, since it is known to promote the graphitization of carbon materials below 2000°C by acting as a cross-linker [29]. Finally sulfur may favor the wetting between the tube and the filling material, in particular through the sulfur atoms which are released after the graphitization process.

The very high sulfur-concentration observed in the filling material can be explained first by its very strong affinity with metals. In the present scheme, the liquid-like metallic particle is therefore rich in sulfur (at least in average, as we will discuss later in the case of Cr). The sulfur atoms which are released by the carbon layers after the graphitization process, also contributes to increase this concentration proportionally to the number of carbon layers.

The final microstructure and chemical composition of the nanowires are governed by the cooling conditions and by the thermodynamics dictated by the phase diagram. The dispersion observed in the microstructures can be explained by different cooling rates (as a function of the position in the reaction chamber for instance): a rapid quench will lead to a microcrystalline filling whereas a slow solidification will lead to a directional solidification front allowing the growth of long single-crystals. Two types of heterogeneities in chemical composition have been observed: either within a given tube for some filling elements or from one tube to the other for other filling elements. We demonstrate below that these two radically different situations correspond to two types of sulfur-metal phase diagrams and that this result can be simply explained following the solidification process of the sulfur-metal particle.

We first consider the schematic phase diagram drawn in Fig. 3(a). It is a simplified representation of the phase diagrams of S-M systems where M is a metal such as Ni, Co, Fe, Pd [30]. Sulfur is soluble in the liquid phase but not in the solid metallic phase, leading to the existence of different sulfides, S1, S2..., depending on the concentration in sulfur. As the solidification starts upon cooling, small crystallites of pure metal start to nucleate within the liquid phase. As the temperature decreases these crystallites grow and the sulfur concentration in the liquid increases, until the eutectic temperature is reached. At this point, the liquid is at a composition close to S1 and almost totally transforms into a S1 solid. This scenario perfectly explains why one finds crystals of pure metal surrounded by a sulfide crystal, and exactly corresponds to the situation illustrated in Fig. 2 in the case of Ni. In the case where the nucleation of metal crystallites in the liquid is too slow, another scheme accounting for this variation of

140

composition has been proposed previously [15]. It is based on the idea of an alternative growth of the M and S1 solids: when the metal grows the sulfur concentration in the liquid increases until it approaches the S1 composition; then S1 crystallizes, but its sulfur concentration being larger than the average sulfur concentration in the liquid, the propagation of the solidification front creates a depletion of sulfur in the liquid close to the liquid-solid interface ; therefore another cycle starts with the crystallization of the metal etc... Note that in the latter scheme a temperature gradient along the tube must be assumed, while the first scheme can be generalized to a homogeneous cooling of the whole tube.

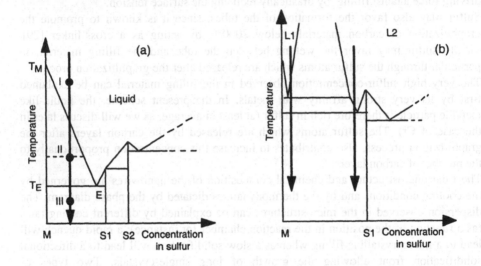

Figure 3: Simplified representation of metal-sulfur phase diagrams and their impact on the filling of tubes. a) : eutectic-type phase diagrams typical of metals like Ni, Co, Fe ; S1 and S2 are definite sulfides. b) : immiscible-liquid type phase diagram typical of metals like Cr or Ge ; the phase separation inside a liquid with an initial composition indicated by the dashed arrow leads to two solidification routes indicated by the vertical arrows.

We now consider the schematic phase diagram drawn in Fig. 3(b), occurring for elements such as Cr and Ge [30]. The difference with the previous case lies in the existence of a miscibility gap in the liquid. There are indeed two liquid phases: an almost pure-metal liquid phase L1 and a S-rich liquid phase L2. These two liquid phases will tend to phase separate. This phase separation within the liquid phase, i.e. before the filling occurs, is very different from the previous situation where the phase separation occurs once the liquid is inside the tube. In the present case the tubes will be filled over very long lengths either by the liquid L1 - which once crystallized yields a filling by the pure metal - or by the liquid L2, with a composition close to a definite sulfide such as S1, which will constitute the main material after solidification. This process nicely accounts for the two kinds of fillings observed in the case of Cr (CrS or pure Cr) in the experiment where the anode was doped by trace amounts of sulfur. In the experiment with the 99.4% graphite, the amount of sulfur is higher so that the L2 liquid phase is dominant and consequently only Cr-S sulfides were detected.

In conclusion we have shown that the presence of sulfur in catalytic quantity is crucial for the arc-discharge production of abundant fillings of nanotubes by metal based nanowires. It is suggested that sulfides are more favourable than pure metals for nanotube fillings because they have a lower solidification temperature and a much lower surface tension. Other elements than sulfur, such as selenium, hydrogen, oxygen, may have a similar effect on the filling process. It is striking that using an hydrogen arc, very similar Ge nanowires have been produced [31]. Furthermore the discussion of the growth mechanism has shown that the "metallurgy" of nanowires can be understood by considering the equilibrium phase diagrams of the corresponding metal-sulfur system and that the microstructures result from a directional solidification of a liquid phase.

4. BN-C nanotubes

4.1 STRUCTURAL ANALYSIS

In the case of the experiments with an HfB_2 anode and a graphite cathode, the most interesting part of the product was the anode deposit. Its analysis is presented in detail in [32]. This deposit contains dominantly nanoparticles encapsulated by a few tens of graphitic layers and some micrometer-long well defined nanotubes. We focus here on the analysis of these nanotubes shown in Fig. 4: they are very straight and well crystallized and their HRTEM images do not allow to distinguish them from pure C or BN nanotubes. Chemical profiles of boron, carbon and nitrogen were extracted from EELS line-spectra recorded along nanotube cross sections. The common feature to all analyses is that B and N profiles are always identical and perfectly correlated, attesting for a B:N ratio close to 1, whereas they are anti correlated with carbon profiles. This indicates a strong phase separation between BN and C and means that nanotubes are made of stacks of layers which are virtually pure C or BN [32]. The presence of ordered BN domains and carbon domains is also attested by the EELS fine structure of the absorption edges of the sp^2 bonds. No evidence of domains corresponding to a ternary compound like BC_2N was found. The number of stacks varies between 3 and 5 from one nanotube to another whereas the number of layers within a given set is typically 2 to 10. The second feature common to all the analyses is that the sets of layers at the free surfaces are always made of carbon. It is therefore concluded from these analyses that the present nanotubes are sandwich $C/BN/C(/BN/C)_n$ coaxial structures.

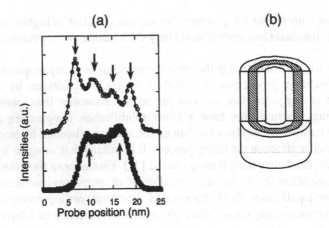

Figure 4: Chemical structure of BN-C nanotubes. a) Chemical profiles of carbon (open circles), boron (filled squares) and nitrogen (diamonds) extracted EELS line spectra [12]. b) Schematic cross-section representation of coaxial structure deduced for the tube: the light areas represent the carbon layers and the dark areas the BN layers.

4.2 DISCUSSION: THE GROWTH MECHANISM

The growth mechanism has to explain both the BN - C phase separation and the radial organisation of the layers with C layers at the free surfaces. The first point is characteristic of a total immiscibility of C and BN graphitic phases. Although the equilibrium phase diagram of the B - N - C system has not yet been experimentally established, the phase diagram calculated using the CALPHAD method predicts a phase separation graphite and h-BN with no intermediate phase [33]. It has been shown experimentally that metastable ternary compounds can be synthesized in certain conditions such as thin films. However the present observations demonstrate that arc-discharge conditions produce structures which are consistent with thermodynamical equilibrium of bulk systems.

The chemical radial organisation of the layers can be driven by energetics or by kinetics. The layer-layer interactions and therefore the surface tension in h-BN and graphite are very weak [34], the surface tension difference is even weaker, and therefore the energy difference between the observed organisation and other configurations is too small to be compatible with phase separation taking place after the growth of the tubular structure. We therefore propose that the sandwich structure results from a self organisation of the layers during their growth which is governed by kinetics and local energetics.

Our model is based on an analogy made between the C/BN/C domain sequence and the lamellar structures of unidirectional eutectic growth patterns. These patterns are found in alloys for which the phase diagram presents an eutectic point of transformation involving a liquid phase and two solid phases of different chemical concentrations, namely α and β. The solidification of the liquid phase within a unidirectional temperature gradient leads to a phase separation at the solidification front and to a lamellar microstructure where the lamellae of each phase alternate according to the sequence α/β/αβ/... . The width of each lamella is governed by the diffusion of the different chemical species and the characteristics of the temperature gradient. The theoretical phase diagram of the B-N-C system gives striking evidence for such an eutectic solidification of a liquid+gas phase into graphite and h-BN phase since the two-dimensional section of the diagram for the BN - C axis (Figure 5.(a)) shows the existence of an eutectic point involving these three phases. Therefore the present observations of self organized nanotubes strongly support a directional solidification from a liquid like particle. The process that we propose is sketched in Fig.5(b) and follows the solidification path indicated by the arrow in Fig.5(a). The number of domains of each phase and the number of layers within each domain are fixed by kinetic arguments (size of the initial liquid droplet, composition and temperature gradient). As a consequence of the asymmetry of the BN - C diagram, the composition of the liquid-like phase certainly lies on the C-rich side of the eutectic point in such a way that the first solid phase to appear upon cooling from the liquid state is graphite. Therefore between the high temperature liquid droplet region (mark I in Fig. 5) and the low temperature tube body region (mark III) an intermediate region exists (mark II) where cylindrical lamellae of BN-rich liquid phase alternate in the radial direction with cylindrical lamellae of graphite. Since the surface energy of the liquid is larger than that of solid carbon it is likely that graphite layers are favoured at free surfaces. This explains why free surfaces, including at the inner core of the tube, are always made of carbon layers.

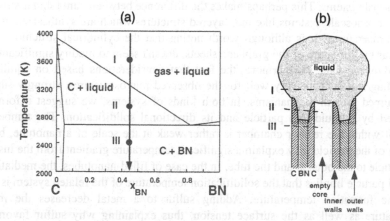

Figure 5: a) Simplified two-dimensional section of the C-B-N phase diagram based on the features of the theoretical phase diagrams of Ref [38]. The vertical arrow indicates the typical quenches which are considered. b) Schematic cross-section representation of the growth mechanism proposed for the formation of a C/BN/C coaxial nanotube mediated by the solidification of a liquid-like droplet.

In conclusion on this part, we have shown that the arc-discharge method modified to provide carbon, boron and nitrogen from separate sources leads to the formation of multiwall nanotubes characterized by a phase separation between BN and C phases and a radial self organization of the layers. All these features can be understood by considering the thermodynamic phase diagrams and the formation of these nanotubes is proposed to be governed by a kinetic path of solidification of a fluid droplet. Since it is based on thermodynamical considerations, the present scheme has a predictive power and could be used for synthesizing tubes with anticipated composition and structure.

5. General discussion: clues for the formation of nanotubes

We have presented in this paper the study of two very different kinds of multi-elements tubular structures, namely nanotubes filled with metallic nanowires and composite BN-C nanotubes, and suggested for both of them a coherent growth scheme to explain their chemical and crystalline structure. It is very striking that, in spite of the obvious differences between the two systems, similar physical processes are found to account for their formation. In both cases, the observed structures have the particularity of being thermodynamically quite stable although these systems are far from being true bulk materials and are grown far from bulk equilibrium conditions. The observed crystalline phases have indeed been identified as known stable phases except in a few cases, and the microstructures are consistent with local energetic minima. This tends to prove that the formation temperatures are high enough to achieve local equilibrium within the evolution time scale and that the systems are large enough to behave as three dimensional systems. This perhaps makes the difference between nanotubular structures and other nanoscale systems like thin layered structures which are synthesized at much lower temperatures. It is although worth noting that the cylindrical structure, and in particular the curvature of the graphene sheets, doesn't seem to modify significantly the thermodynamic data. Furthermore, the fact that considerations based on equilibrium phase diagrams account very well for the observed microstructures, strongly supports the assumed growth mechanisms. In both kinds of systems, we suggest a formation mediated by a liquid like particle and its directional solidification. The temperature gradient within the reactor chamber is rather weak at the scale of a nanotube, but the cooling of the particle may explain a significant temperature gradient from the inside of the particle to the outside and the tube. In the case of filled-nanotubes, the mediation by a liquid particle implies that the solidification temperature of the related system is lower than the formation temperature. Adding sulfur to a metal decreases the melting temperature as well as the surface tension, thus explaining why sulfur favors tube fillings. The self-organisation between BN and C layers in the case of the B-C-N nanotubes studied here is proposed to be governed by a sequential solidification of the two phases, following the characteristics of the BN-C phase diagram. One can in turn consider using these phase diagrams to adjust the synthesis conditions in order to produce other tubular nanostructures.

The similarities of the physical processes driving the formation of multi-element nanotubes emphasized here, make it very tempting to extend the present arguments to the formation and growth of ropes of single wall nanotubes. The later are indeed always obtained by adding a few percent of metals (from one or two elements among transition metals such as Ni, Co, Pd, Pt, Rh and rare earth metals such as Y, La [1, 3 - 5, 35 - 38]) to graphite to form either the cathode for the arc-discharge process or the target for the laser ablation method. The formation and growth of SWNT are therefore also multi-element processes since they involve the metal-carbon binary system. In the following we will show that the ideas behind the growth mechanism proposed initially by Saito [39] are perfectly consistent with the present approach. In this mechanism, liquid metal-carbon droplets form upon condensation from the vapor phase and as the particle is cooled, the solubility of carbon in the metal decreases as indicated by the metal-carbon phase diagram sketched in Fig. 6a. When the saturation threshold is attained, carbon segregates onto the surface; the solidification of carbon then leads to the formation of domes or bubbles rather than to layers for reasons which remain to be understood (Fig. 6 (b)). After this nucleation step, nanotubes will grow from these structures at the surface of the particle by incorporation of carbon atoms coming either from the particle or from the vapor phase by further condensation.

In view of the present study, the solidification path of the metal-carbon particle has to be considered with the help of thermodynamical phase diagrams. In general, carbon has a low solubility in the metallic solid phase and metal-rich metallic carbides are unstable or metastable. As a consequence, metal-carbon phase diagrams display in general an eutectic point between the liquid solution, graphite and the metal [30] (Fig.6(a)). If we consider a slow cooling process, when the surface reaches the solidification temperature - assuming that the composition of the particle is on the carbon-rich side of the eutectic point - carbon will start to crystallize at the surface. The situation may however be closer to a rapid quench from above the solidification temperature to below the eutectic temperature leading to a more complex solidification process involving simultaneously the metal and graphitic carbon. A better understanding of either of this two solidification processes is essential to explain why it could lead to nanotube nuclei at the surface. It is interesting to notice that, as shown recently, in the laser ablation method if the temperature of furnace is increased above the eutectic temperature the abundance of SWNT drops abruptly [37], suggesting that being below the eutectic temperature is essential either for the nucleation or for the growth of nanotubes.

Concerning the growth itself, the root-growth mechanism has theoretically been predicted to be kinetically quite feasible [40] and share some similarities with the formation of catalytic vapor grown fibres [24]. Experimental evidence of this root growth is supported by several observations of sea urchin morphologies [29, 6], where metallic particles are much larger (several tens of nm) than the section of the bundles. The observation of a single bundle emanating radially from a metallic particle as small as a few nm [7] attests that this mechanism operates what ever the size of the particle.

146

The detailed analysis of this mechanism, together with the confrontation with TEM observations of the structures at the surface of metallic particles in the case of successful or unsuccessful SWNT growth, will be the object of a forthcoming paper.

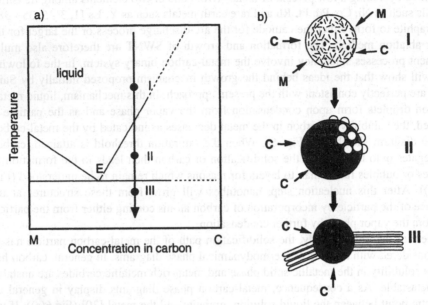

Figure 6: a) Simplified M-C phase diagram, where M is a transition metal like Ni, Co, Pd, Pt ; the vertical arrow indicates the typical solidification path which is considered. b) Schematic growth mechanism for the formation of SWNT bundles [46].

References

1. Thess, A., et al (1996) Crystalline ropes of metallic carbon nanotubes, Science 273, 483-487 ; Journet, C., et al (1997) Large-scale production of single-walled carbon nanotubes by the electric-arc technique, Nature **388**, 756-758.

2. Journet C.(1998) La production de nanotubes de carbone, *PHD thesis*, University of Montpellier.

3. Yudasaka, M., et al (1997) Single-wall carbon nanotube formation by laser ablation using double-targets of carbon and metal, *Chem. Phys. Lett.*, **278**, 102-106; Yudasaka, M., et al (1998) Pressure-dependence of the structures of carbonaceous deposits formed by laser-ablation on targets composed of carbon, nickel, and cobalt, *J. Phys. Chem. B* **102**, 4892-4896; Yudasaka, M., et al (1998) Roles of laser light and heat in formation of single-wall carbon nanotubes by pulsed laser ablation of CxNiyCoy targets at high temperature, *J. Phys. Chem. B* **102**, 10201-10207; Yudasaka, M., et al (1999) Single-wall carbon nanotubes formed by a single laser-beam pulse, *Chem. Phys. Lett.* **299**, 91-96.

4. Saito, Y., et al (1995) Extrusion of single-wall carbon nanotubes via formation of small particles condensed near an arc evaporation source, *Chem. Phys. Lett.* **236**, 419-426; Saito, Y., et al (1995) Single-layered carbon nanotubes synthesized by

catalytic assistance of rare-earths in a carbon-arc, *J. Phys. Chem.* **99**, 16076-16079; Saito, Y., et al (1998) High yield of single-wall carbon nanotubes by arc discharge using Rh-Pt mixed catalysts, *Chem. Phys. Lett.* **294**, 593-598.

5. Kataura, H., et al (1998) Formation of thin single-wall carbon nanotubes by laser vaporization of Rh/Pd-graphite composite rod, *Jpn J. Appl. Phys.* **37**, L616-618.

6. Seraphin, S., and Zhou, D. (1994) Single-walled carbon nanotubes produced at high-yield by mixed catalysts, *Appl. Phys. Lett.* **64**, 2087-2089, Zhou, D., et al (1994) *Appl. Phys. Lett.* **65**, 181.

7. For reviews see: Bernholc, J., et al (1998) Theory of growth and mechanical properties of nanotubes, *Appl. Phys. A* **67**, 39-46 ; Charlier, J.Ch., et al (1999) Microscopic growth mechanisms for carbon and boron-nitride nanotubes, *Appl. Phys. A* **68**, 267-273.

8. Arepalli,S., and Scott, C. D. (1999) Spectral measurements in production of single-wall carbon nanotubes by laser ablation, *Chem. Phys. Lett.* **302**, 139-145; Arepalli, S., et al (2000) Diagnostics of laser-produced plume under carbon nanotube growth conditions, *Appl. Phys. A* **70**, 125-133.

9. Kokai, F., et al (1999) Growth dynamics of single-wall carbon nanotubes synthesized by CO_2 laser vaporization, *J. Phys. Chem. B* **103**, 4346-4351.

10. Kataura, H., et al (2000), *Carbon* **38**, 1691-1687; Ishigaki, T., Suzuki,S., Kataura, H., Krätschmer, W., and Achiba, Y. (2000) Characterization of fullerenes and carbon nanoparticles generated with a laser-furnace technique, *Appl. Phys. A* **70**, 121-124.

11. Puretzki, A.A., Geohegan, D.B., Fan, X., and Pennycook, S.J. (2000) Dynamics of single-wall carbon nanotube synthesis by laser vaporization, *Appl. Phys. A* **70**, 153-160.

12. Journet, C., and Bernier, P. (1998) Production of carbon nanotubes, *Appl. Phys. A* **67**, 1-9.

13. Terrones, M., et al (1999) Nanotubes: a revolution in materials science and electronics, *Topics in Current Chemistry* **199**, (Springer Verlag, Berlin), p.190.

14. Guerret-Piécourt, C., Le Bouar, Y., Loiseau, A., and Pascard, H. (1994) Relation between metal electronic structure and morphology of metal compounds inside carbon nanotubes, *Nature* **372**, 761-765; Loiseau, A., and Pascard, H. (1996) Synthesis of long carbon nanotubes filled with Se, S, Sb and Ge by the arc method, *Chem. Phys. Lett.* **256**, 246-252.

15. Loiseau, A., et al (2000) Filling carbon nanotubes using an arc discharge, in *Science and applications of nanotubes,* D. Tománek and R. J. Enbody (Eds), Kluwer Academic Press, New York, pp. 1-16.

16. Lee, S.T., et al (1999) Oxide-assisted semiconductor nanowire growth, *MRS Bulletin* **24**(8), 36-42; Terrones, M., et al (1999) Advances in the creation of filled nanotubes and novel nanowires, *MRS Bulletin* **24**(8), 43-49.

17. Ugarte, D., et al (1998) Filling carbon nanotubes, *Appl. Phys. A* **67**, 101-105.

18. Suenaga, K., Colliex, C., Demoncy, N., Loiseau, A., Pascard, H., and Willaime, F. (1997) Synthesis of nanoparticles and nanotubes with well-separated layers of boron nitride and carbon, *Science* **278**, 653-655.

19. Loiseau, A., Willaime, F., Demoncy, N., Hug, G., and Pascard, H. (1996) Boron nitride nanotubes with reduced numbers of layers synthesized by arc discharge, *Phys. Rev. Lett.* **76**, 4737-4740. For a review see Loiseau, A., et al (1998) Boron nitride nanotubes, *Carbon* **36**, 743-752.

20. Stephan, O., et al (1994) Doping graphitic and carbon nanotubes structures with boron and nitrogen, *Science* **266**, 1683-1685; Z. Weng-Sieh et al (1995) Synthesis of $B_xC_yN_z$ Nanotubes, *Phys. Rev. B* **51**, 11229-11232; Terrones, M., et al (1996) Pyrolytically grown $B_xC_yN_z$ nanomaterials - nanofibres and nanotubes, Chem.

Phys. Lett. **257**, 576-582; Redlich, Ph., et al (1996) B-C-N nanotubes and boron doping of carbon Nanotubes, *Chem. Phys. Lett.* **260**, 465-470.

21. Tencé, M., Quartuccio, M., and Colliex, C. (1995) PEELS compositional profiling and mapping at nanometer spatial-resolution, *Ultramicroscopy* **58**, 42-54.

22. Audier, M., Oberlin, A., and Coulon, M.J. (1981) Crystallographic orientations of catalytic particles in filamentous carbon: case of simple conical particles, *J. of Cryst. Growth* **55**, 549-556; Baker, R.T.K., Barber, M.A., Harris, P.S., Feates, F.S., and Waite, R.J. (1972) Nucleation and growth of carbon deposit from the nickel catalyzed decomposition of acetylene, *J. Catal.* **26**, 51-62.

23. Demoncy, N., Stéphan, O., Brun, N., Colliex, C., Loiseau, A., Pascard, H. (1998) Filling carbon nanotubes with metals by the arc-discharge method: the key role of sulfur, *Eur. Phys. J. B* **4**, 147-157.

24. Tibbetts, G. (1984) Why are carbon filaments tubular?, *J. Cryst. Growth* **66**, 632-638.

25. Kumikov, V.K and Khokonov, Kh. B. (1983) *J. Appl. Phys.* **54**, 1346.

26. Ip, S.W., and Toguri, J.M. (1993) Surface and interfacial-tension of the Ni-Fe-S, Ni-Cu-S, and fayalite slag systems, *Metallurgical Transactions B* **24**, 657-668.

27. Dujardin, E., Ebbesen, T.W., Hiura, H., and Tanigaki, K. (1994) Capillarity and wetting of carbon nanotubes, *Science* **265**, 1850-1852.

28. Ugarte, D., Châtelain, A., and de Heer, W.A. (1996) Nanocapillarity and chemistry in carbon nanotubes, *Science* **274**, 1897-1899.

29. Oberlin, A. (1984) Carbonization and graphitization, *Carbon* **22**, 521-541; Bourrat, X., Oberlin, A., Escalier, J.C. (1987) *Fuel* **542**, 521.

30. Massalski, T. B. (1990) *Binary Alloy Phase Diagrams*, ASM International.

31. Dai, J.Y., Lauerhaas, J.M., Setlur, A.A., Chang, R.P.H. (1996) Synthesis of carbon-encapsulated nanowires using polycyclic aromatic hydrocarbon precursors, *Chem. Phys. Lett.* **258**, 547-53.

32. Suenaga, K., Willaime, F., Loiseau, A., and Colliex, C. (1999) Organisation of carbon and boron nitride layers in mixed nanoparticles and nanotubes synthesised by arc discharge, *Appl. Phys. A* **68**, 301-308.

33. Kasper, B. (1996) *PhD thesis*, Stuttgart University.

34. Schabel, M.C., and Martins, J.L. (1992) Energetics of interplanar binding in graphite, *Phys. Rev. B* **46**, 7185-7188.

35. Thess, A., et al (1996) Crystalline ropes of metallic carbon nanotubes, *Science* **273**, 483-487; Young Hee, L., Seong Gon, K., Tomanek, D. (1997) Catalytic growth of single-wall carbon nanotubes: An ab initio study *Phys. Rev. Lett.* **78**, 2393-2396.

36. Journet, C., et al (1997) Large-scale production of single-walled carbon nanotubes by the electric-arc technique, *Nature* **388**, 756-758.

37. Kataura, H. et al (2000), *A workshop on nanotubes and fullerenes chemistry*, Elsevier Science Ltd, Oxford.

38. Guo, T., et al (1995) Self-assembly of tubular fullerenes, *J. Phys. Chem.* **99**, 10694-10697; Guo, T., et al (1995) Catalytic growth of single-walled nanotubes by laser vaporization, *Chem. Phys Lett.* **243**, 49-54.

39. Saito, Y., et al. (1994) Single-wall carbon nanotubes growing radially fron Ni fine particles formed by arc evaporation, *Jpn. J. Appl. Phys.* **33**, L526-L529; Saito, Y. (1995) Nanoparticles and filled nanocapsules, *Carbon* **33**, 979-988.

40. Maiti, A., Brabec, C.J., and Bernholc, J. (1997) Kinetics of metal-catalyzed growth of single-walled carbon nanotubes, *Phys. Rev B* **55**, R6097-6100.

FIRST-PRINCIPLES THEORETICAL MODELING OF NANOTUBE GROWTH

JEAN-CHRISTOPHE CHARLIER
Unité de Physico-Chimie et de Physique des Matériaux,
Université Catholique de Louvain, Place Croix du Sud 1,
B-1348 Louvain-la-Neuve, Belgium.

XAVIER BLASE
Département de Physique des Matériaux, U.M.R. n° 5586,
Université Claude Bernard, 43 bd. du 11 Novembre 1918,
F-69622 Villeurbanne Cedex, France.

ALESSANDRO DE VITA
Istituto Nazionale di Fisica della Materia (INFM) and
Department of Material Engineering and Applied Chemistry,
University of Trieste, via Valerio 2, I-34149 Trieste, Italy.

AND

ROBERTO CAR
Department of Chemistry, Princeton University,
107HB Hoyt Lab, Princeton, NJ 08544, USA.

Abstract. The growth of carbon (C) and boron nitride (BN) nanotubes cannot be directly observed and the underlying microscopic mechanism is a controversial subject. Here we report on the results of first-principles dynamical simulations of both single- and double-walled carbon nanotube edges. We find that the open end of carbon single-walled nanotubes (SWNTs) spontaneously closes by forming a graphitic dome in the 2500-3000 K temperature range of synthesis experiments. On the other hand, "lip-lip" interactions, consisting of chemical bonding between the edges of adjacent coaxial tubes, trap the end of the double-walled carbon nanotube into a metastable energy minimum, preventing dome closure. The resulting end geometry is highly chemically active, and can easily accommodate incoming carbon fragments, thus allowing for growth by chemisorption from the vapour phase.

149

L.P. Biró et al. (eds.), Carbon Filaments and Nanotubes: Common Origins, Differing Applications, 149–170.
© 2001 *Kluwer Academic Publishers. Printed in the Netherlands.*

Electron microscopy observations and electron diffraction patterns reveal that B doping considerably increases the length of carbon tubes and leads to a remarkable preferred "zigzag" chirality. These findings are corroborated by first-principles static calculations and dynamical simulations which indicate that, in the "zigzag" geometry, B atoms act as surfactant during growth preventing tube closure. This mechanism does not extend to "armchair" tubes suggesting a helicity selection during growth.

The growth mechanisms of boron nitride SWNTs are studied as well and compared to the case of pure carbon tubes. In the experimental conditions of temperature, the behavior of growing BN nanotubes strongly depends on the nanotube network helicity. In particular, we find that open-ended "zigzag" tubes close rapidly into an amorphous like tip, preventing further growth. In the case of "armchair" tubes, the formation of squares traps the tip into a flat cap presenting a large central even-member ring. This structure is metastable and able to revert to a growing hexagonal framework by incorporation of incoming atoms. These findings are directly related to frustration effects, namely that B-N bonds are energetically favored over B-B and N-N bonds.

1. Introduction

About ten years after their discovery in 1991 [1, 2], carbon nanotubes are attracting much interest for their potential applications in highly performing nanoscale materials [3] and electronic devices [4,5]. Synthesis techniques for carbon nanotubes have achieved high production yields as well as good control of the tubes multiplicity, shape and size [6]. Carbon nanotubes typically grow in an arc discharge at a temperature of ~3000 K, explaining that the mechanisms of nanotube formation and growth under such extreme conditions remain unclear [6]. The earliest models [7,8] for growth of multi-walled nanotubes (MWNTs) were based on topological considerations and emphasised the role of pentagon and heptagon rings to curve inside or outside the straight hexagonal tubular network. The most debated issue, in later works, was whether these nanotubes are open- or close-ended during growth. In favour of the closed-end mechanism, it was proposed that tubes grow by addition of atoms onto the reactive pentagons present at the tip of the closed structure [9, 10]. However, further experimental studies suggest an open-end growth mechanism [11,12]. Moreover, there is controversy on whether in MWNTs the inner or the outer tubes grow first [8,13] or if different tubes may grow together [14,15]. In addition, the growth of SWNTs requires the presence of metal catalysts, contrarily to the multi-walled case [16,17]. Electron microscopy observations and electron diffraction patterns also reveal that B doping considerably increases the length of carbon tubes and the metallic character of the tube [18]. Experimental studies [19,20] have also shown that the morphology of BN nanotubes differ significantly from the one of their carbon analogs. BN nanotubes tend to have amorphous-like tips or closed-flat caps, which are less frequently observed in carbon systems.

Although to date no single model seems to be capable of explaining all the experimental evidence mentioned above, quantum molecular dynamics is a very accurate and powerful tool to investigate the growth of nanoscale systems at the atomic level. In the next sections, we will study the microscopic mechanisms underlying the growth of pure carbon (section II), boron-doped (section III) and boron-nitride (section IV) nanotubes by performing first-principles molecular dynamics (MD) simulations on single- and double-walled tubes [21–23]. In this approach [24], the forces acting on the atoms are derived from the instantaneous electronic ground state, which is accurately described within density functional theory in the local density approximation. In such a framework, the instantaneous electronic ground state is given within a formulation in which only valence electrons are explicitly taken into account. The interaction between valence electrons and nuclei plus frozen core electrons is described using norm-conserving pseudopotential for carbon, boron and nitrogen [25] and for hydrogen [26]. Periodic boundary conditions are adopted, using a supercell size for which the distance between repeated images is larger than 5 Å. This is sufficient to make negligible the interaction between images. The electronic wavefunctions at the Γ-point of the supercell Brillouin zone are expanded into plane waves with a kinetic energy cutoff of 40 Ry, which gives well converged ground-state properties of carbon and BN systems [27].

2. Microscopic growth mechanisms for carbon nanotubes

In the present section, we perform calculations on finite tubular C nanotubes terminated on one side by an open end and on the other side by hydrogen atoms which passivate the dangling bonds. We consider two 120 carbon atoms systems representing a (10,0) "zigzag" (0.8 nm of diameter) (Figs.1.a-1.b) and a (5,5) "armchair" (0.7 nm of diameter) (Figs.2.a-2.b) SWNTs, and a 336 carbon atoms system representing a (10,0)@(18,0) "zigzag" (0.8 nm and 1.4 nm of diameter) double-walled nanotube (Figs.3.a-3.b) [28]. *Ab initio* calculations with such large systems and for a total simulation time of ~30 ps are made possible by massively parallel computing. All nanotubes are initially relaxed to their open-end equilibrium geometries before being gradually heated up to 3500 K by performing constant temperature simulations [29].

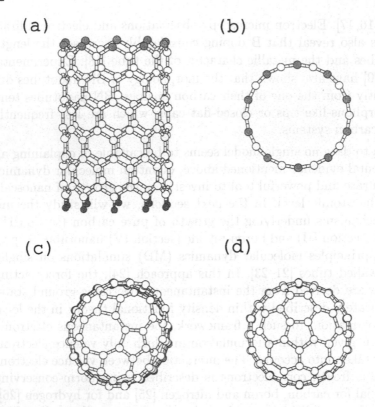

Figure 1. Single-walled nanotube clusters. The tubes belong to the zigzag geometry and have a diameter of 0.8 nm. Each cluster contains 120 carbon atoms (large white spheres when the atomic coordination is 3: sp^2 bonding) and 10 hydrogen atoms (small dark grey spheres). Atoms which coordination is below 3 (sp^1 bonding and dangling atoms) or higher than 3 (sp^3 bonding) are represented with large light grey and black spheres respectively. The hydrogen atoms and the attached ring of carbons are kept fixed during the simulations. (a) Side view of the open-ended starting configuration at 0K, where 10 dangling bonds (light grey spheres) are present on the top edge of the zigzag geometry. (b) Top view of the same system illustrating the cylindrical geometry. (c) Top view of the close-ended configuration, after a thermalization period of 5 ps at 3500 K. The final cap topology is composed of 7 hexagons, 5 pentagons, 1 heptagon, 1 octagon, and 2 squares, verifying Euler's theorem. (d) Top view of a perfect capping topology (hemi-fullerenes) where only isolated pentagons and hexagons are used to close the structure.

2.1. SELF-CLOSURE OF ZIGZAG AND ARMCHAIR C SWNTS

The choice of the (10,0) single-walled system is consistent with a dominant peak observed in the diameter frequency histogram of synthesised mono-layer nanotubes [16]. In this case, the first observed structural rearrange-ment occurs at 300 K: it is a reconstruction of the tip edge characterised by the formation of triangles. This eliminates most of the dangling bonds and induces an initial inside-bending of the nanotube edge. At ~1500 K, the first pentagon is created from one of the top hexagonal rings, and leads to more substantial inside-bending. The extra carbon atom, dangling over the tube edge, moves onto the nearest trimer to form a square. The formation of two more pentagons with the same mechanism is observed at temperatures lower than 2500 K. Finally at temperatures of ~3000K, a global reconstruction of the tip develops, with the nanotube edge completely closing into a structure with no residual dangling bonds (Fig.1.c). This gives an energy gain of $\simeq$ 18 eV with respect to the initial open-end geometry. The closed structure obtained by molecular dynamics is still 4.6 eV higher in energy than the "ideal" C_{60}-hemisphere cap, containing only hexagons and isolated pentagons (as proposed in Fig.1.d). However, the timescale of our simulations prevents us from studying the further evolution of the tip geometry.

A similar calculation for a (5,5) armchair SWNT [28] also leads to tip closure (Fig.2). In this case the open nanotube edge consists of dimers (Figs.2.a-2.b), rather than single atoms as in Figs.1.a-1.b, and its chemical reactivity is lower. This is presumably the reason why we observe the first atomic rearrangement only at ~3000 K, with the formation of a pentagon plus a dangling atom (Fig.2.c). Due to the armchair symmetry no squares are formed, at variance with the zigzag case. On the other hand, a second pentagon is often created by the connection of the extra dangling carbon with a neighboring carbon dimer (see Fig.2.c). Successive processes of this kind lead to the self-closure of the nanotube (Fig.2.d) into a hemi-C_{60} (energy gain $\simeq$15 eV). We expect that similar self-closing processes should also occur for other nanotubes having a similar diameter, as it is the case for most SWNTs synthesised so far [16]. We notice that the reactivity of closed nanotube tips is considerably reduced compared to that of open end nanotubes. It is therefore unlikely that SWNTs could grow by sustained incorporation of C atoms on the closed tip. This is in agreement with the finding that C atoms are not incorporated into C_{60} [30]. Although several mechanisms have been proposed to account for the growth mechanisms of SWNTs, the key role played by the metal catalyst at the microscopic level is still a highly debate issue.

2.2. LIP-LIP INTERACTION IN C MWNTS

In another set of calculations, we consider a (10,0)@(18,0) zigzag bilayer system as a prototypical example of MWNTs. It consists of the (10,0) nanotube previously studied plus a concentric (18,0) tube (Figs.3.a-3.b). The interlayer distance is 3.3 Å, in good agreement with typical experimental

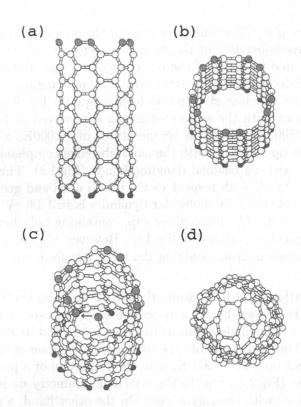

(a) (b)

(c) (d)

Figure 2. The carbon nanotubes belong to the armchair geometry and have a diameter of 0.7 nm. Each cluster contains 120 carbon atoms and 10 hydrogen atoms. The black and white scale is defined in Fig.1. The hydrogen atoms and the attached ring of carbons are kept fixed during the simulations. (a) Side view of the open-ended starting configuration at 0K, where 10 dangling bonds coupled in dimers (light grey spheres) are present on the top edge of the armchair geometry. (b) Top view of the same system. (c) At 3000K, the first pentagon is created from one of the top hexagonal rings. Nine dangling bonds remain among which one extra dangling C atom. The latter is going to move aside, forming a second pentagon as indicated by the arrow. (d) Top view snapshot of the close-ended configuration at 3500 K. The final cap topology is composed of pentagons and hexagons only. Some pentagons are sharing the same edge, and are not isolated like in the topology described in Fig.1.d.

values [2]. In order to study the effect of the second shell on edge stability, we carry out simulations at temperatures ranging from 0 K to 3000 K. At 300 K, the topmost atoms of the inner and outer tube edges rapidly move towards each other forming several bonds to bridge the gap between adjacent edges. At the end, only two residual dangling bonds are left (Fig.3.c). The resulting picture is one in which the open-ended bilayer structure is stabilized by "lip-lip" interactions as first proposed in Ref. [31], which inhibit the spontaneous dome closure of the inner tube observed in the previous simulations. Notice that the non-perfect reconstruction that we observe is

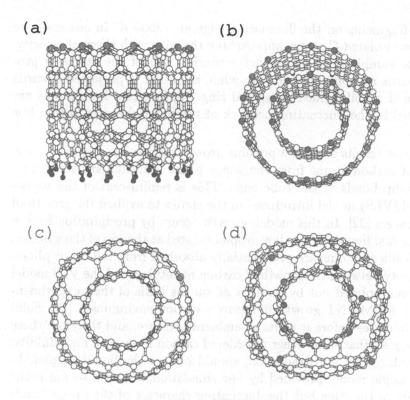

Figure 3. Double-walled nanotube clusters. The tubes belong to the zigzag geometry and have diameters of 0.8 nm (inner shell) and 1.4 nm (outer shell), respectively. Each cluster contains 336 carbon atoms and 28 hydrogen atoms. The black and white scale is defined in Fig.1. (a) Side view of the open-ended starting bilayer system at 0K, where 28 dangling bonds are present on the two top edges. (b) Top view of the same carbon system. (c) Top view of "lip-lip" interactions between the two concentric shells, reducing the number of dangling bonds by the creation of 10 inter-shell and 3 intra-shell covalent bonds (triangles) at ~300 K. (d) Top view snapshot of the fluctuating edge of the bilayer system at ~3000 K, illustrating the high chemical reactivity of this region. This edge exhibits 12 dangling bonds, 2 pentagonal rings positioned side-by-side, and 1 four-fold coordinated atoms (black sphere), presenting the metastability of the nanotube top edge.

favored by the incommensurability of the two graphene networks. During the short time span of this simulation (1.5 ps), we also observe isolated exchanges between atoms at the tube edge, suggesting low diffusional barriers. We report in Fig.3.d a snapshot of a typical configuration at ~3000 K.

2.3. GROWTH MECHANISM FOR C MWNTS

Since the residual dangling bonds and the continuous bond-breaking processes should provide active sites for absorption, we investigate the impact of carbon fragments on the fluctuating edge at ~3000 K. In one case, we consider two isolated C atoms approaching the edge with thermal velocity. In order to sample different impact scenarios, one of the atoms is projected towards a dangling bond site, while the other is projected towards the middle of a well formed hexagonal ring. We find that both atoms are incorporated in the fluctuating network of the nanotube edge within less than 1 ps.

The above results suggest a possible growth mechanism in which incorporation of carbon atoms from the vapor phase is mediated by the fluctuating lip-lip bonds of the tube edge. This is reminiscent of the vapor-liquid-solid (VLS) model introduced in the sixties to explain the growth of silicon whiskers [32]. In this model, growth occurs by precipitation from a super-saturated liquid-metal-alloy droplet located at the top of the whisker, into which silicon atoms are preferentially absorbed from the vapor phase. The similarity between the growth of carbon nanotubes and the VLS model has also been pointed out by Saito *et al.* on the basis of their experimental findings for MWNT growth in a purely carbon environment [14]. Solid carbon sublimates before it melts at ambient pressure, and therefore these authors suggest that some other disordered carbon form with high fluidity, possibly induced by ion irradiation, should replace the liquid droplet. In the microscopic model provided by our simulation, it is not so much the fluid nature of the edge but the fluctuating character of the lip-lip bonds that makes possible a rapid incorporation of carbon fragments.

3. Boron-mediated growth of long helicity-selected carbon nanotubes

To understand the remarkable increase in length of carbon nanotubes under B doping, we investigate by means of static and dynamical *ab initio* calculations the growth mechanisms of such systems. We first study the relative energies of structures obtained by substituting a C atom with a B atom at different locations at the open end of a (9,0) zigzag nanotube (Fig.4.a). After atomic relaxation, we find that B atoms preferentially substitute the two-coordinated topmost C atoms at the tube end.

Substitutional B atoms located one, two and three bonds away from this location into the tube body are found to be less stable by 0.9, 0.4, and 1.5 eV respectively. Clearly, the stabilization of B atoms at the zigzag edge is due to the removal of dangling bonds at the two-coordinated atomic sites. Similar results are obtained for a graphitic zigzag edge (Fig.5), indicating that the energy ordering is independent of the curvature and thus of the radius of the nanotube. These results suggest that B atoms will preferentially remain on the "lip" of the nanotubes, thus acting as surfactant during the growth.

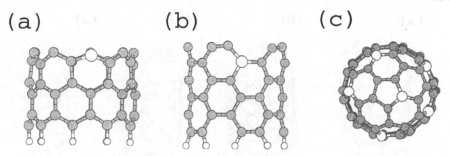

Figure 4. Symbolic ball-and-stick representation of various B-doped carbon nanotube lips. In (a) and (b), C(9,0) and C(5,5) edges, respectively, are represented. In (c), the B-doped hemi-C_{60} caped C(5,5) nanotube used for the MD run is seen from the top.

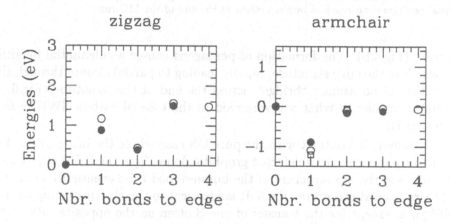

Figure 5. Substitutional energy for B atoms at various sites in the zigzag and armchair cases. The reference of energy is taken to be the "edge" site. In both cases, filled and empty circles indicate the results for respectively the model nanotubular systems (represented in Figs.4.a-4.b) and a planar graphene sheet edge. The empty squares indicate calculations on a C_{24} flake presenting an armchair rim.

We now address the role of surfactant B atoms in the dynamical behavior of zigzag nanotubes at synthesis temperature by means of finite temperature first-principles molecular dynamics. As a test case, we consider a C(9,0) nanotube in an open-end configuration terminated by a complete ring of two-coordinated B atoms [33]. Starting from this "ideal" cleaved B-terminated geometry, we gradually raise the temperature up to 2500 K, which is typical of the experimental synthesis conditions. At a relatively low temperature (~ 500 K), the only observed phenomenon is that B atoms transiently dimerize (Fig.6.a). At ~ 2500 K, B trimers are created, followed by the appearance of B-B-B-C-C and C pentagons at the tip of the nan-

158

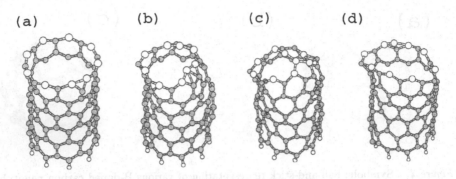

Figure 6. Symbolic ball-and-stick representation of a B-doped C(9,0) nanotube. In (a), the ground state geometry at T=0 K is represented. In (b), (c), and (d) three different snapshots for the nanotube at 2500 K are given. Snapshot (d) represents the final configuration reached by the system at the end of the MD run.

otube (Fig.6.b). The formation of pentagons causes a substantial bending inwards of the edge structure, rapidly leading to partial closure through the creation of an atomic "bridge" across the end of the nanotube (Fig.6.c). This is similar to what was observed in the case of carbon SWNTs (see section II).

However, in contrast with the pure CN case where the bridging quickly led to tip closure into a defected graphitic dome, here the B-B bonds easily break, and the tip structure of the boron-doped tube eventually re-opens. The re-opened structure (Fig.6.d) is identical to the starting configuration (Fig.6.a) except for the transfer of one B atom on the opposite side of the tube edge. In contrast with the CN case, the presence of a C pentagon (front right of Figs.6.b-6.d) is not sufficient to drive the system to full tip closure. Further, we verify that C pentagons located at the edge of an open-ended nanotube reopen spontaneously to form a hexagon upon arrival of a C atom from the vapor phase. This indicates that a C pentagon which does not succeed in closing the tube tip is short lived at the growing edge.

It appears therefore that B atoms located at the top of growing zigzag CNs can act against tube closure. This would explain the increased length of zigzag B-doped carbon nanotubes. We note that while the present simulations were performed on a SWNT, similar conclusions also apply to the case of MWNTs. Since it opposes the closure of each individual shell, B doping is expected to favor the growth of long MWNTs. The ability of growing long tubes is by itself an important result, given the importance attached to the production of long ballistic one-dimensional nanoconductors, or for large aspect ratio fibers for composite materials. It is well-known, however, that the electronic properties of a nanotube depends strongly on the tube helicity. Therefore, a further important development in the nanotube production techniques would be the ability of selectively growing long structures with a given helicity.

We now turn to the helicity selection revealed by the electron diffraction experiments [23]. The same kind of mechanism discussed above for a zigzag tube may *a priori* apply to armchair nanotubes. We therefore perform the same type of analysis for armchair boron-doped tubes. In contrast with the zigzag case, we find that the preferred substitutional site is not on the lip of the open nanotube but is located instead one bond away from the edge (see Fig.4.b and Fig.5). In this most stable site, the B atom is threefold coordinated, which makes it much less mobile (and, thus, less likely to act as a surfactant) than it would be as an edge atom. In the case of a $C(5,5)$ nanotube, this configuration is found to be 0.9 eV more stable than the B-C dangling-dimer configuration and 0.75 eV more stable than a configuration in which the B atom is three bonds away from the lip. Once more, similar results (Fig.5) are obtained at the armchair edge of a graphitic sheet and for a C_{24} flake with a rim presenting dangling dimers only. We note that while dangling C_2 dimers can be stabilized by dangling bond pairing, the formation of an "unbalanced" B-C pair seems to significantly reduce the efficiency of this hybridization, explaining therefore the relative unstability of the dangling BC dimer. Since we are left with no mechanism to place B mobile atoms at the very edge of the growing structure, it seems difficult at this stage to generalize to the armchair case the results obtained earlier for the zigzag tube. In a further simulations, starting from a $C(5,5)$ tube closed by the B-doped hemi-C_{60} represented in Fig.4.c, we did not observe the cap reopening during 10 ps of simulations at 3000 K [34]. Comparing this simulation with the one described above in the case of the zigzag tube, it seems likely that maximizing the number of B-B bonds upon attempted closure is crucial for observing the "re-opening" of the tip. Since in the armchair case B atoms tend to "sink" into the graphitic network, B-B bonds will form much less frequently (if at all) and will thus not be available to play the catalytic role observed in the zigzag case. This can explain why B atoms, acting as surfactant, can selectively catalyze the growth of long zigzag nanotubes. We note, however, that the present simulations do not rule out the growth of MWNTs of "standard" length with chiral or armchair helicities.

4. Frustration effects and growth mechanisms for BN nanotubes

While carbon nanotubes [1] have attracted a lot of attention in the last few years, boron-nitride (BN) nanotubes [19,20,35] have been studied much less extensively. BN nanotubes are insulating with a ~ 5.5 eV band gap [36].

More generally, it has been shown that controlling the (x,y,z) stoichiometry of composite $B_xC_yN_z$ nanotubes can be used as a mean to tailor their electronic properties [37]. In addition, the lack of chemical reactivity displayed by BN nanotubes leads to the idea that BN nanotubes could be used as "protecting cages" or "molds" for any material encapsulated within [38]. These potential applications provide a strong motivation for pursuing the experimental and theoretical effort of better understanding BN tubular systems.

Recent experimental studies [19,20] have shown that the morphology of BN nanotubes differ significantly from the one of their carbon analogs. BN nanotubes tend to have amorphous-like tips or closed-flat caps, which are very rarely observed in carbon systems. In order to understand the origin of these differences, we have studied the growth mechanism of BN nanotubes by means of first-principles molecular dynamics simulations which was shown in the previous section to elucidate the microscopic processes of carbon nanotube growth. We find that the high energy cost of "frustrated" B-B and N-N bonds strongly affects the modalities of growth. In particular, pentagons and other odd-member rings are not stable at the growing tube edge at experimental temperatures. A metastable "open" tip structure with even-member rings only and no frustrated bonds is created when this is compatible with the tube network helicity (armchair nanotubes). The calculations indicate that SWNTs with this structure may grow uncatalyzed by chemisorption from the vapor phase. On the contrary, if the network helicity imposes the presence of frustrated or dangling bonds (zigzag nanotubes), these bonds are unstable, breaking and forming during the simulations. This leads to an amorphous tip structure, preventing growth. The results provide evidence in support of the topological models proposed in Ref. [19,20] for final BN tip geometries.

4.1. STABILITY OF BN NANOTUBES

Before studying the growth mechanism of BN nanotubes, and compare to the case of their carbon analogs, we first study the stability and geometry of infinite BN nanotubes on the basis of the present *ab-initio* approach. By minimizing both stress and Hellman-Feynman forces, we determine first the equilibrium geometry for (n,0) and (n,n) tubes with diameters ranging from 4 to 12 Å (index notations for the tubes refer to the convention of Ref. [28] as defined for graphitic nanotubes). The main relaxation effect is a buckling of the boron-nitrogen bond, together with a small contraction of the bond length ($\sim$ 1%). In the minimum energy structure, all the boron atoms are arranged in one cylinder and all the nitrogen atoms in a larger concentric one.

Due to charge transfer from boron to nitrogen, the buckled tubular structure forms a dipolar shell. The distance between the inner "B-cylinder" and the outer "N-cylinder" is, at constant radius, mostly independent of tube helicity and decreases from 0.2 a.u. for the (4,4) tube to 0.1 a.u. for the (8,8) tube. As a result of this buckling, each boron atom is basically located on the plane formed by its three neighboring nitrogen atoms so that the sp^2 environment for the boron atom in the planar hexagonal structure is restored (at most, the $\widehat{NBN}$ angles differ from 120^0 by 0.2% for the smallest tube). This tendency for three-fold coordinated column III atoms to seek 120° bond angle is extremely strong. For example, it explains the atomic relaxation of the (110) and (111)-2x2 surfaces of GaAs and other III-V compounds. On the other hand, the $\widehat{BNB}$ angles approach the value of the bond angle of the s^2p^3 geometry. Buckling and bond length reduction induce a contraction of the tube along its axial direction by a maximum of 2% for the smallest tube studied.

Energies per atom for the relaxed tubes are plotted in Fig.7. as a function of the tube diameter. The zero of energy is taken to be the energy per atom of an isolated hexagonal BN sheet. On the same graph, the energy per carbon atom above the graphite sheet energy is represented. As for graphitic tubes, BN tube energies follow the classical $1/R^2$ strain law, where R is the average radius of each tube. However, for the same radius, the calculated strain energy of BN nanotubes is smaller than the strain energy of graphitic tubes. This is related mostly to the buckling effect which reduces significantly the occupied band energy in the case of the BN compounds. Therefore, it is energetically more favorable to fold a hexagonal BN sheet onto a nanotube geometry than to form a carbon nanotube from a graphite sheet.

We also address the question of stability of a small tube versus opening into a strip of planar hexagonal BN. We performed total energy calculations for the strip corresponding to the small tube (6,0), allowing complete relaxation of the strip geometry. As shown on Fig.7 (filled square), the strip is less stable than the corresponding tube. As for carbon nanotubes, BN nanotubes with a radius larger than 4 $\mathring{A}$ are stable with respect to a strip.

4.2. FRUSTRATION ENERGIES

We now turn to the study of the growth mechanisms of BN nanotubes. Firstly, we evaluate the energy cost associated with a frustrated N-N or B-B bond as compared to a B-N bond. Starting from an isolated BN hexagonal sheet, we exchange two neighboring B and N atoms, creating 2 B-B bonds and 2 N-N bonds. The computed energy cost of this anti-site defect is 7.1 eV, after atomic relaxation. This corresponds to an average of 1.8 eV

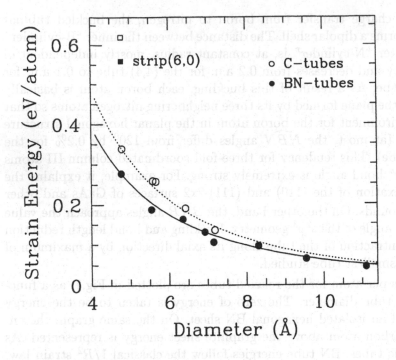

Figure 7. Total energy of nanotubes in eV/atom as a function of tube diameter (in Å). The black circles represent the BN nanotube energies above the energy of an isolated hexagonal BN sheet. The opened circles represent the graphite nanotubes energies above the energy per atom of an isolated graphite sheet. The solid and dashed lines are guides to the eye. The energy of the strips corresponding to the (6,0) BN and carbon tubes are given respectively by the filled and empty square.

per frustrated bond. Secondly, we study various isomers of small C_{24} and $B_{12}N_{12}$ clusters. In particular, we compare the energies of two fullerene-like closed structures, one with pentagons and hexagons only (labeled [5,6]), the other with squares and hexagons (labeled [4,6]). In the case of carbon, we find that $C_{24}[5,6]$ is more stable than $C_{24}[4,6]$ by 1.5 eV, while in the case of BN compounds, the $B_{12}N_{12}[5,6]$ isomer is less stable than $B_{12}N_{12}[4,6]$ by 7.8 eV [39]. Assuming that elastic energies are equivalent for both C and BN compounds, we can estimate the cost of the 6 frustrated bonds present in the $B_{12}N_{12}[5,6]$ isomer to be 9.3 eV, that is ~ 1.6 eV per frustrated bond. This is consistent with the energy found above for a planar geometry.

4.3. THE ARMCHAIR BN SWNTS

We now turn to the study of the dynamics and growth mechanisms of BN nanotubes. We consider first the case of a (5,5) armchair nanotube. Starting from an "ideal" cleaved geometry (Fig.8.a), we gradually raise the temperature up to 3000 K which is typical of the experimental synthesis conditions. At relatively low temperature ($\sim$ 1500 K), the top-most BN hexagons bend inside connecting each other. This creates 5 squares and a decagon (Fig.8.b). In this configuration, all atoms are three-fold coordinated and no frustrated bonds are present. The energy of this BN [4, 6, 10] configuration at T=0 K is 5.7 eV lower than the energy of the cleaved starting configuration and 2.0 eV lower than the energy of an hypothetical BN(5,5) nanotube capped with a perfect hemi-C_{60} structure (Fig.8.c). Brought to 3000 K for more than 5 ps, the squares and the decagon never open and no further reorganization is observed. This suggests that tip structures containing several squares and a "large" even-member ring are quite stable in the case of BN based systems. An analogous molecular dynamics study has been performed for a BN(4,4) nanotube. Similarly to what was found before, this leads to the creation of a [4,6,8] tip containing 4 squares and an octagon.

Further, we test the stability of the obtained caps against arrival of incoming B and N atoms from the plasma. We send a B atom at 3000 K thermal velocity along the axis of the BN(4,4) nanotube onto the [4,6,8] tip structure.

We find that the impinging atom is rapidly incorporated by re-opening a square (Fig.9.a) to create a pentagon. We repeat the same "experiment" with a BN dimer and, again, the dimer is rapidly incorporated by a square in order to form a perfect hexagon (Fig.9.b). The picture which emerges therefore from the present set of simulations is that armchair BN nanotubes can be stabilized into an intermediate "semi-opened" metastable cap, consisting of squares and a large even-member ring, which is able to incorporate incoming B or N atoms (or BN dimers) from the plasma phase. This contrasts significantly with the case of armchair carbon nanotubes where the formation of pentagons leads to closure onto a hemi-C_{60} cap which does not incorporate any incoming atoms, preventing further growth.

4.4. THE ZIGZAG BN SWNTS

We consider now the case of a (9,0) zigzag nanotube. In the chosen starting configuration, the final ring of atoms is made of B atoms only and important frustration effects can be expected in an attempt to close this configuration. For temperatures lower than 3000 K, the only observed process is a dimerization of some B atoms at the top (Fig.10.a). However, around 3000 K, a remarkable "diffusion" process at the tip of the nanotube

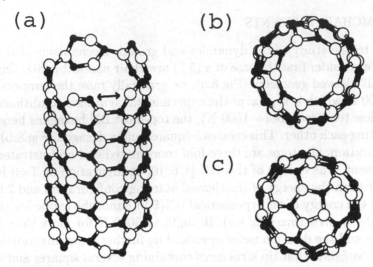

Figure 8. Symbolic ball-and-stick representation of the (5,5) BN nanotube. Large white and small black circles are respectively B and N atoms. In (a), the initial configuration of the molecular dynamics simulation is shown (0K). In (b), the final configuration at 3000 K exhibits the [4,6,10]-rings geometry discussed in the text. In (c), the hypothetical C_{60}-like capping is shown. The arrows indicate the motion of the atoms.

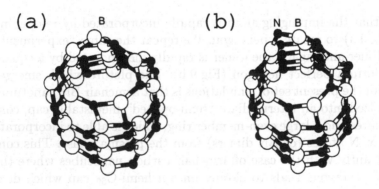

Figure 9. Symbolic ball-and-stick representation of a (4,4) BN nanotube after incorporation of (a) a B atom and (b) a BN dimer. The incorporated atoms have been labeled by their chemical element symbol.

brings N atoms above the terminal B-ring (Fig.10.b) in order to create a BN square (Fig.10.c).

The formation of two more squares by the same mechanism is subse-

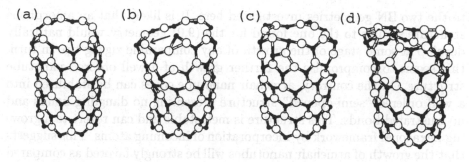

Figure 10. Symbolic ball-and-stick representation of the (6,0) BN nanotube. In (a), the low-temperature dimerized configuration is shown. In (b) and (c), the flipping over of a N atom followed by the formation of a square is depicted. As a result, a frustrated pentagon has been "buried" inside the hexagonal network. In (d), we represent the final configuration reached by the system at the end of our simulation run. The arrows indicate the motion of the atoms.

quently observed in the simulation. This first reorganization is followed by the formation of a "bridging" bond across the open end of the nanotube. However, contrarily to the armchair case, the system does not evolve towards a stable minimum energy configuration but samples several tip structures. We plot in Fig.10.d the configuration reached by the system after 3 ps of simulation at 3000 K.

4.5. GROWTH MECHANISMS FOR BN SWNTS

The zigzag BN nanotube open edge is therefore very unstable, evolving rapidly towards an amorphous-like tip. The creation of triangles in Fig.10.a indicates that the energy cost of two dangling bonds is larger than the cost associated with an homopolar bond. As a result, $\sim$ 4-5 homopolar bonds are formed at the tip of our (9,0) system. At given "constant" frustration cost, we may expect that pentagons are preferred over triangles, driving the system towards tip closure. The rearrangements observed in Figs.10.b-10.c which lead to the replacement of triangles by pentagons and squares is indeed very similar to what was observed in the carbon case. However, rearrangements at the zigzag BN nanotube egde are not as stable as in the case of carbon nanotubes. This is because homopolar bonds are much weaker than B-N bonds. In the simulations, we observe these bonds to break and form frequently at temperatures $\sim$ 3000 K. As a result, the zigzag BN edge does not reach any stable configuration as long as these frustrated bonds are present.

We can now compare the growth mechanisms which may be attributed to the two BN geometries investigated here. It is likely that an amorphous structure similar to the one found for the (9,0) geometry would naturally develop at some stage of the growth of any comparable zigzag tip. In turn, this seems to compromise the further growth of a well ordered nanotube structure. On the contrary, armchair nanotube edges can be stabilized into a well ordered "semi-opened" structure presenting no dangling bonds and no frustrated bonds. This structure is metastable and can revert to a growing hexagonal framework by incorporation of incoming atoms. This suggests that the growth of armchair nanotubes will be strongly favored as compared to the one of zigzag nanotubes. Further, the present simulations lead to the idea that armchair BN SWNTs may grow uncatalyzed, in great contrast with the carbon case. This conclusion is supported by the recent experimental observations of BN SWNTs without any trace of catalytic particle incorporated in (or protruding from) their tips [20].

We note that in the case of larger (n,n) nanotubes, the bending inside of the dangling dimers cannot lead to the formation of exactly n squares as in the previous cases. Indeed, the length of the BN bonds shared by the squares and the central large even-member ring will increase with increasing tube radius. However, local fluctuations of curvature at 3000 K can allow for the creation of one or several squares through the same mechanism described above. Since these squares do not survive the arrival of impinging atoms, the final nanotube closure will only occur under the unlikely event that such squares are incorporated in the hexagonal network. This is consistent with the topological models proposed in Ref. [19, 20] which show a strong reduction of the local radius of curvature where the squares are present. As emphasized in these works, the presence of squares in the hexagonal network induces flat tips of the kind observed experimentally. On the basis of our simulations, we find that the number of squares in the final tip may be larger than three. This is an important result since it implies that large even-member rings such as octagons and decagons will be present at the tip of BN nanotubes. This possibility was pointed out in Ref. [19] on the basis of Euler's theorem for even-member ring systems. To explore the consequences of the presence of such large rings, we have performed static total energy calculations which show that Li atoms can diffuse without any barrier through the center of octagons and decagons. This suggests that BN nanotubes are potential candidates for a very pure metallic doping, in contrast with the case of carbon nanotubes for which re-opening of the cap by oxidation is necessary.

4.6. GROWTH MECHANISMS FOR BN MWNTS

In the case of carbon MWNTs, lip-lip interactions have been shown to stabilize the open edge of each nanotube by forming bridging bonds between concentric shells. These bridging bonds easily break to accomodate incoming atoms from the vapor phase, thus allowing for further growth. We expect that such a mechanism would apply also to BN concentric nanotubes, with the additional constraints introduced by frustration effects. As we saw above the open edge of individual armchair tubes is already metastable and lip-lip interactions are expected to further favour the growth. In the case of the zigzag geometry, lip-lip interactions may contribute to reduce the number of frustrated bonds (e.g. if a B-terminated edge is adjacent to a N-terminated edge). We recall however that previous simulations on carbon double-wall zigzag nanotubes have shown that, at 3000 K, triangles are dynamically created at the growing shell edge of each constituting nanotube. We expect that the "trimer-to-pentagon" switching process observed in Figs.10.a-10.c could also drive towards amorphization the zigzag "lip-lip edge", preventing the further growth of a well ordered MWNT. This suggests that most of the observed BN concentric nanotubes may be of the armchair type. Finally, we note that lip-lip interactions should favour the growth of BN MWNTs in which all the concentric shells have the same helicity, since this is the best strategy to minimize the number of frustrated bonds during growth.

BN nanotubes with a large aspect ratio and an amorphous-like tip are frequently observed [19, 20]. This indicates that, even in the case of BN nanotubes which can sustain growth (e.g. armchair nanotubes), a sudden non-stoichiometric fluctuation of plasma phase composition during growth would induce the formation of a large number of frustrated bonds at the growing edge, thus leading to amorphization, as shown above. We propose that arc discharge synthesis based on the used of stoichiometric BN electrodes, as compared to other techniques where B and N atoms are introduced separatly in the reaction chamber (see Ref. [20]), should enhance the homogeneity of the plasma. In this respect, an experimental study of the plasma composition for different experimental approaches [19,20,35] would be most helpful in optimizing the production of well formed BN nanotubes.

5. Conclusion

Quantum molecular dynamics is shown to be a valuable and very accurate tool used to elucidate microscopic growth mechanisms for pure and doped carbon tubes, as well as BN nanotube systems.

Our results for a double-walled nanotube should be contrasted with those obtained for SWNTs. In the carbon SWNTs case, dome closure rules out uncatalyzed growth. In contrast, for the double-walled carbon tube, the existence of a metastable edge structure can explain why MWNTs can grow by a non-catalytic process. More generally, the lip-lip interaction favors a growth in terms of tube pairs, which is consistent with the relative abundance of even-numbered walls observed in TEM images [1, 11, 40].

Our simulations do not allow us to address directly the issue of MWNT closure. However, on the basis of our results for SWNTs, we can speculate that multiple dome closure may initiate when two pentagons form simultaneously at the growing edge of two adjacent walls. The rather low probability of such an event may explain why carbon MWNTs tend to grow long.

Experimental evidence that B doping of carbon nanotubes catalyses the growth of long ($\sim$ 5-100 μm) and well graphitized nanotubes are corroborated by first-principles static calculations and dynamical simulations which support a model of surfactant B atoms catalyzing the growth of long nanotubes. The model only applies to the zizag tube geometry, implying that this non-chiral geometry should be favoured in B-doped nanotubes with a large aspect ratio, which is consistent with the helicity preference revealed by the experimental electron diffraction patterns.

In the case of BN nanotubes, different growth mechanisms can be expected for BN nanotubes with different chiralities. In particular, it appears that armchair BN SWNTs may grow uncatalyzed while zigzag tubes rapidly evolve into an amorphous-like structure. This difference originates in the $\sim$ 1.6 eV average frustration energy associated with N-N or B-B bonds as compared to B-N bonds. BN nanotubes eventually close through the permanent insertion of squares in the bond network, inducing flat tips and the presence of large even-member rings such as octagons or decagons. The role played by frustration effects and that of square rings in determining disordered or ordered tube-edge structures are in agreement with the observed morphology of BN nanotube tips and with the helicity selection evidenced experimentally in the case of BN MWNTs.

6. Acknowledgments

J.C.C. acknowledges the National Fund for Scientific Research [FNRS] of Belgium for financial support. This work has been partly funded by the interuniversity research project on reduced dimensionality systems (PAI P4/10) of the Belgian Office for Scientific, Cultural, and Technical affairs. This work is carried out within the framework of the EU Human Potential - Research Training Network project under contract N^o HPRN-CT-2000-00128 and within the framework of the specific research and technological development EU programme "Competitive and Sustainable Growth" under contract G5RD-CT-1999-00173.

References

1. Iijima, S. (1991) Helical microtubules of graphitic carbon, Nature **354**, 56-58.
2. Ebbesen, T.W., and Ajayan, P.M. (1992) Large-scale synthesis of carbon nanotubes, Nature **358**, 220-222.
3. Treacy, M.M.J., Ebbesen, T.W., and Gibson, J.M. (1996) Exceptionnaly high young's modulus observed for individual carbon nanotubes, Nature **381**, 678-680.
4. De Heer, W.A., Châtelain, A., and Ugarte, D. (1995) A carbon nanotube field-emission electron source, Science **270**, 1179-1180.
5. Tans, S.J., Devoret, M.H., Dai, H., Thess, A., Smalley, R.E., Geerligs, L.J. and Dekker, C. (1997) Individual single-wall carbon nanotubes as quantum wires, Nature **386**, 474-477; Tans, S.J., Verschueren, A.R.M., and Dekker, C. (1998) Room-temperature transistor based on a single carbon nanotube, Nature **393**, 49-52.
6. Ebbesen, T.W. (1996) Carbon nanotubes, Physics Today **49**, 26-32.
7. Iijima, S., Ichihashi, T., and Ando, Y. (1992) Pentagons, heptagons and negative curvature in graphite microtubule growth, Nature **356**, 776-778.
8. Iijima, S. (1993) Growth of carbon nanotubes, Mater. Sc. Eng. B **19**, 172-180.
9. Endo, M., and Kroto, H.W. (1992) Formation of carbon nanofibers, J. Phys. Chem. **96**, 6941-6944.
10. Saito, R., Dresselhaus, G., and Dresselhaus, M.S. (1992) Topological defects in large fullerenes, Chem. Phys. Lett. **195**, 537-542.
11. Iijima, S., Ajayan, P.M., and Ichihashi, T. (1992) Growth model for carbon nanotubes, Phys. Rev. Lett. **69**, 3100-3103.
12. Kiang, C.-H., and Goddard III, W.A. (1996) Polyyne ring nucleus growth model for single-layer carbon nanotubes, Phys. Rev. Lett. **76**, 2515-2518.
13. Zhang, X.F., Zhang, X.B., Van Tendeloo, G., Amelinckx, S., Op de Beeck, M., and Van Landuyt, J. (1993) Carbon nanotubes; their formation process and observation by electron microscopy, J. Crystal Growth **130**, 368-382.
14. Saito, Y., Yoshikawa, T., Inagaki, M., Tomida, M., Hayashi, X. (1993) Growth and structure of graphitic tubules and polyhedral particles in arc discharge, Chem. Phys. Lett. **204**, 277-282.
15. Gamaly, E.G., and Ebbesen, T.W. (1995) Mechanism of carbon nanotube formation in the arc discharge, Phys. Rev. B **52**, 2083-2089.
16. Iijima, S., and Ichihashi, T. (1993) Single-shell carbon nanotubes of 1-nm diameter, Nature **363**, 603-605.
17. Bethune, D.S., Klang, C.H., de Vries, M.S., Gorman, G., Savoy, R., Vazquez, J., and Beyers, R. (1993) Cobalt-catalysed growth of carbon nanotubes with single-atomic-layer walls, Nature **363**, 605-607.
18. Carroll, D.L., Redlich, Ph., Blase, X., Charlier, J.-C., Curran, S., Ajayan, P.M., Roth, S. and Rühle, M. (1998) Effects of nanodomain formation on the electronic structure of doped carbon nanotubes Phys. Rev. Lett. **81**, 2332-2335.
19. Terrones, M., Hsu, W.K., Terrones, H., Zang, J.P., Ramos, S., Hare, J.P., Castillo, R., Prassides, K., Cheetham, A.K., Kroto, H.W. and Walton, D.R.M. (1996) Metal particle catalyzed production of nanoscale BN structures, Chem. Phys. Lett. **259**, 568-573.
20. Loiseau, A., Willaime, F., Demoncy, N., Hug, G., and Pascard, H. (1996) Boron nitride nanotubes with reduced numbers of layers synthesized by arc discharge, Phys. Rev. Lett. **76**, 4737-4740.
21. Charlier, J.-C., De Vita, A., Blase, X., and Car, R. (1997) Microscopic growth mechanisms for carbon nanotubes, Science **275**, 646-649.
22. Blase, X., De Vita, A., Charlier, J.-C., and Car, R. (1998) Frustration effects and microscopic growth mechanisms for BN nanotubes, Phys. Rev. Lett. **80**, 1666-1669.

170

23. Blase, X., Charlier, J.-C., De Vita, A., Car, R., Redlich, Ph., Terrones, M., Hsu, W.S., Terrones, H., Carroll, D.L., and Ajayan, P.M. (1999) Boron-mediated growth of long helicity-selected carbon naanotubes, Phys. Rev. Lett. 83, 5078-5081.

24. Car, R., and Parrinello, M. (1985) Unified approach for molecular dynamics and density functional theory, Phys. Rev. Lett. 55, 2471.

25. Troullier, N., and Martins, J.L. (1991) Efficient pseudopotentials for plane-wave calculations, Phys. Rev. B 43, 1993.

26. Bachelet, G.B., Hamann, D.R., and Schlüter, M. (1982) Pseudopotentials that work: from H to Pu, Phys. Rev. B 26, 4199.

27. The calculated intraplanar lattice constant of graphite, 2.46 Å, and the C-C bond lengths in C_{60}, 1.40 Å and 1.47 Å, agree with experiment within 1%.

28. in the notation of : Hamada, N., Sawada, S.-I., and Oshiyama, A. (1992) New one-dimensional conductors: graphite microtubules, Phys. Rev. Lett. 68, 1579-1581. With standard notations, (n,0) and (n,n) nanotubes will be called respectively zigzag and armchair. Cutting these nanotubes perpendicularly to their axis will create two-coordinated "dangling" atoms in the first case and dangling dimers in the second one.

29. Nosé, S. (1984) A unified formulation of the constant temperature molecular dynamics methods, J. Chem. Phys. 81, 511-519; Hoover, H.G. (1985) Canonical dynamics: equilibrium phase-space distributions, Phys. Rev. A 31, 1695-1697.

30. Eggen, B.R., Heggie, M.I., Jungnickel, G., Latham, C.D., Jones, R., Briddon, P.R. (1996) Autocatalysis during fullerenes growth, Science 272, 87-89.

31. Guo, T., Nikolaev, P., Rinzler, A.G., Tománek, D., Colbert, D.T., and Smalley, R.E. (1995) Self-assembly of tubular fullerenes J. Phys. Chem. 99, 10694-10697.

32. Wagner, R.S., and Ellis, W.C. (1964) Vapor-liquid-solid mechanism of single crystal growth, Appl. Phys. Lett. 4, 89-90.

33. The studied nanotube contains 117 atoms, including 9 B and 9 H atoms. The H atoms are used to passivate the nanotube end opposite to the "growing lip". They are maintained fixed during the simulation. The total simulation time is of ∼ 10 ps.

34. The initial cap was prepared assuming closure without formation of B-B bonds. We note such a tip exhibits a rather large ∼20 % doping (6 B atoms out of 30 C atoms), much larger than what is measured for the body of the nanotube. This implies that the effect of isolated B atoms on the electronic properties (see Ref. [18]) of the nanotube body or cap is not sufficient to enhance the growth of carbon nanotubes.

35. Chopra, N.G., Luyken, R.J., Cherrey, K., Crespi, V.H., Cohen, M.L., Louie, S.G., and Zettl, A. (1995) Boron nitride nanotubes, Science 269, 966-967.

36. Blase, X., Rubio, A., Louie, S.G., and Cohen, M.L. (1994) Stability and band gap constancy of BN nanotubes, Europhys. Lett. 28, 335-340; Blase, X., Rubio, A., Louie, S.G., and Cohen, M.L. (1995) Quasiparticle band structure of bulk hexagonal BN and related systems, Phys. Rev. B. 51, 6868-6875.

37. Blase, X., Charlier, J.-C., De Vita, A., and Car, R. (1997) Theory of composite BxCyNz nanotube heterojunctions, Appl. Phys. Lett. 70, 197-199, and references therein.

38. Rubio, A., Myamoto, Y., Blase, X., Cohen, M.L., and Louie, S.G. (1996) Theoretical study of one-dimensional chains of metal atoms in nanotubes, Phys. Rev. B 53, 4023-4026.

39. Similar results have been found at the level of MP2/DZP calculations: Jensen, F., and Toftlund, H. (1993) Structure and stability of C24 and B12N12 isomers, Chem. Phys. Lett. 201, 89-96. We refer the reader to this reference for figures of the studied clusters.

40. Seraphin, S., Zhou, D., and Jiao, J. (1994) Extraordinary growth phenomena in carbon nanoclusters, Acta Microscopica 3, 45-64.

CONTROLLED PRODUCTION OF TUBULAR CARBON AND BCN ARCHITECTURE

Mauricio Terrones

Max-Planck-Institut für Metallforschung, Seestr. 92, D-70174 Stuttgart, Germany

Abstract: Pure carbon filaments and nanotubes are of interest in materials science because of their remarkable electronic and mechanical properties. In this context, efficient self assembly pyrolytic routes to large arrays of aligned carbon nanotubes and CN_x nanofibers are presented. These routes include hydrocarbon decomposition over preformed metal-cluster substrates and pyrolysis of organometallic compounds. During the thermolytic process, the metal particles are responsible for carbon agglomeration and subsequent tube axial growth. The generation of aligned Fe-filled carbon nanotube films is also described in detail. Nanowire arrangements of this novel kind should find applications in the fabrication of high density magnetic recording devices, as well as fine particle magnets in magnetic inks and toners in xerography. Electronic properties and state-of-the-art electron microscopy studies on these nanostructures are presented. In addition, pyrolytic routes to $B_xC_yN_z$ materials (hybrids of h-BN and graphite) are also described. Finally, the synthesis and recent electronic calculations of WS_2 and MoS_2 nanotubes are briefly discussed.

1. Introduction

The discovery of a third carbon allotrope Buckminsterfullerene (C_{60}) in the mid 1980's by Kroto, Curl and Smalley [1] initiated a novel, round and curved carbon nanocosmos. As a result, in the early 1990's, elongated cage-like carbon structures (known as nanotubes) were identified [2]. This provided the impetus for an international, multidisciplinary field of research, which has proved to be highly productive since novel electronic and material technologies are now envisaged.

171

L.P. Biró et al. (eds.), Carbon Filaments and Nanotubes: Common Origins, Differing Applications, 171–185.
© 2001 *Kluwer Academic Publishers. Printed in the Netherlands.*

Fullerenes consist of closed (sp^2 hybridized) carbon networks, organized on the basis of 12 pentagons and any number of hexagons except one. This definition encompasses nanotubes [3]. In 1992, a few months after Iijima's publication [2], Ebbesen and Ajayan described the bulk synthesis of nanotubes [4] formed as an inner core cathode deposit generated by arcing graphite electrodes in an inert atmosphere; a similar procedure to that used for fullerenes. Nowadays, carbon nanotubes can be produced by diverse techniques such as arc discharge [5], pyrolysis of hydrocarbons over catalysts [5-8], laser vaporization of graphite [5, 9], and by electrolysis of ionic salts using graphite electrodes [10]. The products exhibit different degrees of crystallinity and various morphologies (e.g. straight, curled, hemitoroidal, branched, spiral, helix-shaped, etc.).

Nanotube research - both theory and experiment - accelerates, and it is likely that the proportion of publications devoted to applications will continue to increase. Recent examples [11] include use of nanotubes as: (i) gas storage components for Ar and H_2; (ii) STM probes and field emission sources; (iii) high power electrochemical capacitors; (iv) chemical sensors; (v) electronic nanoswitches; (vi) magnetic storage devices (*e.g.* Fe-filled nanotubes), etc (see Ref. 11 and references therein).

2. Carbon Filaments and Nanotubes from hydrocarbon decomposition

Pyrolysis of hydrocarbons (e.g. benzene, acetylene, naphthalene, ethylene, etc.) in the presence of catalysts (e.g. Co, Ni and Fe deposited on substrates such as silicon, graphite or silica), provides an additional route to fullerenes and carbon nanotubes. Prior to the discovery of fullerenes in 1985, pyrolytically grown nanofibers/nanotubes had actually been observed and structurally identified by several groups [12].

Various authors have proposed mechanisms to account for the pyrolitic formation of carbon filaments (which can be applied to nanotubes). At present, several mechanisms have been widely recognised [12].

2.1 ALIGNED NANOTUBES

In 1994, Ajayan and co-workers generated aligned nanotube arrays by cutting thin slices (50-200 nm thick) of a nanotube-composite polymer [13]. The tubes appeared to be well separated as a result of mechanical deformation suffered by their being embedded in the polymer. However, the alignment strongly depended upon the thickness of the composite slice and the process becomes impractical for larger areas [13]. Therefore, in order to generate large areas of aligned nanotubes, alternative bulk growth methods should be considered.

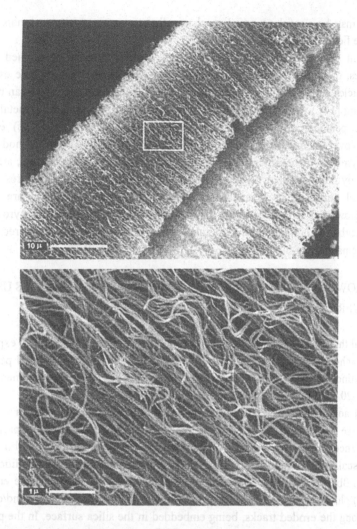

Figure 1. SEM images of pyrolytically grown carbon nanotubes: (Top) double bundles at low magnification, in which nanotubes appear aligned; (Bottom) higher magnification of these bundles showing aligned nanotubes of uniform length (20 μm) and diameter (30-50 nm).

In this context, laser etching of cobalt, nickel and iron thin films provides a novel route to catalysts which, in conjunction with the pyrolysis of organic precursors (e.g. 2-amino-4,6-dichloro-s-triazine, triaminotriazine, *etc.*) yields aligned nanotubes [8, 14-15]. These tubes are of uniform length (≤ 100 μm) and diameter (30-50 nm od) and grow perpendicularly to the catalytic substrate 'only' in the etched regions (Fig. 1). The orientation of the catalyst particles, ablated during etching, appears to be crucial, nanotubes with preferred helicity being favored (*e.g.* armchair 25-30 %). It is possible that matrices consisting of aligned nanotube bundles may be useful as novel mechanically strong composite materials and as ultra-fine field emission sources.

Recent reports also describe the large scale synthesis of aligned carbon nanotubes, of uniform length and diameter by: (a) passage of acetylene over iron nanoparticles embedded in mesoporous silica [16], in which nanotubes can be up to 2 mm long [17] ; (b) thermolysis of metal-containing precursors (*e.g.* metallocenes and iron pentacarbonyl) in conjunction with hydrocarbons [18].; (c) ethylene pyrolysis over chemically patterned Fe-Si substrates [19]. The latter methods, based on the pyrolysis of organic precursors over templated/catalysts supports, are by far superior by comparison with plasma arcs, since other graphitic structures such as polyhedral particles, encapsulated particles and amorphous carbon are notably absent. However, it is noteworthy that the degree of crystallinity of pyrolytically grown carbon filaments and nanotubes can be lower when compared to arc discharge nanotubes.

2.2 GROWTH MECHANISM FOR ALIGNED NANOTUBE ARRAYS USING LASER ETCHING

Metal thin films are deposited on silica substrates. The plates are then exposed to air and etched with single laser pulses (Nd:YAG 266nm or 355nm; 5 mJ per pulse) using cylindrical lenses (65 mm focal length; Fig 2), thus creating linear tracks (width 1-20 mm; length ≤ 5mm) across the substrates.

SEM and AFM studies reveal a marked roughness within the silica substrates along these tracks (Fig. 3). AFM profilometry measurements show the presence of protuberances at the track edges of *ca.* 100-170 nm below the metal film surface, which usually increased further towards the center of the track by an additional 100 nm (Fig. 3b). It is believed that during this severe etching process, the energized (ablated) clusters may condense and recrystallize uniformly, as metal and/or metal oxide along the eroded tracks, being embedded in the silica surface. In the presence of gaseous carbon, these clusters (≤ 50nm OD) must act as nucleating seeds being responsible for nanotube growth. It is noteworthy that the creation of a rough surface and the embedding of the nanoparticles, prevents coalescence of the metal clusters during substrate heating, thus encouraging nanotube formation.

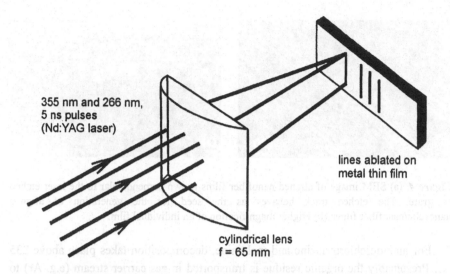

Figure 2. Etching technique using a Nd:YAG (266 nm or 355 nm) laser and cylindrical lens (65 mm focal length).

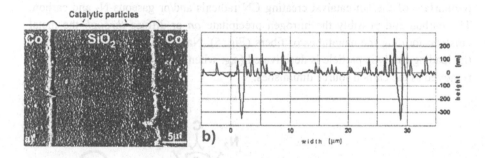

Figure 3. (a) AFM image of an etched nanotrack showing how the uniform channel is made; (b) AFM profilometry profile of a track exhibiting depths at the edges of ca. 100-170 nm below the metal film surface [17].

Pyrolysis experiments are usually carried out in silica tubes fitted in 1 or two furnace systems in the presence of flowing gases such as Ar, H_2, He, *etc*. The catalytic substrate is placed in the high temperature furnace region (800-1050 °C). After theromolysis, nanotubes/nanofibers grow only on the laser etched regions (Fig. 4).

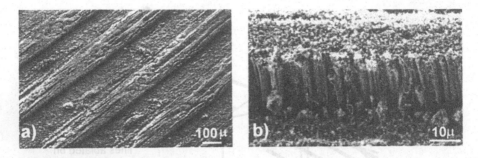

Figure 4. (a) SEM image of aligned nanofiber films grown perpendicular to the laser etched substrate. The etched track behaves as the seed for the generation of aligned nanotube/nanofiber films; (b) Higher magnification of an individual film.

For aminodichlorotriazine and melamine, decomposition takes place above 235 °C. Presumably the organic residue is transported in gas carrier stream (e.g. Ar) to the pyrolysis zone containing the catalytic film at 950-1050 °C. The liberated HCl (in the case of aminodichlorotriazine) reacts with the metal catalyst, generating metal chloride in the gas phase (detected by mass spectrometry studies). For both aminodichlorotriazine and melamine, organic residues possibly fragment further on the surface of the hot catalyst creating CN radicals and/or gaseous N_2 and carbon. The carbon and possibly the nitrogen precipitate *on* or diffuse *through* the metal clusters, forming the nanotubes or fibers (Fig. 5). Nanotube growth continues until the leading catalytic particle is deactivated (*e.g.* when the temperature decreases), or removed as the carbon source diminishes.

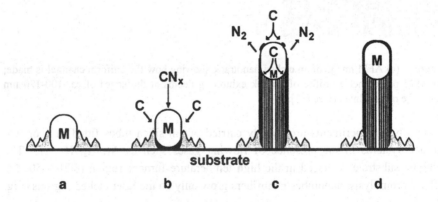

Figure 5. Mechanism for the formation of carbon nanotubes by pyrolysis of triazine compounds over metal particles. The catalytic particles are though to be embedded in etched silica substrate due to the laser etching process. M denotes metal and C denotes carbon.

3. Metal Filled Nanotubes

In the case of hydrocarbon decomposition over metal catalysts, it is envisaged that high catalyst concentrations (*ca.* > 10-20%) may lead to encapsulation of the metal within carbon filaments and multiwalled nanotubes (MWNT's). In particular, pyrolysis of organic compounds in the presence of Co [14], Ni (Fig. 6) [15] and Fe [16] generates nanotubes containing these metals. Interestingly, C_{60} (a source of pure carbon) can be used in conjunction with a metal catalyst, resulting in the formation of tapered MWNT's (25-40 % yield; 2-5 µm in length; 10-100 nm OD) fully-filled with Ni and exhibiting a high degree of 'graphitization' [20]. From these studies it was observed that Ni single crystals are present in the inner core of large diameter (ca. 10-50 nm) nanotubes. A possible growth scenario has been proposed [20] involving: (a) reaction of C_{60} (or hydrocarbon) with the metal catalyst (e.g. Ni) to form a metal carbide (e.g. Ni_3C) coated clusters; (b) precipitation of graphitized carbon in a tube-like fashion containing Ni droplets, caused by temperature increase or exothermic reactions; (c) agglomeration and/or subsequent insertion of additional metal/carbide (e.g. Ni_3C/Ni; present in the gas phase) in the growing open tube; (d) secondary growth related to a carbon tube thickening, due to subsequent extrusion of carbon from the metal carbide and (e) inhibition of growth if the carbon supply is exhausted or the temperature becomes unstable.

Figure 6. Ni-filled nanotube grown by pyrolysis of C_{60}/Ni substrates (interlayer spacing, ca. 0.34 nm).

Fe-coated nanowires (Fe-filled MWNT's) have been obtained in high yield by a simple method involving pyrolysis of ferrocene under a vacuum or low Ar pressure [18]. However, the degree of order exhibited by the carbon walls is not very high. Fortunately, films (< 2 mm^2; <40 µm thick) of aligned Fe-filled carbon nanotubes (Fig. 7), usually composed of single Fe crystals (5-40 nm OD and < 10 µm in length) have been recently produced by pyrolysing ferrocene/C_{60} mixtures in an Ar

178

atmosphere. In this case, the carbon tubes which coat the wires are highly crystalline with diameters of *ca.* 20-70 nm and are < 40 μm in length [21].

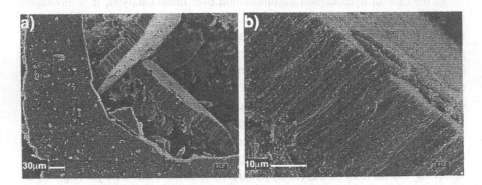

Figure 7. SEM images of flake-like particles (films) consisting of aligned Fe-filled nanotubes at lower (a) and higher (b) magnification. The material was grown by pyrolysing Ferrocene/C_{60} mixtures (e.g. 1:1, 2:1, 3:1, etc.) at 900-1050 °C in an Ar flow.

The near edge EELS fine structure of the carbon K shell (*ca.* 284 eV) confirms that the material is highly 'graphitic', the L-edge (*ca.* 708 eV) being characteristic of metallic Fe [21]. HREELS line-scans across individual filled tubes reveal that the C and Fe concentration profiles anticorrelate, indicating that pure Fe is indeed encapsulated within the C layers (Fig. 8). These measurements also confirm that other elements, such as oxygen and sulphur, are not present to any significant extent. The material exhibits coercivities greater than those associated with Ni and Co nanowires [22].

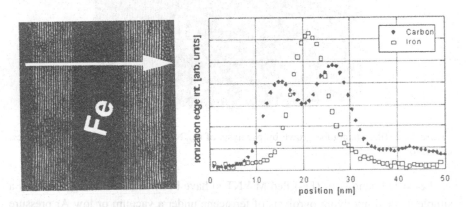

Figure 8. Left HRTEM image (interlayer spacing ca. 0.34 nm). *Right* Concentration profiles of C and Fe across the Fe-filled tube derived from EELS line-scans (white arrow left image).

These nanowire arrays may have significant potential in magnetic data storage devices (Quatized Magnetic Disks) due to their size and anisotropic behaviour, which permit the use of a smaller bit size (one per nanowire), thus increasing the attainable recording density. Film structures of this kind, based on single-domain elements, exhibit the best available storage densities (*e.g.* 65 Gb/in^2) [23].

4. CN$_x$ aligned Nanofibers

The synthesis of crystalline C$_3$N$_4$ [24-25] and CN nanotubes [26], predicted to be super-hard and metallic respectively, remains a challenge for the future. The generation of aligned C$_{13}$N$_x$ (x < 1) nanofibers (< 100 nm OD; < 60 μm in length) in high yield can be achieved by pyrolysing melamine over laser etched Fe substrates (Fig. 4). Self-assembly processes are also used to create large arrays (< 400x400 μm^2) of aligned C$_9$N$_x$ (x ≤ 1) nanofibers (< 100 nm OD, 60 μm length) in a *single* step process: the pyrolysis of ferrocene/melamine mixtures at 1050 °C in Ar atmospheres.

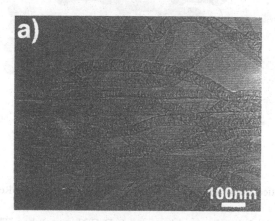

Figure 9. TEM images of a typical region showing densely packed hollow nanofibers. Their morphology consists of compartmentalized "stacked cones".

High Resolution Transmission Electron Microscopy (HRTEM) and HREELS studies show that: (a) the hollow nanofibers possess unusual somewhat irregular stacked-cone corrugated morphologies, and (b) the degree of tubular perfection of the structure decreases as a result of the nitrogen incorporation (Fig. 9). EELS spectra of the nanofibers revealed the presence of ionization edges at *ca.* 284.5 eV and 400 eV corresponding to the C and N K-shells. XPS analyses on the N1s signal reveal binding energies at 398.7 eV and 400.9 eV. The latter energies correspond to highly coordinated N atoms replacing C atoms in the graphene sheets (*ca.* 401 - 403 eV), and pyridinic nitrogen (*ca.* 399 eV). We note that it is difficult to generate highly ordered structures in which large concentrations of N are incorporated into the carbon network.

5. $B_xC_yN_z$ Tubular Nanostructures

$B_xC_yN_z$ materials constitute another possible layer-by-layer system, composed of h-BN and graphite hybrids. They may be semiconductors, and be used as photoluminescent materials (light sources), transistors working at high temperatures, lightweight electrical conductors or high temperature lubricants [27]. These applications depend not only on the composition but also on the arrangement of the B, C and N atoms. In this context, BC_2N nanotubes, have been proposed theoretically (Fig. 10) [28] and are predicted to be semiconducting.

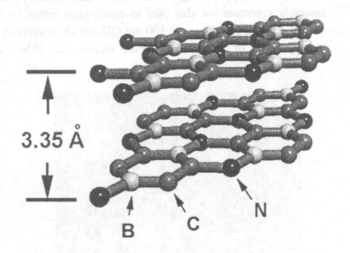

3.35 Å

B   C   N

Figure 10. Theoretical model of homogenous BC_2N layered material (see Ref. 28).

Pyrolytic methods can be used to generate $B_xC_yN_z$ nanotubes. These techniques, when compared to arc discharge methods, produce a rather uniform range of $B_xC_yN_z$ nanostructures, preserving stoichiometry [29]. In particular, pyrolysis of $CH_3CN \cdot BCl_3$ over cobalt powder at 950-1000 °C leads to ca. $[BC_2N_z]_n$ (z = 0.3-1.0) nanofibers and nanotubes (Fig. 11) possessing different morphologies following the reaction

$$CH_3CN + BCl_3 \rightarrow BC_2N + 3\ HCl.$$

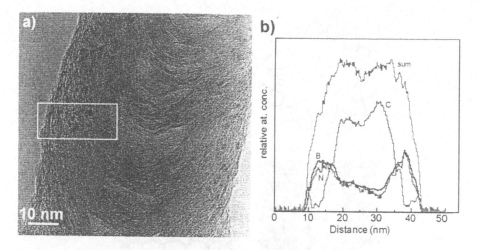

Figure 11. (a) TEM image of a typical BC_2N nanofibres with stacked-cone morphology: (b) Concentration profiles of the B, C and N across the fiber, showing a C double peak with the BN crest in between, signature of BN and C segregation.

Their internal structures were investigated, at the nanometer level, using high spatial resolution electron energy-loss spectroscopy. Line scans along the fiber axis were recorded in order to verify that the observed C-BN-C sandwich structure is not only a feature of the outer walls (Fig. 11b). It was found that B, C and N are not homogeneously distributed within the nanostructures, but are separated into pure C and BN domains [30]. Novel calculations of these segregated BCN materials need to be performed, since they may prove useful as high temperature operation nanoscale electronic devices.

6. MoS₂ and WS₂ Nanotubes

Tenne et. al. [31, 32] have successfully produced inorganic fullerene-like and tubular structures from WS_2 and MoS_2. In the synthetic process, H_2S is passed over MO_3 (M = W, Mo) nanoparticles at 800-900° C and MS_2 (M = W, Mo) nanotubes, together with polyhedral particles are collected from the rear of the reaction zone. It has also been shown that WO_x nanowhiskers behave as efficient templates of long WS_2 nanotubes [33] after sulfidisation. In addition, it has been recently demonstrated that polyhedral particles of WS_2 and MoS_2 appear to be exceptional lubricants [34].

182

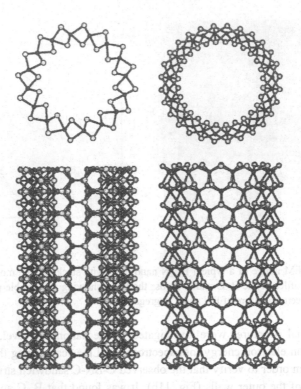

Figure 12. Arm chair (8,8) WS$_2$ nanotube (left) and zig-zag (14,0) WS$_2$ nanotube (right). Light atoms are S, dark atoms are Mo.

The most stable forms of bulk MoS$_2$ and WS$_2$, consist of a metal layer (Mo or W) sandwiched between two sulphur sheets. These "triple layers" are stacked together, similarly to graphite, by van der Waals interactions and are separated by c.a. 6.2 Å. Therefore, tubular structures can be constructed, as in carbon nanotubes, by rolling up the sheets (Fig. 12). Recent Density-Functional-based Tight-Binding (DFTB) calculations [37-39] on MoS$_2$ and WS$_2$ nanotubes predict that all nanotubes observed experimentally (>20Å OD) should be semiconductors with a small band gap (lower than the bulk material) depending on diameter and chirality. For MoS$_2$ and WS$_2$, it has been found that the band gap tends to vanish for diameters < 10 Å.

In order to bend layered WS$_2$ or MoS$_2$ to form cages, square-like "defects" must be introduced, thus obtaining cages exhibiting the T$_d$ symmetry.

Semiconducting dichalcogenide nanotubes may be ideal for constructing electronic nanodevices and robust composites. Further experimental and theoretical research of intercalated MS$_2$ (M = W, Mo, Vn, Nb, *etc*.) nanotubes need to be explored in detail, thus opening up new avenues in the challenging field of nanotechnology.

ACKNOWLEDGEMENTS

I am in debt to H.W. Kroto, D. Walton, M. Endo, H. Terrones, N. Grobert, Ph. Redlich, G. Seifert, P. M. Ajayan, J.C. Charlier, K. Kordatos, W.K. Hsu, Y. Q. Zhu, M. Yumura, A.J. Pidduck, D. Wallis and C.L. Reeves for stimulating discussions and valuable assistance in some of the work presented here. I am grateful to the Alexander von Humboldt Stifftung, EPSRC, Royal Society, and CONACYT-México (grant J31192U) for financial support.

REFERENCES

1. Kroto, H. W., Heath, J. R., O'Brien, S. C., Curl, R. F., Smalley, R. E. (1985) C_{60}: Buckminsterfullerene, *Nature* 318, 162-163.
2. Iijima, S. (1991) Helical microtubules of graphitic carbon, *Nature* 354, 56-58.
3. Endo, M., Kroto, H.W. (1992) Formation of carbon nanofibres, *J. Phys. Chem.* 96, 6491.
4. Ebbesen, T.W., Ajayan, P.M. (1992) Large-scale synthesis of carbon nanotubes, *Nature* 358, 220-221.
5. Dresselhaus, M.S., Dresselhau, G., Eklund, P.C. (1996) Science of Fullerenes and Carbon Nanotubes, Academic Press, New York.
6. Endo, M., Takeuchi, K., Igarashi, S., Kobori, K., Shiraishi, M., Kroto, H. W. (1993) The production and structure of pyrolytic carbon nanotubes (PCNT's), *J. Phys. Chem. Solids* 54, 1841-1843.
7. Amelinckx, S., Zhang, X.B., Bernaerts, D., Zhang, X.F., Ivanov, V., Nagy, J.B. (1994) A formation mechanism for catalytically grown helix-shaped graphite nanotubes, *Science* 265, 635-638.
8. Terrones, M., Grobert, N., Olivares, J., Zhang, J.P., Terrones, H., Kordatos, K., Hsu, W.K., Hare, J.P., Townsend, P.D., Prassides. K., Cheetham, A.K., Kroto, H.W., Walton. D.R.M. (1997) Controlled production of aligned-nanotubes bundles, *Nature* 388, 52-55.
9. Thess, A., Lee, R., Nikolaev, P., Dai, H., Petit, P., Robert, J., Xu, C., Lee, Y.H., Kim, S.G., Rinzler, A.G., Colbert, D.T., Scuseria, G.E., Tománek, D., Fischer, J.E., Smalley, R.E. (1996) Crystalline ropes of metallic carbon nanotubes, *Science* 273, 483-487.
10. Hsu, W.K., Hare, J.P., Terrones, M., Harris, P.F.J., Kroto, H.W., Walton, D.R.M. (1995) Condensed-phase nanotubes, *Nature* 377, 687.
11. Terrones, M., Hsu, W. K., Kroto, H. W., Walton, D. R. M. (1998) Nanotubes: A revolution in material science and electronics". In *Fullerenes and Related Structures;* Topics in Chemistry Series, Ed. A. Hirsch (Springer-Verlag), vol. 199, ch. 6, pp.189-234.
12. Dresselhaus, M.S., Dresselhaus, G., Sugihara, K., Spain, I.L., Goldberg, H.A. (1988) Synthesis of graphite fibers and filaments, In *Graphite Fibers and Filaments*, vol 3, Springer, Berlin New York London.
13. Ajayan, P.M., Stephan, O., Colliex, C., Trauth, D. (1994) Aligned carbon nanotube arrays formed by cutting a polymer resin nanotube composite, *Science* 265, 1212-1214.

184

14. Terrones, M., Grobert, N., Zhang, J. P., Terrones, H., Olivares, J., Hsu, W. K., Hare, J. P., Cheetham, A. K., Kroto, H. W. and Walton, D. R. M. (1998) Formation of aligned carbon nanotubes catalysed by laser-etched cobalt thin films, *Chemical Physics Letters* **285**, 299-305.

15. Grobert, N., Terrones, M., Trasobares, S., Kordatos, K., Terrones, H., Olivares, J., Zhang, J.P., Redlich, Ph., Hsu, W.K., Reeves, C.L., Wallis, D.J., Zhu, Y.Q., Hare, J.P., Pidduck, A.J., Kroto, H.W., Walton, D.R.M. (2000) A novel route to aligned nanotubes and nanofibres using laser patterned catalytic substrates, *Applied Physics A* **70** 175-183.

16. Li, W.Z., Xie, S.S., Qian, L.X., Chang, B.H., Zou, B.S., Zhou, W.Y., Zhao, R.A., Wang, G. (1996) Large scale synthesis of aligned carbon nanotubes, *Science* **274**:1701-1704.

17. Pan, Z.W., Xie, S.S., Chang, B.H., Wang, C.Y., Lu, L., Liu, W., Zhou, M.Y., Li, W.Z. (1998) Very long carbon nanotubes, *Nature* **394**, 631-632.

18. Rao, C.N.R. Sen, R., Satishkumar, B.C., Govindaraj, A. (1998) Large aligned-nanotube bundles from ferrocene pyrolysis, *J. Chem. Soc. Chem. Commun.* **15**, 1525-1526.

19. Fan, S.S., Chapline, M.G., Franklin, N.R., Tombler, T.W., Cassell, A.M., Dai, H.J. (1999) Self-oriented regular arrays of carbon nanotubes and their field emission properties, *Science* **283**, 512-514.

20. Grobert, N., Terrones, M., Osborne, O.J:, Terrones, H., Hsu, W.K., Trasobares, S., Zhu, Y.Q., Hare, J.P., Kroto, H.W., Walton, D.R.M. (1998) Thermolysis of C_{60} thin films yields Ni-filled tapered nanotubes, *Appl. Phys. A* **67**, 595-598.

21. Grobert N., Terrones, M., Redlich, Ph., Terrones, H., Escudero, R., Morales, F., Hsu, W.K., Zhu, Y.Q., Hare, J.P., Rühle, M., Kroto, H.W., Walton, D.R.M. (1999) Enhanced Magnetic Coercivities in Fe Nanowires *Applied Physics Letters* **75** 3366-3368.

22. Whitney, T. M., Jiang, J. S., Searson, P. C., Chien, C. L. (1993) Fabrication and magnetic-properties of arrays of metallic nanowires, *Science* **261**, 1316-1319.

23. Chou, S. Y. (1997) Patterned magnetic nanostructures and quantized magnetic disks, *Proceedings of the IEEE* **85**, 652-671.

24. Liu, A. Y., Cohen, M. L. (1989) Prediction of new low compressibility solids, *Science* **245**, 841-842.

25. Teter, D. M., Hemley, R. J. (1996) Low-compressibility carbon nitrides, *Science* **271**, 53-55.

26. Miyamoto, Y., Cohen, M. L., Louie, S. G. (1997) Theoretical investigation of graphitic carbon nitride and possible tubule forms, *Solid State Commun.* **102**, 605-608.

27. Kawaguchi, M., Bartlett, N., (1995) In: Nakajima T (eds) *Fluorine-carbon and fluoride-carbon materials*, Marcel Dekker Inc, New York, pp. 187-195.

28. Miyamoto,Y., Rubio, A., Cohen, M.L., Louie, S.G. (1994) Chiral tubules of hexagonal BC_2N, *Phys. Rev. B* **50**, 4976-4979.

29. Terrones, M., Benito, A.M., Manteca-Diego, C., Hsu, W.K., Osman, O.I., Hare, J.P., Reid, D.G., Terrones, H., Cheetham, A.K., Prassides, K., Kroto, H.W., Walton, D.R.M. (1996) Pyrolytically grown $B_xC_yN_z$ nanostructures: nanofibres and nanotubes, *Chem. Phys. Lett.* **257**, 576-582.

30. Kohler-Redlich, Ph., Terrones, M., Manteca-Diego, C., Hsu, W.K., Terrones, H., Rühle, M., Kroto, H.W., Walton, D.R.M. (1999) Stable BC_2N nanostructures: Low temperature production of segregated C/BN layered materials, *Chem. Phys. Lett.* **310**, 459-465.

185

31. Tenne, R., Margulis, L., Genut, M., Hodes, G. (1992) Polyhedral and cylindrical structures of tungsten disulfide, *Nature* **360**, 444-446.
32. Margulis, L., Saltra, G. Tenne, R., Tallanker, M. (1993) Nested fullerene-like structures, *Nature* **365**, 113-114.
33. Zhu, Y.Q., Hsu, W.K:, Grobert, N., Chang, B.H., Terrones, M., Terrones, H., Kroto, H.W. Walton, D.R.M., Wei, B. Q. (2000) In-situ production of WS_2 nanotubes, *Chemistry of Materials* **12**, 1190-1194.
34. Rapoport, L., Bilik, Y., Feldman, Y., Homyonfer, M., Cohen, S.R., Tenne, R. (1997) Hollow nanoparticles of WS_2 as potential solid-state lubricants, *Nature* **387**, 791-793.
35. Porezag, D., Frauenheim, Th, Köhler, Th., Seifert, G., Kaschner, R. (1995) Construction of tight-binding-like potentials on the basis of density-functional theory - application to carbon, *Phys. Rev. B* **51**, 12947-12957.
36. Seifert, G., Terrones, H., Terrones, M., Jungnickel, G., Frauenheim, T. (2000) On the electronic structure of ws_2 nanotubes, *Solid State Commun.* **114**, 245.
37. Seifert, G., Terrones, H., Terrones, M., Jungnickel, G., Frauenheim, T. (2000) Structure and electronic properties of MoS_2 nanotubes, *Phys. Rev. Lett.* in press.

LARGE SCALE PRODUCTION OF VGCF

MAX L. LAKE
Applied Sciences, Inc.
P.O. Box 579, Cedarville, OH 45314, USA
www.apsci.com

Abstract. The phenomenon that is the basis for synthesis of vapor grown carbon nanofibers (VGCF) has been observed for many years. In particular VGCF has been the subject of relatively intense research over the past twenty-five years due to the promise of achieving physical properties approaching single crystal graphite in the form of an inexpensive carbon filament. Production of VGCF on a commercial scale has lagged behind expectations for a material with such a desirable combination of properties that is synthesized in a simple, low-cost process. Barriers to commercial production are due to the unique properties of VGCF. In fact, VGCF are discontinuous fibers, surface state modifications are required for many attractive applications and there remain many unknowns in the manufacturing technologies required to support use in specific applications. These latter unknowns include processing methods to achieve the ideal surface states, the form of the fiber most likely to enable preservation and translation of the desired property, and handling methods to achieve the appropriate forms. Recently, the development of the requisite technologies to address these barriers has been undertaken, through quantitative analysis of the fibers morphology and surface state generated for various combinations of the production parameters. This was done by seeking manufacturing analogies within the carbon black, carbon fiber, and glass fiber industries, and by development of new techniques for fiber synthesis, modification, and handling where no existing methods were suitable. The results of this effort, and prospects for future availability of a family of VGCF products will be reported in this chapter.

1. Introduction

Vapor grown carbon nanofibers (VGCF) have a unique method of production, leading to unique properties. These unique features are similar to the so-called "bookkeeping equation" in accounting – the assets equal the liabilities. In the present instance, the differences in the simple fiber production process and the fiber properties represent advantages over conventional PAN-based and pitch-based carbon fiber, but the same differences represent barriers to insertion into many of the existing markets for carbon reinforcements and additives.

Carbon fibers derived from PAN or pitch are available as continuous filaments with diameters of five to fifteen microns. PAN is the dominant precursor for high strength carbon fibers, while mesophase pitch is the dominant precursor for ultra-high modulus fibers. The

187

L.P. Biró et al. (eds.), Carbon Filaments and Nanotubes: Common Origins, Differing Applications, 187–196.
© 2001 *Kluwer Academic Publishers. Printed in the Netherlands.*

mechanical properties of these two types of carbon fibers are directly related to the textile processing methods used to produce them. Due to precursor impurities and defects introduced during melt-spinning, these property values are only a fraction of that which could be theoretically obtained from a single crystal graphite filament. Other limitations of composites derived from textile technology are the limited range of available fiber diameters, susceptibility of continuous fibers to damage during composite processing, and processing complexity required to achieve desired fiber architectures in composite articles and components. These limits aside, a mature industry exists for design and fabrication of composite articles having predicted properties competitive to structural metals, albeit at a cost which precludes widespread use in automobiles and other consumer goods.

In contrast, VGCF is a discontinuous fibrous reinforcement produced in a relatively simple process directly from a pyrolyzed hydrocarbon. The purity of the carbon source and the mechanics of growth result in a highly graphitic fiber, providing physical properties approaching those of single-crystal graphite. The enhanced mechanical properties may enable engineered composites to be considered where previously only metal structures or advanced composites could be contemplated.

Because VGCF are produced in a process very similar to that of carbon black, directly from a simple hydrocarbon source, the economies of scale are not tied to the production costs associated with spun textiles. In equivalent production volumes, VGCF are projected to have a cost comparable to E-glass on a per-pound basis, yet possess properties which far exceed those of glass and are equal to or exceed those of much more costly PAN and pitch-based carbon fibers. Being a discontinuous reinforcement, VGCF can be incorporated into commercially available thermoplastics, thermosets and elastomers and can be used directly in existing high volume molding processes without any significant new manufacturing development, thus allowing for engineered composites with an inherently low processing cost.

Because of their extraordinary intrinsic properties, particularly the elastic modulus, VGCF are expected to enable a reduction in the material required to produce a given stiffness, thus providing net weight and cost savings.

And yet, key technologies for utilization of carbon nanofibers are only now beginning to emerge, roughly twenty-five years after the first flurry of reports in the literature. The first of these technologies aims at the modification of the surface of the fibers to enable wetting of and adhesion by the desired matrix materials, using industrial scale processes. While principles for surface modification using etchings and sizing are available from the continuous carbon fiber and glass fiber industries, different methods will be necessary to achieve the desired surface states on discontinuous reinforcements.

A second body of technology required is the development of fiber forms analogous to carbon fiber multifilament tow and glass fiber roving to enable incorporation into a composite. Carbon nanofibers are produced in a state of low bulk density, on the order of 3.2×10^{-4} gms/cm^3 (0.02 lb/ft^3). To be usable by a composite compounder, the material must be de-bulked and incorporated into a form usable in the compounding or molding process. Alternative forms such as pellets, felts, paper, and yarn are required as standard product forms for VGCF.

One of the primary advantages of advanced composites is the ability to design materials with anisotropic physical properties optimized for a given application. Methods of alignment of discontinuous fiber are required to enable the full exploitation of the outstanding array of physical properties available from carbon nanofibers.

Further barriers to the industrial application of carbon nanofibers are health and safety concerns related to the use of a new and unknown material. Because of the similarity of the geometric domain of carbon nanofibers to asbestos, the possibility exists that this material may represent health risks associated with other respirable fines. Hence, appropriate consideration must be applied to the handling and utilization of the material to ameliorate these risks.

2. Interface Technology

Establishing appropriate fiber-matrix adhesion is the key to imparting good mechanical properties to a fiber-reinforced polymer matrix composite. Understanding of the chemistry and physics of fiber-matrix adhesion and its effect on composite mechanical properties has advanced significantly in recent years. Modern theory explains adhesion behavior in terms of a microscopic layer existing at the interface with properties different from those of either the fiber or the matrix. This region is known as the interface, and it is now considered possible to optimize the interface and the level of fiber-matrix adhesion for virtually any polymer matrix.

The principles of controlling the buildup of the interface are now generally well established. One must first remove any loose surface residues or coatings, which are not chemically attached to the fiber, and then modify the surface chemistry and microtopography. Chemical surface treatment should make the fiber surface somewhat reactive with the intended matrix.

For the case of PAN fiber/epoxy composites, a well-accepted treatment of adhesion describes the phenomenon as a combination of chemical and physical bonding. Chemical bonding of the matrix to the surface of conventional carbon fiber involves only about 3% of the available fiber surface sites, but accounts for 25% of the adhesion. Another 25% of the adhesion is attributed to the surface microtopography at the 10-50nm level. It is not useful to maximize the fiber matrix adhesion in an attempt to maximize mechanical properties. When adhesion is too high, the composite becomes very brittle and loses fracture toughness. Rather one should attempt to tailor the adhesion to produce an optimum set of mechanical properties.

These details and methods of imparting the desired surface properties are well developed for continuous PAN-derived fibers. In the case of carbon nanofibers, the optimized surface chemistry, optimized bonding configurations, and methods of surface modification generally remain to be developed. It is clear, however, that in order to maintain the lowest possible processing costs, the need exists to act as much as possible in the gas phase process by which the fibers are produced, in contrast to adding subsequent processing steps to modify their surface. As a beginning, studies have been conducted which are directed at enhancing the surface energies and surface areas of the carbon nanofibers, through the addition of gas phase oxidizing agents such as H_2O and CO_2 in the production process.

The ability to influence the surface area by this approach is demonstrated in Table 1. Surface modifications can have a positive impact on the mechanical properties, as observed for different composites of polypropylene (PP) reinforced with carbon nanofibers.

TABLE 1. Range of Pyrograf™-III nanofiber surface properties

Process designation	Surface area, m^2/g		Surface energy, mJ/m^2
	Total	External	
PR-18	15	10	33
PR-1	14	10	48
PR-1A	34	16	57
PR-1A	123	46	79
PR-11-2	83	28	106
PR-11-5	26	15	156
PR-11-7	91	53	248
PR-11-0	191	57	470

Figure 1 provides further evidence of the beneficial effect of optimizing the surface state of the fibers for selected polymers. A series of eight PP based composites were prepared utilizing a 15 volume % loading of fibers with a range of surface areas and surface energies, as listed in Table 1. The figure presents a contour plot of the corresponding tensile strength vs. surface area and energy.

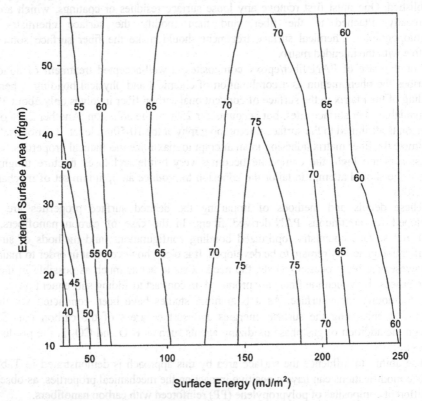

Figure 1. Contour plot of the tensile strength (in MPa) of polypropylene composites as a function of the surface properties of the nanofibers employed

It is readily apparent that a maximum tensile strength is achieved within a specific range of these parameters. This is attributed to an increasing level of adhesion between the fiber and matrix as the surface energy increases. The lack of still higher tensile strengths as the surface energy is increased beyond a certain level is thought to be due to the probable occurrence of fiber damage as more aggressive means are used to increase the surface energy.

The tensile moduli of these composites range from 3.2 to 4.2 GPa relative to a value of 1.6 GPa for the neat polymer. This property reflects the stiffness of the fiber and is not impacted by fiber-matrix adhesion to the same extent as tensile strength.

3. Fiber de-bulking and Forms

In order for carbon nanofibes to be successfully incorporated into engineering composites, they must be available in forms which composite fabricators are equipped to handle. If the fabricators are forced to radically change their procedures and equipment to use these fibers, then the economic advantage of low cost synthesis is lost. Essentially what is required is technology for performing suitable operations on the fibers to reduce them to the appropriate form for the intended application. These post-reactor operations may include de-bulking, application of sizing or other surface treatment, drying, pelletizing, sheeting, size reduction, and alignment.

3.1 PELLETS

The as-grown high-density fibers are de-bulked by dispersing in an aqueous solution, filtering, and drying to form a filter cake. This cake is then broken up into a relatively loose powder. Sizings compatible with the aqueous media may be applied at this time. High shear mixing is a suitable method to de-bulk the fibers into a more loosely flowing powder with controlled aspect ratio, allowing for size reduction and partial alignment through injection molding.

3.2 PAPER

Work has begun to demonstrate the feasibility of using paper processing techniques as a method for creating sheets of carbon nanofibers, which could be subsequently incorporated into composites having approximately isotropic in-plane properties. Carbon nanofiber papers have been made at the Paper Science and Engineering Department at Miami University (Oxford, OH). The fibers were first incorporated into a slurry with water and a small amount of polyvinyl alcohol, intended to act as a binder, and then drawn onto filter paper to create round sheets, 15 cm in diameter. Each sheet was then compressed between the rollers of a calendar to increase the density. The sheets were strong enough to support their own weight and could be easily handled. Such papers or non-woven mats are suitable for infiltration with a resin and may then be either wound or cut to size for ply lay-up. Work is ongoing to explore the efficacy of other binders selected on the basis of their compatibility with intended matrix materials.

4. Fiber Alignment

Unlike conventional pitch and PAN carbon fibers, nanofibers are non-continuous, and usually produced as a randomly oriented and entangled mass. Therefore, in order to take maximum advantage of the reinforcement properties, it is desirable to develop techniques to induce and control their alignment in composites.

Recent models of composite reinforcement by short fibers suggest that even nanofibers can provide mechanical reinforcement similar to that of longer fibers, provided that they are aligned and their aspect ratio is sufficiently high. A relatively simple model to explain short-fiber reinforcement, developed first by Cox [1] and later extended by Baxter [2], estimates stress transfer to fibers of length, l, diameter, d, modulus, E_f, and volume fraction, V, in a matrix of modulus, E_m, and Poisson's ratio, υ, with the factor

$$\beta = \frac{l}{d}\sqrt{\frac{E_m}{(1+\upsilon)E_f \cdot \ln\,(\pi/4V)}} \tag{1}$$

Combining Eq. 1 with the rule of mixtures yields a composite modulus E_c, of

$$E_c = (1 - V)E_m + \alpha\,(1 - \tanh\beta/\beta)VE_f \tag{2}$$

The pre-factor, α, is a function of the degree of fiber alignment and is equal to 1 for one-dimensionally oriented fibers, 1/2 for two-dimensionally oriented fibers and 1/6 for fibers randomly oriented in three dimensions.

In Eq. 1, we see that the value of β scales as the aspect ratio of the fibers, l/d. As a result, when the aspect ratio is large, the term $\tanh\beta/\beta$ in Eq. 2 tends to zero, and Eq. 2 reduces to the regular rule-of-mixtures form, modified only by the orientation factor, α. Thus, according to this theory, short fibers equal the reinforcement performance of arbitrarily long fibers as long as their aspect ratio is sufficiently large. In practice, any aspect ratio over 30 should be large enough to gain virtually all the benefits of the theory. Baxter's findings imply that if carbon nanofibers, which have very high aspect ratio, can be restricted in orientation to two dimensions, the resulting composite could be up to nine times as stiff as glass reinforced composites. A panel with only 30 vol. % fibers could exhibit stiffness equivalent to aluminum. Baxter's model also supplies theoretical tensile strengths by assembling three different fracture mechanisms, longitudinal fiber failure, matrix failure, and interfacial failure.

Some early experiments in the processing of nanofiber filled polymers, suggest that fiber alignment in one or two dimensions can be obtained. Kuriger and Alam [3] have carried out some preliminary tests on extruding polypropylene containing nanofibers. These efforts have resulted in thin polymer strands that have some degree of nanofiber alignment and increased strength characteristics. The ultimate tensile strength of the strands was determined by using a Tinius Olsen 1000 Benchtop Testing Machine. Prior to each test, the cross-sectional area of each strand was accurately determined by optical microscopy. Two measurements of the strand's diameter at right angles to each other were taken using the calibrated eyepiece. The cross-sectional area was then calculated by using the area formula for an ellipse:

$$Area = \frac{\pi x a x b}{4} \qquad (3)$$

Where, a is the smallest span across the section and b is the largest. This provided the information needed to calculate the ultimate tensile strength.

The values of the ultimate strength of the extruded polymer composite strands are summarized below in Figure 2. The mechanical properties were significantly improved by reinforcing with nanofibers (indicated as Pyrograf™-III in the figure). A 56% increase in strength in the extrudates was observed with only 10% fiber mass content. More modest results of 39% and 33% occurred in the 5% and 2% fiber mass mixtures, respectively. For comparison, a larger (more expensive) Mitsubishi brand fiber ('REPLARK', Type 30) was chopped to about 3 mm lengths and extruded with the polymer to produce a polymer composite strand for comparison. This strand also demonstrated strength improvement.

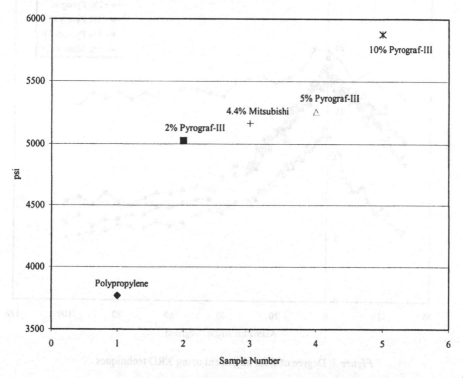

F

Figure 2. Tensile strength for different carbon fiber mass concentration

A more direct measurement of alignment of the carbon nanofibers inside the polymer strand was made using an X-ray diffraction technique. Figure 3 shows the intensity versus azimuthal angle for each specimen. The intensity at different angles indicates the

194

distribution of fiber orientations in the strand. Ideally, the intensity at high angles should be zero if the background had been eliminated properly and there are no diffracting planes at those angles. If the fiber loading is very low, the latter assumption may be in error due to the polypropylene having a diffracting plane near the fiber peak that extends to higher azimuthal angles. This error is apparent in the 2% specimen, where the intensity does not show a sharp peak. It can be seen from this figure that, in the strand with 5% chopped Mitsubishi fibers, the carbon fibers are oriented within ± 18 degrees. In the strand with 5% nanofibers, the fibers are oriented within approximately ± 20 to 25 degrees of the axis. Since the Mitsubishi fibers are much longer (in fact longer than the nozzle diameter), the enhanced alignment is to be expected. It should be noted that the two curves for the 5% were obtained for two different strands. 5% Pyrograf-R refers to a specimen that was produced by chopping up a 5% Pyrograf strand and extruding it a second time through the nozzle die. The two strands do not show any significant difference in the fiber orientation.

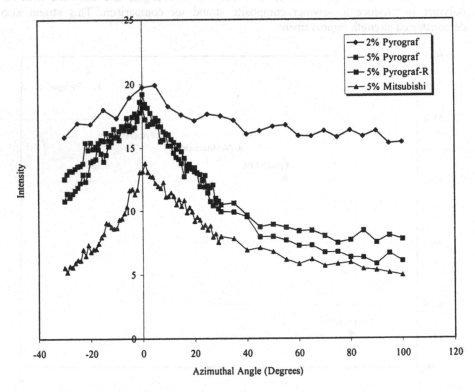

Figure 3. Degree of fiber alignment using XRD techniques

While these experiments clearly demonstrate the feasibility of improving fiber alignment and physical properties though extrusion, there remains plenty of room for improvement in the alignment by refinement of the die design. There is also a need to quantify the degree of alignment that can be achieved at higher loading, and as a function of fiber aspect ratio.

5. Health and Safety Issues

Airborne particles with diameters less than 1 micron, as in the case of asbestos, are potentially respirable. Therefore, the manufacture of any submicron diameter carbon particles such as carbon black, fullerenes, and carbon nanofiber includes a responsibility to ensure that no health hazards are present in the production or use of such material. Additionally, as in the manufacture of carbon black, various hydrocarbons can be formed during VGCF production, which is a health concern.

At present, there are no known risks attendant to the production and handling of carbon nanofibers other than those associated with condensed polycyclic aromatic hydrocarbons (PAH) which may include known carcinogens. Such PAH compounds are also generated in the production of carbon black, and other industrial pyrolytic processes. The PAH distributions and concentrations have been measured using gas chromatography/mass spectroscopy [4]. With respect to the formation of unwanted PAH compounds in the production process, it has been established that conditions can be maintained where such formation is negligible with respect to EPA and OSHA standards for the handling of carbon black. As production volumes are increased, it will be incumbent on manufacturers to maintain a set of operating parameters, which produce an environmentally benign product. Current information regarding the process for fiber formation reveals no barriers to accomplishing this

In contrast to asbestos, there are no documented health risks, which accrue from exposure to carbon nanofibers. Asbestos is composed of mineral fibers having different chemical compositions, which in turn have considerably different propensity to increase health risks [5]. Thus particle size alone is not a predictive factor pertaining to health risks. Furthermore, longitudinal studies of workers in the carbon black industry indicate no statistically significant health risks associated with long-term exposure to carbon black. Finally, a recent preliminary report on fullerene exposure using dermatological testing and effects on bronchoalveolar environment and pulmonary function indicates no measurable allergic response or inflammation in the respiratory system of guinea pigs [6]. These observations support the premise that health risks from carbon nanofibers exposure cannot be automatically inferred from the evidence relating to asbestos.

The problem of carbon nanofibers being within a respirable range, can be dealt with by continuous containment. This concept implies the containment of the fibers from the point of formation inside the reactor to permanent entrainment in the polymer material of choice. As currently produced, this type of fiber is entangled, or bird nested, and made in large agglomerations of fiber resembling cotton (except for the color). The degree of entanglement is so complete that periodic air sampling of the exhaust from a laboratory reactor has revealed no evidence of dispersed individual fibers. Higher volume production rates may impact this condition. Higher production rates may also require more rigorous collection systems, followed by application of sizing, pelletization, paper formation, or other de-bulking process, similarly leaving the fiber in a state of agglomeration and containment. Finally, the production process may be completed by entrainment of the fiber in a binder or polymer matrix material before packaging and shipping to compounders or other users. Thus exposure to individual fibers is anticipated to be an extremely rare exception to anticipated normal handling operations.

6. Conclusions

The remarkable properties of vapor grown carbon fibers (VGCF) have been reported in the scientific literature for approximately twenty-five years, with the expectation that this novel graphitic fiber would ultimately be used as a low cost reinforcement for numerous engineering applications. Technical barriers embodied in large volume production of a material suitable for fabrication of useful components have impeded the realization of this expectation. Today many of the critical advances have been captured in processing hardware for large volume production of fibers having tailored geometries and surface properties for selected applications. The availability of such "designer nanofibers" is enabling an ever-expanding number of organizations to conduct composite synthesis trials directed at uses ranging from lightweight composite structures through applications in chemical and biomedical processing and electronics.

Acknowledgements

This discussion was prepared with the benefit of the significant contributions of researchers at Applied Sciences, Inc., including R. Alig, D. Burton, G. Glasgow, J. Guth, J. Hager, R. Jacobsen, E. Kennel, C. Tang and J-M. Ting, whose efforts and service are gratefully acknowledged. This work was sponsored in part by a NIST Advanced Technology Program under Agreement No. 70NANB5H1173, and the Ohio Coal Development Office under Grant No. CDO/D-96-3.

References

1. Cox, H.L. (1952) Elasticity and Strength of Paper and other Fibrous Materials, *British Journal of Applied Physics* **3**, 72-79.
2. Baxter, W.J. (1992) An analysis of the Modulus and Strength of PYROGRAF/Epoxy Composites, Report No. PH-1717, General Motors Research Laboratories, Warren, MI.
3. R.J. Kuriger and M. K. Alam (2000) The Influence of Extrusion Conditions on Properties of Vapor Grown Carbon Fiber Reinforced Polypropylene, to be published in Polymer *Composites Journal*.
4. Pederson, T.C., Powell, C.A., Santrock, J., Rosenbaum, L., Siak, J., Tibbetts, G.G. and Alig, R.L. (1991) Analysis of Polycyclic Aromatic Hydrocarbons on Vapor Grown Carbon Fibers, in R.C. Brown (ed.), *Mechanisms in Fibre Carcinogenisis*, Plenum Press, New York, pp. 199-212.
5. Gibbs, G.W., Valic, F. and Brown, K. (1994), Health Risks Associated with Chrysotile Asbestos, *Ann. Occup. Hyg.* **38**, no. 4, 399-426.
6. Huczko A. et. Al (2000) On Some Aspects of the Bioactivity of Fullerene Nanostructures, to be published in *Fullerene Sci. and Technology*.

DIFFRACTION BY MOLECULAR HELICES
The range of morphologies of sp² carbon and the basic theory of diffraction by an atomic helix

A. A. LUCAS, PH. LAMBIN and F. MOREAU
Physics Department, Facultés Universitaires
Notre-Dame de la Paix
61, rue de Bruxelles, B5000 Namur, Belgium

1. Introduction

In its most abundant natural form, pure carbon crystallizes in the graphitic structure which consists of *planar* honeycomb lattices of sp^2-bonded atoms called graphene, loosely piled up at a regular distance a = 0.34 nm (see Figure 1a on the next page).

In the last ten years, a great variety of similar but *nonplanar* sp^2 carbon morphologies have been discovered. Figure 1 sketches the many globular, tubular or other curved forms presently known [1]. The graphene sheets are rolled up into spheroidal, cylindrical or other curved shapes, always maintaining the regular interlayer spacing a.

The most common form of cylindrically "curved graphite" and the first to be discovered [1c] is the straight, hollow nanotube (Fig.1c). This variety perhaps constitutes the ultimate mesoscopic core of some of the macroscopic carbon filaments discussed in other chapters of this book.

The detailed atomic structure of single wall (SWNT) and multiple wall (MWNT) nanotubes (NT) will be the subject of this chapter. Like for practically all other large molecular assemblies, this structure has been determined by a combination of high resolution microscopies and wave diffraction methods. The present chapter will focus on the interpretation of Electron Microscopy (EM) and Electron Diffraction (ED) results (and extensive review can be found in [1f]).

In this part, we derive the kinematic amplitude for the wave diffraction by a regular, circular monoatomic helix, the basic atomic assembly from which any NT can be built up.

L.P. Biró et al. (eds.), Carbon Filaments and Nanotubes: Common Origins, Differing Applications, 197–204.
© 2001 *Kluwer Academic Publishers. Printed in the Netherlands.*

198

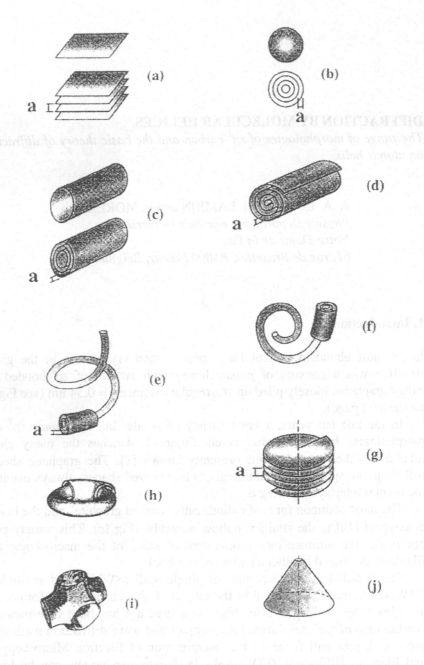

Figure 1. Various forms of sp^2 carbon [1a]. (a) refers to ordinary planar graphite and graphene. Curved morphologies (b) to (j) are described in references [1b] to [1j].

2. Mathematical ingredients

To establish the scattering form factor of a regular, circular atomic helix, a few fundamental mathematical tools need to be recalled.

2.1. SCALAR HELMHOLTZ EQUATION

Consider the scalar wave equation

$$\Delta \Psi + q^2 \Psi = 0 \tag{1}$$

where $\mathbf{q} = (q_x, q_y, q_z)$ is a real 3-D wave vector.

This equation is separable in eleven curvilinear coordinate systems [2]. Two of them are the cartesian and the cylindrical coordinates for which the elementary solutions take the forms

$$\Psi_{\mathbf{q}}(\mathbf{r}) = e^{i\mathbf{q}.\mathbf{r}} \tag{2}$$

$$\Psi_{q_\perp, q_z, n}(\mathbf{r}) = J_n(q_\perp \rho) e^{-in\varphi} e^{iq_z z} \tag{3}$$

where $q_\perp = (q_x, q_y)$, $\rho = (x, y)$ and $J_n(q_\perp \rho)$ is a Bessel function of integer order n. Both famillies of functions (2) and (3) being complete sets of eigenfunctions of the hermitian operator $\Delta + q^2$, any (well-behaved) function can be expressed in terms of these sets as a Fourier or a Fourier-Bessel transform, respectively.

2.2. FOURIER SERIES

In particular, the functions of set (2) can be expressed in terms of the functions of set (3) (and vice versa). The way to do this is to notice that the exponential function $exp(i\mathbf{q}.\mathbf{r})$, where $\mathbf{r} = (\rho\cos\varphi, \rho\sin\varphi, z)$, is a periodic function of the cylindrical azimuthal angle φ and can therefore be expanded in a Fourier series :

$$e^{i\mathbf{q}.\mathbf{r}} = \sum_m C_m e^{-im\varphi} e^{iq_z z} \tag{4}$$

The expansion coefficients C_n are obtained, in the usual way, by multiplying (4) by $exp(in\varphi)$ and integrating over φ :

$$C_n = \frac{1}{2\pi} \int_0^{2\pi} e^{iq_\perp \rho \cos\theta} e^{in\varphi} d\varphi \tag{5}$$

In (5), the angle between $q_\perp$ and ρ has been written as $\theta = \varphi - \psi_q$ where $\psi_q = \tan^{-1}(q_y/q_x)$ is the azimuthal angle of $q_\perp$.

Apart from phase factors, (5) is just the integral representation of the Bessel function [2] :

$$J_n(x) = \frac{i^{-n}}{2\pi} \int_0^{2\pi} e^{i(x\cos\theta+n\theta)} d\theta \tag{6}$$

where $x = q_\perp \rho$.

The end result is known as the Jacobi-Anger expansion of a plane wave [2] :

$$e^{i\mathbf{q}\cdot\mathbf{r}} = \sum_n e^{in(\psi_q+\frac{\pi}{2})} J_n(q_\perp\rho)e^{-in\varphi} e^{iq_z z} \tag{7}$$

2.3. DELTA FUNCTION

The last ingredient which we shall need is the sum of plane waves evaluated at the atomic positions of a monoatomic chain. This involves the geometrical series

$$\sum_{n=-N}^{+N} e^{ixn} = \frac{\sin\left[(N+\frac{1}{2})x\right]}{\sin(\frac{1}{2}x)} = 2\pi\delta_N(x) \tag{8}$$

$\delta_N(x)$ is clearly a periodic function of x with period 2π and constitutes a sequence of functions approaching the Dirac delta function for large N [2]. For N going to infinity one can thus write

$$\sum_{n=-\infty}^{+\infty} e^{ixn} = 2\pi \sum_{m=-\infty}^{+\infty} \delta(x - 2\pi m) \tag{9}$$

3. Diffraction by an atomic helix

We are now in a position to combine the results above to construct the scattering factor of a monoatomic helix.

3.1. CROSS SECTIONS

In the kinematic (first Born) approximation, the differential elastic cross section for the scattering for a wave scattered from the initial $\mathbf{k}_i$ to the final $\mathbf{k}_f$ wave vectors by a collection of atoms at $\mathbf{r}_j$ is the square modulus of a complex amplitude $S(\mathbf{q})$

$$\frac{d\sigma}{d\Omega} = |S(\mathbf{q})|^2 = \left| \sum_j f_j(\mathbf{q}) e^{i\mathbf{q}.\mathbf{r}_j} \right|^2 \tag{10}$$

where $\mathbf{q} = \mathbf{k}_f - \mathbf{k}_i$ is the scattering wave vector and where $f_j(\mathbf{q})$ is the atomic scattering factor. The latter is the Fourier transform of an atomic property which can be either the electron density (X-rays), the screened Coulomb potential of the nucleus (fast electrons), or the bare nucleus density (neutrons). The behavior of $f_j(q)$ vs $q=|\mathbf{q}|$ for the light elements B, C, N is shown in Figure 2 for X-rays and for fast electrons [3]. With neutrons $f_j(q)$ is a constant for values of q up to inverse nuclear sizes. $f_j(q)$ has the dimension of a length. For X-rays the natural length scale is the classical electron radius $r_e = e^2/4\pi\varepsilon_0 mc^2 = 2.8 \ 10^{-15}$ m. whereas for neutrons it is the nuclear radius, i.e. of the same order of magnitude as for X-rays. For fast electrons, the length scale is the Bohr radius, $0.5 \ 10^{-10}$ m. Hence the scattering power of atoms for electrons is 8 to 9 orders of magnitude larger than for X-rays or neutrons. This explains why Electron Microscopy or Diffraction are able to work with one single nanoscopic object such as a NT, while X-rays or neutrons require macroscopic samples.

3.2. CCV FORMULA

It will be shown in the second part that a carbon NT can be constructed from a set of regular, circular, atomic helices. A first step toward the theory of diffraction by a NT is therefore to establish the scattering amplitude of a monoatomic helix.

$f(q)$ being the isotropic scattering factor of carbon, the amplitude to compute is

$$S(q) = f(q) \sum_j e^{i\mathbf{q}.\mathbf{r}_j} \tag{11}$$

The positions $\mathbf{r}_j = (r, \varphi_j, z_j)$ of atoms sitting on a continuous (right-handed) helix of pitch P around the z-axis are given by the cylindrical coordinates $\rho_j = r$, $\varphi_j = \varphi_0 + 2\pi(z_j-z_0)/P$ and $z_j = z_0 + jp$ where p is the regular axial repeat inter atomic distance. Here $\mathbf{r}_0 = (r, \varphi_0, z_0)$ is the position of an arbitrary atom taken as origin.

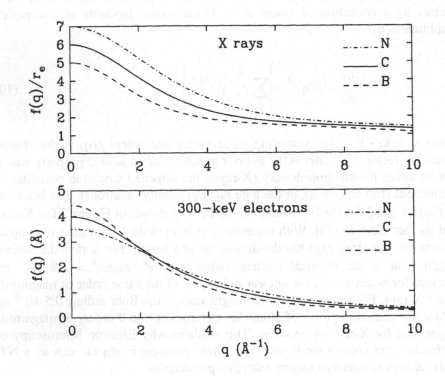

Figure 2.
Atomic structure factors of B, C, N for X-rays and for fast electrons.

Using now the Jacobi-Anger formula (7) for the exponentials in (11), one encounters a summation such as (9) which causes the quantization of the axial component q_z of the transfer wave vector. The end result is

$$S(q) = \frac{2\pi}{p} f(q) \sum_{n,m} J_n(q_\perp r) e^{in(\psi_q - \varphi_0 + \pi/2)} e^{iq_z z_0} \delta(q_z - n\frac{2\pi}{P} - m\frac{2\pi}{p}) \qquad (12)$$

Due to the delta function, the scattered intensity is nonzero only in directions given by the circular intersections, in reciprocal space, of the Ewald sphere with layer planes perpendicular to the z-axis. These planes are equidistant at heights

$q_z = l2\pi/P_a$ where P_a is the true axial repeat period of the (discrete) atomic helix and where the integer l is given by the selection rule

$$\frac{l}{P_a} = \frac{n}{P} + \frac{m}{p} \tag{13}$$

On a distant screen parallel to the helix axis, the scattered directions project as hyperbolae, called "layer lines", labelled by l. The concept of layer line was introduced in the early use of X-ray diffraction to investigate the structure of fibrous materials [4]. The layer line structure of diffracted intensity is characteristic of any linear polymer having a regular repeat period P_a in one dimension, not just of helical ones. For high-energy electrons, the Ewald sphere is very nearly planar. Then the layer lines, along which the intensity is nonzero, appear as equidistant, straight, horizontal lines on either side of the $l = 0$ equatorial line.

These beautiful results were first derived by Cochran, Crick and Vand (CCV) in 1952 for the interpretation of the X-ray diffraction pictures of helical polypeptide molecules [5]. In addition, the CCV formula (12) was to play a crucial role in the intellectual path toward the discovery in 1953 of the double helical structure of DNA by Watson and Crick. We will come back to this fascinating application at the end of the second part.

Acknowledgements

The authors are grateful to prof. S. Amelinckx and Antwerp colleagues for fruitful discussions. We are also grateful for the support of the Fund for the Industrial and Agricultural Research FRIA. This work has been supported in part by the Belgian Ministry of Research in the framework of the PAI-IUAP program P4/10.

References

1. a) Terrones M., Hsu W.K., Hare J.P., Kroto H.W., Terrones H. and Walton D.R.M. (1996), Graphitic structures: from planar to spheres, toroids and helices, *Phil. Trans. R. Soc. Lond. A.* **354**, 2025-2054.
b) Kroto H.W., Heath J.R., O'Brien S.C., Curl R.F. and Smalley R.E. (1985) *Nature* **318**, 162-163; Iijima S. (1987), The C60-Carbon Cluster has been revealed, *J. Phys. Chem.* **91**, 3466-3467.
c) Iijima. S. (1991), Helical microtubules of graphitic carbon, *Nature* **354**, 56-58; Iijima S. and Ichihashi T. (1993), Single-shell carbon nanotubes of 1 nm

diameter, *Nature* **363**, 603-605; Bethune D.S., Kiang C.H., de Vries M.S., Gorman G., Savoy, R., Vasquez J. and Beyers R. (1993), Cobalt-catalyzed growth of carbon nanotubes with single-atomic-layer walls, *Nature* **363**, 605-606.

d) Amelinckx S., Zhang X.B., Bernaerts D., Zhang X.F., Ivanov V. and B.Nagy J. (1994), A formation mechanism for catalytically grown helix shaped graphite nanotubes, *Science* **265**, 635-636.

e) Zhang X.B., Zhang X.F., Bernaerts D., Van Tendeloo G., Amelinckx S., Van Landuyt J., Ivanov V., B.Nagy J., Lambin Ph. and Lucas A.A. (1994), The texture of catalytically grown coil-shaped carbon nanotubes, *Europhys. Letters* **27**, 141-146

f) Amelinckx S., Lucas A.A. and Lambin Ph. (1999), Electron diffraction and microscopy of nanotubes, *Rep. Prog. Phys.* **62**, 1-54.

g) Amelinckx S., Luyten W., Van Tendeloo G. and Van Landuyt J. (1992), Conically, helically wound, graphite whiskers : a limiting member of the fullerenes, *J. Cryst. Growth* **121**, 543-558.

h) Liu J., Dai H., Hafner J.H., Colbert D.T., Smalley R.E., Tans S.J. and Dekker C. (1997), Fullerene "crop circles", *Nature* **385**, 780-781.

i) Plumber nightmare morphology (undiscovered).

j) Krishan A., Dujardin E., Treacy M.M.J., Hugdahl J., Lynum S. and Ebbesen T.W. (1997), Graphitic cones and the nucleation of curved carbon surfaces, *Nature* **388**, 451-54.

2. Arfken B. and Weber H. (1995), *Mathematical Methods for Physicists*, Academic Press, New York.

3. Doyle P.A. and Turner P.S. (1968), Relativistic Hartree-Fock X-ray and electron scattering factors, *Acta Cryst.* **A24**, 390-397.

4. Polanyi M. (1921), *The X-ray fiber diagram*, Z. Phys. **7**, 149-180.

5. Cochran W., Crick F.H.C. and Vand V. (1952), The structure of synthetic polypeptides. I. The transform of atoms on a helix, *Acta Cryst.* **5**, 581-585.

DIFFRACTION BY MOLECULAR HELICES

Transmission electron microscopy and diffraction by nanotubes and other helical nanostructures

A. A. LUCAS, PH. LAMBIN and F. MOREAU

Physics Department, Facultés Universitaires
Notre-Dame de la Paix
61, rue de Bruxelles, B5000 Namur, Belgium

1. Introduction

In the present part, we construct the diffraction amplitude of SWNT and MWNT from the form factor of a monoatomic helix obtained in the previous part. This will enable us to simulate numerically and interpret the observed nanotube (NT) electron diffraction (ED) patterns.

In order to deepen our understanding of such patterns we perform, with student participation, experiments which simulate the diffraction phenomenon optically. By using specially designed diffraction gratings and a coherent source of light such as a laser pointer, all the essential features of the observed ED patterns can be displayed in the visible range.

As a further, related application of great historical interest, the basic helix form factor derived in the first part, along with other suitable optical diffraction gratings, will also be used to interpret the X-ray diffraction picture produced by the celebrated DNA double helix.

2. Constructing Nanotubes from Helices

We summarize here the essential points in the construction of a SWNT from a flat graphene sheet (details can be found e.g. in [1]).

Consider a piece of graphene oriented, as in Figure 1a, with one of its three families of atomic zig-zag lines placed in the horizontal direction. An origin (0, 0) and a lattice vector $C = na_1 + ma_2$ ending on the lattice point (n, m) are chosen. Let's assume $0 \leq m \leq n$ (the general case is considered in [1]). Let's roll up the piece into a circular cylinder in the convex sense, i.e. the (n, m) point is rotated clockwise around the normal to C until the point (n, m) is made to coïncide with the origin $(0, 0)$. The result is a (n, m) SWNT shown in Fig.1b.

In general, a SWNT with $n \neq m \neq 0$ is a chiral object, i.e. it cannot be superposed to its mirror image. The angle θ between the graphene C and the horizontal zig-zag line is called the chiral angle of the NT. C becomes the

L.P. Biró et al. (eds.), Carbon Filaments and Nanotubes: Common Origins, Differing Applications, 205–217.
© 2001 *Kluwer Academic Publishers. Printed in the Netherlands.*

circumference of the NT in Fig.1b. The normal to C from the origin necessarily meets a lattice point at a finite distance T which is the true axial period of the NT [2]. The zig-zag line of atoms at 60° from the horizontal in the graphene sheet (highlighted in Fig.1a) is seen to wind up as a right-handed double helix around the NT axis in Fig.1b. The entire NT can be generated by just n such diatomic helices separated from each other by a constant screw operation $(\Delta z, \Delta\varphi)$ entirely determined by the pair of indices (n, m) [1]. This is very useful for computing the NT scattering factor from that of the zig-zag helix, as we now show.

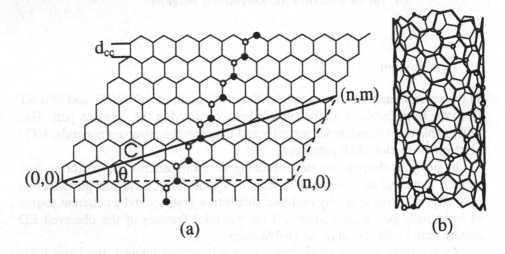

Figure 1. (a) Defining the NT indices (n, m) on a piece of graphene; (b) Rolled up graphene sheet to produce the (n, m) NT.

3. Form Factor of a SWNT

3.1. ACHIRAL SINGLE WALL NANOTUBE

Let us illustrate the method with the simple case of a "parallel" SWNT, i.e. an achiral one of the type $(n, 0)$ for which the axial period is $T = 3d_{cc}$. The reader will treat the equally simple case of a "perpendicular" achiral NT (n, n) as an exercise.

From Fig.1, one sees that the upper helix of the zig-zag pair can be obtained from the lower one by the pure axial translation $(\Delta z = d_{cc}, \Delta\varphi = 0)$. Then if $S_1(q)$

is the diffraction amplitude of the lower monoatomic helix, the amplitude of the zig-zag pair is

$$S_2(q) = S_1(q)(1 + e^{iq_z d_{cc}}) \tag{1}$$

If we now subject the zig-zag pair to $n - 1$ successive pure axial translations ($\Delta z = 3d_{cc}$, $\Delta \varphi = 0$) we generate the entire $(n, 0)$ NT. The total NT scattering amplitude is then obtained by multiplying (1) by the function

$$\sum_{m=0}^{n-1} e^{imq_z 3d_{cc}} = \frac{1 - e^{inq_z 3d_{cc}}}{1 - e^{iq_z 3d_{cc}}} \tag{2}$$

representing the interferences between the amplitudes of the n diatomic helices.

Now recall that S_1 in (1), as given by the CCV formula (see Eq.(12) in the first part), implies that the axial transfer wave vector q_z is quantized to the layer line values $q_z = 2\pi l/P_a$ where, for a parallel NT, the zig-zag helix period is $P_a = nT = 3d_{cc}n$. Hence the phase factor (2) simplifies to

$$\frac{1 - e^{i2\pi l}}{1 - e^{i2\pi l/n}} = \sum_s n\delta_{l,sn} \tag{3}$$

The Kronecker delta suppresses all layer lines of the single zig-zag double helix safe those for which l is a multiple of n, i.e. when $q_z = s2\pi /3d_{cc}$. This indeed is the layer line quantization value for any polymer of period $T = 3d_{cc}$.

The final diffraction amplitude of a parallel SWNT $(n, 0)$ is thus

$$S(q) = \frac{4\pi n}{3d_{cc}} f(q) \sum_l e^{il2\pi z_0 /3d_{cc}} F_l \delta(q_z - l2\pi /3d_{cc}) \tag{4}$$

where

$$F_l = \sum_{s,t} J_{sn}(q_\perp r) e^{isn(\varphi_q - \varphi_0 + \frac{\pi}{2})} (1 + e^{i(s+2t)\frac{2\pi}{3}}) \delta_{s+2t,l} \tag{5}$$

3.2. CHIRAL SINGLE WALL NANOTUBE

The method just described has been fully worked out for an arbitrary chiral SWNT (n, m) in [1]. It will not be repeated here since, apart from a somewhat more cumbersome algebra, the procedures are just the same as for the achiral case.

4. Multiwall nanotube and Nanotube Bundle.

The diffraction amplitude of an arbitrary collection of SWNT can be written by summing up the individual amplitudes $S(\mathbf{q})$ for each of them. The scattered intensity is the square modulus of the coherent sum.

SWNT pertaining to the same MWNT are coaxial and differ only by their chiral indices (n, m) and by the positions of their original atoms $\mathbf{r}_0$.

For bundles of NT (also called "ropes") parallel to the z-axis, a global phase factor $exp(i\mathbf{q}.\rho)$ must be applied to each NT amplitude where ρ is the lateral position of the NT axis normal to the (x, y) plane [1].

5. Numerical Simulations

We now present a few examples of diffraction patterns calculated on the basis of the kinematic theory outlined previously.

5.1. SINGLE WALL NANOTUBES

Figure 2b shows the result for the (20, 0) SWNT of Fig.2a seen at normal incidence. The distribution of intensities along horizontal, equidistant layer-lines determines a hexagonal pattern of spots elongated horizontally. The first order hexagon of spots is oriented parallel to the NT honeycomb hexagons while the second order hexagon is rotated by 30°. Each streaking spot replaces what would be a sharp dot in the diffraction pattern of a flat graphene sheet. Notice that the streaks are sharp in the vertical direction which, by virtue of the rolled up construction of the NT, remains a direction of unchanged linear cristalline order. It is the curved, cylindrical order in the horizontal directions which causes the streaking phenomenon. The incident plane wave sees a shrinking honeycomb lattice parameter towards the edges of the NT : smaller lattice constants lead to larger diffraction angles, hence the streaking. The intensity modulation along a streak is due to the interference between the amplitudes scattered from the two left and right halves of the NT.

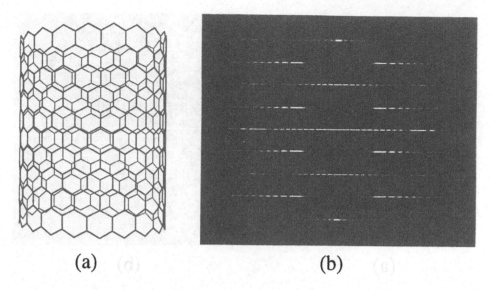

Figure 2. (a) Single wall nanotube (20,0); (b) Calculated diffraction pattern of (a)

Mathematically the streak structure reflects the variations of the Bessel functions in (5). The Bessel functions $J_n(x)$, where $x = q_\perp r$ is a measure of the distance to the meridian axis in Fig.2b, are vanishingly small when $x < n \neq 0$, which is why the central part of certain layer lines has no intensity. Other layer lines, such as the equatorial line, are controlled primarilly by J_0 and are therefore most intense on the meridian line of the pattern. This explains the formation of a hexagonal pattern of streaking spots. Since the zeros of $J_n(x)$ rapidly become equidistant, with separation π, the modulation of intensity along a layer line has period π/r as in the Fraunhofer diffraction by a slit of width $2r$, the diameter of the NT.

Figure 3b is the calculated diffraction pattern of a chiral (19, 2) SWNT shown in Fig.3a. The hexagon of each diffraction order has split into two hexagons rotated in opposite directions by the chiral angle $\theta = 8°$. The left-rotated hexagon is produced by diffraction from the front half of the NT and the right-rotated one from the back half.

Unfortunately, there are very few experimental ED measurements on SWNT with which to compare the above computations. SWNT tend to "evaporate" under the electron beam while recording their weak ED pattern [3,4].

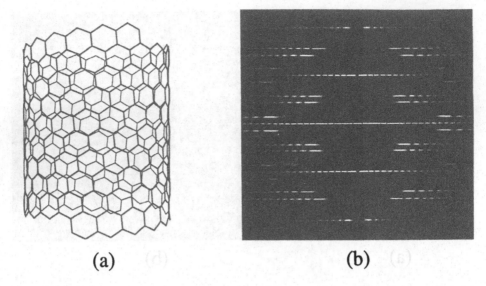

Figure 3. (a) Single wall nanotube (19,2); (b) Calculated diffraction pattern of (a)

5.2. MULTIPLE WALL NANOTUBES

The situation is more favorable for MWNT which resist radiation damage better. Figure 4 compares the observed and calculated ED pictures from a 3-layers MWNT [5]. Although diffraction amplitudes rather than intensities are additive, the interpretation of an ED pattern such as Fig.4b can usefully start by assuming that the layers scatter independently. From the ED pattern, one reads the number of different helicities and the corresponding chiral angles by just counting the number of pairs of hexagons on a diffraction circle and measuring their anglular separation. Thus inspection of Fig.4b reveals that the three layers have three different chiral angles of about $\theta = 4°$, 8° and 26°. However it is presently not feasible to decide, among the 6 different permutations of these three numbers, which is the actual sequence of chiralities in the actual NT. This is because the computed patterns of all permuted sequences do not differ sufficiently to attempt a best match to the observed ED pattern (while the geometry of the pattern in Fig.4b is reliable, the relative spot intensities are not). Fig.4c shows the simulated pattern of the sequence (38, 30) (inner NT), (62, 11) (middle NT) and (73, 6) (outer NT) corresponding to the observed radii in Fig.4a and the chiral angles given above.

The "sequencing" of the chiralities of a MWNT is however not completely hopeless. Various semi-quantitative methods are reviewed in [6].

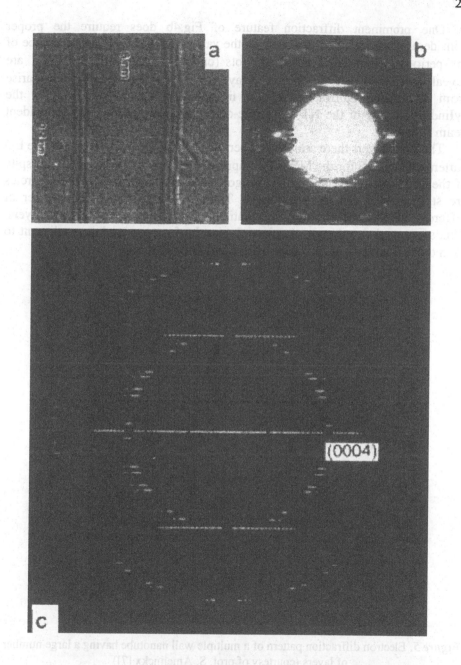

Figure 4. (a) Electron micrograph og a triple wall nanotube; (b) Electron diffraction pattern of (a); Computer simulated electron diffraction pattern of (a).

One prominent diffraction feature of Fig.4b does require the proper consideration of interferences between the various layers. It is the emergence of the periodic set of broad (0,0,2m) spots (only the spots with $m = \pm 1$ are revealed in Fig.4b) on the equatorial layer line separated by $2\pi/a$. These arise from the coherent diffraction by the linear-looking gratings defined by the cylindrical layers of the NT seen edge-on in the plane normal to the incident beam (Fig.4a).

The more layers there are, the sharper the (0,0,2m) spots, as shown in the ED pattern of Figure 5 for the MWNT comprising as many as 15 layers [7]. In spite of the large number of layers, the hexagonal sets of streaks on the Bragg circles are still resolved and even countable. Thus in large MWNT, the number of different chiralities is generally substantially smaller than the number of layers. This suggests that a degree of epitaxial growth of successive layers adjacent to each other must have occurred during the synthesis process [7].

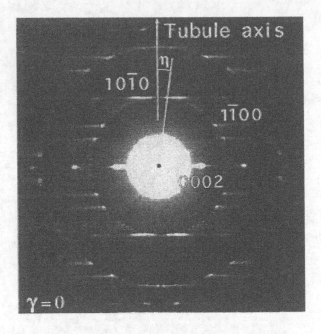

Figure 5. Electron diffraction pattern of a multiple wall nanotube having a large number of layers (courtesy of prof. S. Amelinckx [7])

6. Optical Simulations

6.1. DIFFRACTION BY NANOTUBES

In order to gain a better understanding of the diffraction process, we now present optical simulation experiments which bring out, in the visible range, the essential features of the observed ED patterns.

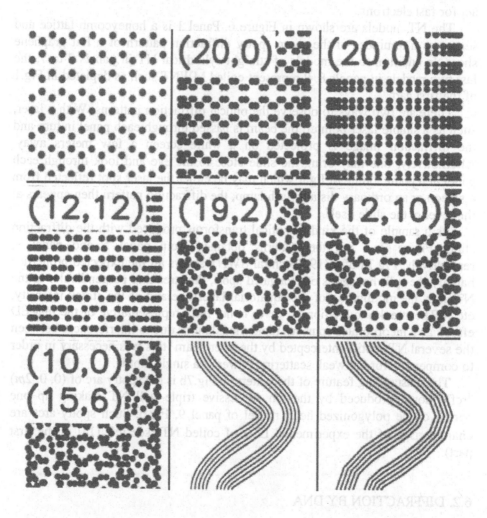

Figure 6. Diffraction motifs used to simulate optically electron diffraction of nanotubes

A set of nine diffraction gratings has been laid down on a slide which serves as a little diffraction laboratory. The motif of a grating is a parallel projection of the atom distribution in a NT on a plane which may be either perpendicular or inclined at an angle to the projection direction. Each motif is first drawn and copied many times in parallel but randomly separated to occupy a single A4 sheet. The nine A4 panels are then mounted together on a board and photographically reduced to the standard size of a 5cm x 5cm slide. The atoms are represented as circular dots which act for visible photons as screened nuclei act for fast electrons.

The NT models are shown in Figure 6. Panel 1 is a honeycomb lattice and serves as a reminder of the sharp spotty diffraction pattern of a flat graphene sheet. In addition to the models of straight cylindrical NT in panels 2 to 7, the last two models in panels 8, 9 represent coiled MWNT such as depicted in Fig.1 of the first part.

There are two ways to observe the optical diffraction patterns. With a laser, such as a simple laser pointer, the beam is passed through each panel in turn and the diffraction image is projected on a white screen a few meters away. Alternatively, one may bring the slide close to the eye and look through each panel at a point source of light which may be either the laser spot reflected from a screen or an ordinary distant flash lamp; the diffraction pattern then appears at the level of the slide itself.

As a sample of the kind of optical transforms produced with the diffraction slide, Figure 7a,b are the patterns obtained from panel 5 and panel 9, respectively. Fig.7a of a (19, 2) chiral SWNT shows all the expected properties namely the split hexagons of elongated spots, the streaking perpendicular to the NT axis, the modulation of the equatorial layer line and of the streak intensity, etc... The streak modulation however is not fully representative of the real ED effect, as it incorporates spurious rapid modulations due to interferences between the several NT motifs intercepted by the laser beam (this was necessary in order to compensate for the weak scattering power of a single motif).

The outstanding feature of the pattern in Fig.7b is the spotty arc of $(0, 0, 2m)$ "reflections" produced by the ten successive triple-wall NT making up one period of the polygonized helix model of panel 9, Fig.6. Such spotty arcs are characteristic of the experimental ED's of coiled NT (see ref. [1e] in the first part).

6.2. DIFFRACTION BY DNA

The CCV diffraction theory described in the first part was developed in the early 1950's for the interpretation of the X-ray diffraction patterns of helical proteins (see ref. [5] in the first part). But it's greatest triumph has been in helping Crick

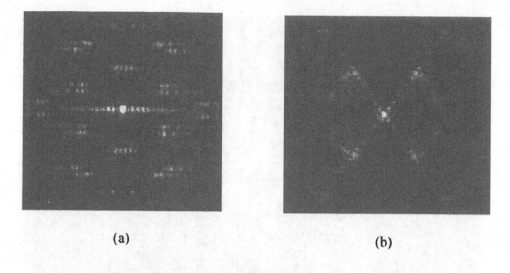

(a)   (b)

Figure 7. (a) Optical transform of panel 5 (Fig.6) representing the planar projection of a (19,2) nanotube; (b) Optical tronsform of panel 9 (Fig.6) representing a polygonized coiled nanotube.

and Watson to recognize the structural significance of several crucial features in the X-ray diffraction diagram of DNA [8]. This critical insight led them to the momentous discovery of the double helical structure of DNA [9].

We want to end the present part by optical simulation experiments similar to the one above, which not only illustrate directly the CCV formula but also provide a deeper understanding of the structural content of the famous X-ray diagram [8].

A nine-panel diffraction slide dedicated to DNA has been realized along the same principles as those of the NT slide. Figure 8 shows the nine diffraction motifs along with the diffraction patterns they produce. The rather straighforward logic of the slide has been explained at length in a recent didactic paper [10] to which the reader is refered for a complete demonstration of the diffraction experiment.

The DNA diffraction slide as well as the NT slide are available on demand from the authors.

216

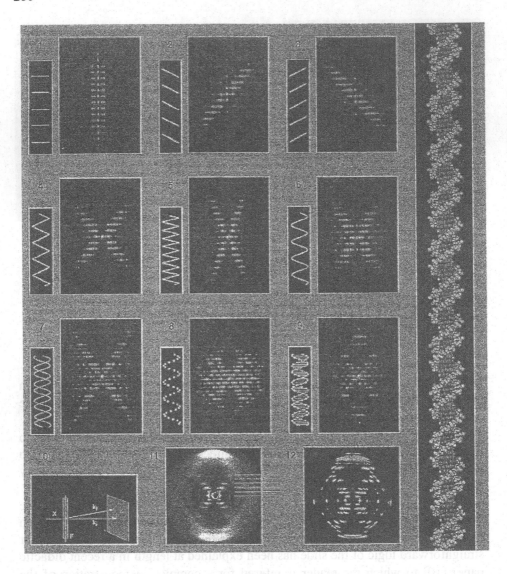

Figure 8. Panels 1 to 9: optical transforms of nine periodic motifs simulating DNA. Panel 9 sketches the experimental X-ray scattering geometry. Panels 11 and 12 reproduce the observed X-ray pictures of a B-DNA fiber [8]. A model double helix is shown on the right

Acknowledgements

We are grateful for the support of the following agencies : the Belgian National Science Foundation, the Fund for the Industrial and Agricultural Research FRIA, the Ministry of Sciences, the European Commission TMR program and the Walloon Government. We thank Prof. S. Amelinckx for allowing us to reproduce Fig.5. We also thank Mr M. Mathot for helping us in the realization of the optical transforms.

References

1. Lambin Ph. and Lucas A.A. (1999), Quantitative theory of diffraction by carbon nanotubes, *Phys. Rev.* **B66**, 3571-3574.
2. Dresselhaus S., Dresselhaus G. and Saito R. (1992), Carbon fibers based on C60 and their symmetry, *Phys. Rev.* **B45**, 6234-6246.
3. Iijima S. and Ichihashi T. (1993), Single-shell carbon nanotubes of 1-nm diameter, *Nature* **363,** 603-605.
4. Henrard L., Loiseau A., Journet C. and Bernier P. (2000), Study of the symmetry of single-wall nanotubes by electron diffraction, *Eur. Phys. J.* **B13**, 661-669.
5. Iijima S. (1994), Carbon nanotubes, *MRS Bulletin* **19**, 43-46.
6. Lambin Ph., Lucas A.A., Amelinckx S. and Dekker C. (2000), Determination of the helicity of the nanotubes, to appear in *"Nanotubular structures. Characterization and simulation at the atomic scale"*, ed. Loiseau A., Rubio A. and Willaime F., Gordon and Breach, New York.
7. Zhang X.B., Zhang X.F., Amelinckx S., Van Tendeloo G.,and Van Landuyt J. (1994), The reciprocal space of carbon nanotubes : a detailed interpretation of the electron diffraction effects, *Ultramicroscopy* **54**, 237-249.
8. Franklin R.E. and Gosling R.G. (1953), Molecular configuration of sodium thymonucleate, *Nature* **171**, 740-741.
9. Watson J.D. and Crick F.H.C. (1953), A structure for deoxyribose nucleic acid, *Nature* **171,** 737-738.
10. Lucas A. A., Lambin Ph., Mairesse R. and Mathot M. (1999), Revealing the backbone structure of B-DNA from laser optical simulations of its X-ray diffraction diagram, *J. Chem. Educ.* **76**, 378-383.

STM INVESTIGATION OF CARBON NANOTUBES

L. P. BIRÓ and G. I. MÁRK

Research Institute for Technical Physics and Materials Science
H-1525 Budapest, P. O. Box 49, Hungary, e-mail: biro@mfa.kfki.hu

Abstract. The general principles of scanning tunneling microscopy (STM) will be presented and their application to image interpretation will be discussed. The particular problems that may arise from the three dimensional nature and from the complexity of the tunneling system in the case of supported carbon nanotubes will be considered. An overview of the milestones of STM and scanning tunneling spectroscopy (STS) experiments performed on carbon nanotubes will be given, with particular emphasis on the questions related with atomic resolution imaging, the influence of the two tunneling gaps (tip/nanotube; nanotube/support) and point contacts during imaging. Experimental STS results on multiwall carbon nanotubes and rafts of nanotubes will be discussed and compared to computer simulations based on wave packet dynamics.

1. Introduction

1.1 CARBON NANOTUBES

A single-wall carbon nanotube (SWCNT) is constituted of a single graphene layer wrapped into a perfect cylinder. The experimentally found typical diameter values of SWCNTs are in the range of 1 - 2 nm. A multi-wall carbon nanotube (MWCNT) is composed of several coaxial SWCNTs with increasing diameters, in a way that the distance between the walls of two consecutive nanotubes is kept at the value of 0.34 nm, close to the inter-layer distance along the c axis in bulk graphite. The diameter of MWCNTs may range up to 100 nm. On the basis of their electronic structure the SWCNTs can be divided in two groups [1,2,3]: *semiconducting nanotubes*, these have a vanishing density of states (DOS) at the Fermi energy, and *metallic nanotubes*, with a finite DOS at the Fermi energy. The strong relation between the atomic structure and the electronic structure makes it a necessity to investigate isolated nanotubes, to resolve their atomic structure, and to measure the electronic structure of the very same nanotube. The only experimental tool which was suitable for achieving these tasks simultaneously was the scanning tunneling microscope (STM) invented in 1981 by Binnig and Rohrer [4] rewarded by the Nobel Prize for their invention in 1986.

1.2. SCANNING TUNNELING MICROSCOPY

In principle, the concept of an STM is very simple: an atomically sharp, metallic tip is brought within a distance of a few tenths of a nanometer to a conducting surface; due to their quantum mechanical behavior, electrons may tunnel from the tip to the surface and

L.P. Bíró et al. (eds.), Carbon Filaments and Nanotubes: Common Origins, Differing Applications, 219–231.
© 2001 *Kluwer Academic Publishers. Printed in the Netherlands.*

220

vice versa. A comprehensive overview of the models used to describe the tunneling through a one dimensional potential barrier is given by Wiesendanger [5]. Two major conclusions arise from the various models:

1) The transmission T of electrons through a one dimensional potential barrier, depends exponentially on the width s of the barrier:

$$T \propto e^{-2\chi s}. \tag{1}$$

1) The decay rate χ, characterizing the decrease of the probability to find the electron inside the barrier is:

$$\chi = \frac{2\pi[2m(V_0 - E)]^{1/2}}{h}, \tag{2}$$

where V_o is the average height of the barrier, E is the energy of the tunneling electron. Independently of the exact shape of the barrier, the strong exponential dependence of T with the barrier width s and the square root of the effective barrier height, $(V_o - E)^{1/2}$ is typical of tunneling. The exponential dependence makes that the tunneling channel will be very narrow, thus making possible the atomic resolution.

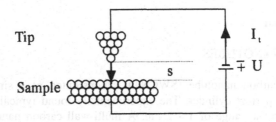

Figure 1. Principle of operation of the STM

Unless a bias V is applied between the tip and the surface the two electron fluxes: surface → tip, and tip → surface, will be equal, and will cancel out each other in equilibrium. When an external bias is applied, depending on the polarity of the bias, one of the tunneling directions is made preferential, therefore a net electronic current can be measured in the circuit, Figure 1. Usually the bias is of the order of 1 V, which yields currents in the 1 nA range. The tunnel current flowing between the tip and the sample when a bias V is applied, will be:

$$I(V) \propto \int_0^{eV} \rho_s(E)\rho_t(E - eV)T(E)dE, \tag{3}$$

where ρ_s is the sample electronic DOS and ρ_t is the DOS of the tip. One may note that formula (3) offers the opportunity to acquire information regarding the DOS of the sample. If the tip DOS is flat and the transmission coefficient T may be taken as constant, than Eq. (3) reduces to:

$$I(V) \propto \int_0^{eV} \rho_s(E)dE.$$ (4)

Than, the quantity $dI/dV(V)$ α $\rho_s(V)$. However, T may be regarded constant only in the limit of small voltages, it gives deviation from this simple dependence at large bias values.

The real case of an STM measurement is different from the simple case of tunneling through a one dimensional potential barrier. First of all, in the case of a sharp tip, one has a three dimensional potential barrier instead of a one dimensional one. The real potential barriers may strongly differ from the rectangular shape assumed in deducing the formulas (1), and (2). The most frequently used model for the interpretation of STM experiments is the model given by Tersoff & Hamann [6]. In this theory the tip is treated as a single *s orbital* with a constant DOS and the tunnel current is proportional to sample DOS at the Fermi energy at vanishingly small bias.

In practical STM instruments, the positioning and scanning of the STM tip is achieved by piezoelectric actuators. The width of the STM gap is controlled by a feedback loop which keeps the value of the tunneling current at a setpoint value selected by the operator. An STM can operate in several regimes, the two most important ones are as follows:

- Topographic (constant current) imaging: the feedback loop is on, the image is generated from the values of the voltage applied to the piezo-actuator to maintain a constant value of the tunneling current. Provided the electronic structure at the sample surface is homogeneous, the topographic profile of the surface will be generated.
- Current-voltage spectroscopy, frequently called scanning tunneling spectroscopy (STS). The scanning, and the feedback loop are switched off, the value of the tunneling gap is fixed, and the bias voltage is ramped from $-U$ to $+U$, and the corresponding current variations are recorded. The function dI/dV gives information about the local DOS of the sample.

2. STM investigation of carbon nanotubes

The first STM experiment proving that atomic resolution is possible on a carbon nanotube was reported by Ge & Sattler [7]. They investigated MWCNTs produced in-situ by the condensation of evaporated carbon on a highly oriented pyrolitic graphite (HOPG) substrate. Superimposed on the atomic lattice, a periodicity of 16 nm was found. This was interpreted as arising from the misorientation of the two outer layers of the MWCNT, analogously with the generation of Moiré patterns known in geometric optic. The STM observation of Moiré patterns on HOPG [8] is well known experimentally, however, no clear theoretical description has been given yet.

The earliest STS measurement on carbon nanotubes was reported by Olk & Heremans [9]. The measurements were carried out in air on MWCNTs grown by the electric arc method transferred onto an Au substrate by ultrasonication in ethanol. Both semiconductor and metallic carbon nanotubes were found. The comparison of measured energy gap and diameter values with gap values predicted by theory [1,2] showed an increasing deviation with decreasing tube diameter. The source of this deviation may be the unavoidable error introduced in the measured diameter by tip/sample convolution effects and by the more complex structure of the tunneling region than in the well known case of a flat, homogeneous sample [10, 11]. These effects can be separated in three classes:

- effects arising from the complexity of the system through which the tunneling takes place,
- effects of pure geometric origin,
- effects arising from the different electronic structure of the nanotube and its support.

In Figure 2 the two cases: a) an STM tip over a flat surface, is compared to b) the case of a tip over a carbon nanotube floating on the Van der Waals potential over a support. While in case a) the current flowing through the STM gap is determined only by R_g, in case b) there are three resistances, two tunneling gap resistances, R_{g1} and R_{g2}, and the resistance of the nanotube itself R_t. Additionally, in case b) there are two interfaces through which the electron can travel only by tunneling. This situation is frequently called resonant tunneling [12].

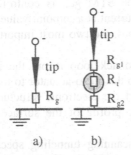
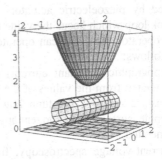

Figure 2. Schematic representation of the STM tip when tunneling directly into the substrate a); and when tunneling through a carbon nanotube b).

Figure 3. Model system used in the tunneling probability calculation. All dimensions are in nm.

The effects arising from the complex structure of the tunneling interface - a nanoscopic object sandwiched between two tunneling gaps – were investigated using a wave packet dynamical computer code [11] for the simulation of the electron tunneling process through a supported carbon nanotube, Figure 3. The quantum mechanical tunneling probability was calculated from the time dependent scattering of a wave packet on the effective potential modeling the system. Figure 4 shows snapshots of the tunneling process, one may note that the time evolution of the tunneling shows significant differences when the tip is situated exactly over the carbon nanotube as compared with the case when the tip is situated over the support. The detailed simulation shows that the nanotube is "charged" during the tunneling, the fraction of the wave packet, which tunneled into the tube, will form a standing wave pattern and the tunneling from the tube into the substrate will be intermittent.

The effects of geometric origin are the tip/sample convolution effects [11, 13]. As a general rule, it can be formulated that in scanning probe microscopy, independently of which object was chosen as tip, always the sharper object (with the smaller radius of curvature) will generate the image. This is illustrated in Figure 5. For an STM tip, the radius of curvature which has to be taken into account is an effective radius of curvature composed from the geometric radius of curvature, to which one has to add the value of the tunneling gap [13]. The value of the tunneling gap may vary from the support to the nanotube.

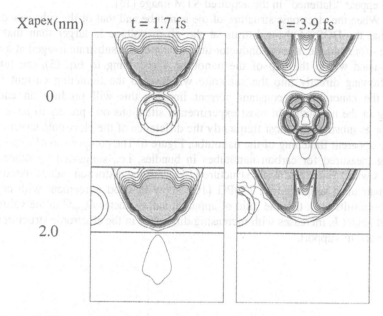

Figure 4. Probability density of the scattered wave packet during the tunneling process through a carbon nanotube (top); and directly into the substrate (bottom). X^{apex} gives the horizontal distance between the apex of the STM tip and the axis of the nanotube, t values indicate the time elapsed since the launching of the wave packet from the bulk of the STM tip [11].

Due to the fact that the typical diameter of a MWCNT is in the range of 10 nm, it is extremely difficult to create experimental conditions in which the apex of the tip may be regarded over a length of 10 nm, as having a negligible width ($d_{tip}/2 < d_t/10$) compared with the diameter of the nanotube. Such a tip would be extremely unstable mechanically. The importance of this effect increases with decreasing tube diameter [13].

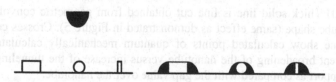

Figure 5. Illustration of distortion arising due to tip-sample convolution effects for objects with different shapes. When the object becomes sharper than the tip, the object will generate the image of the tip.

The effects arising from the difference of the electronic structure of the carbon nanotube and its support will be superimposed on the geometric effects and those arising from the complexity of the tunneling interface. As already discussed, the convolution effects make that the nanotube will have a larger apparent diameter than the geometric one. The existence of the two tunneling gaps will make that the transfer probability of electron through the nanotube will be smaller than in a direct tunneling process from the tip into the substrate, this will yield a smaller tube height than the geometric one. As a result the tube will appear "flattened" in the acquired STM image [13].

When the electronic structure of the nanotube and that of the substrate differ in a way that the DOS of the substrate at the given voltage is larger than that of the nanotube - for example: a semiconductor tube on a metallic substrate imaged at a voltage value situated within the gap of the nanotube – according to Eq. (3), the tunneling current flowing directly into the substrate will exceed the tunneling current flowing through the nanotube. In constant current imaging this will produce an additional flattening of the nanotube. In most experimental situations one prefers to have a good conductor as substrate, so most frequently the difference of the electronic structures will cause the apparent flattening of the nanotube, Figure 6. The comparison of experimental flattening measured for carbon nanotubes in bundles, i.e., supported by other carbon nanotubes with similar electronic structure [14], with the distortion values measured for carbon nanotubes supported on HOPG [10], show in good agreement with computer simulation results [11], that the ratio of apparent half diameter $D_{app}/2$ to the value of the measured height h, increases with increasing difference in the electronic structures of the nanotube and its support.

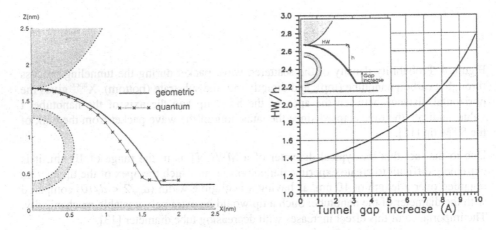

Figure 6. left) Thick solid line is line cut obtained from geometric convolution of tip shape with tube shape (same effect as demonstrated in Figure 5). Crosses connected by thin solid line show calculated points of quantum mechanically calculated line cut. Right) apparent broadening of the nanotube versus increase of the tunneling gap width above the support as compared with the gap value over the nanotube.

Current imaging tunneling spectroscopy measurements (CITS) - in this operation mode, for each pixel point of the image, immediately after the acquiring of the topographic information, the spectroscopic information is acquired, too - of carbon nanotubes on HOPG show that in the vicinity of the Fermi level in a voltage range of   -1.5 to 1.5 V, the electronic structure of large diameter MWCNT (d_t > 50 nm) is practically identical with that of the HOPG, Figure 7. In this case the observed flattening is produced only by the tip/sample convolution and by the second tunneling gap. For the larger diameter tube seen in Figure 7, the ratio $(D_{app}/2)/h$ is 2.81; the geometrically correct ratio, free of convolution effects and neglecting the complexity of the tunneling, should be 0.5. For a SWCNT with identical electronic structure as its support, and a similar radius as the radius of curvature of the STM tip, the simulation [11] gives a ratio of 1.35. This shows that due to the existence of several layers, the tunneling through a MWCNT is more complex.

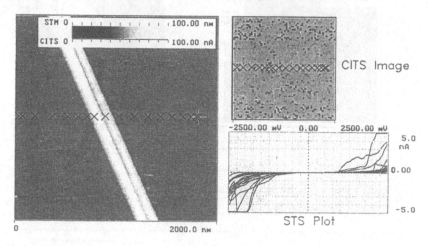

Figure 7. Topographic and CITS image of two multiwall carbon nanotubes on HOPG. The CITS image does not reveal differences in electronic structure. In the STS plot the I-V curves recorded in the points marked by crosses are shown, in the range of -1 to 1 V all curves are practically coincident.

As it was pointed out above, the measured flattening has two components: the broadening due to convolution and the smaller apparent height due to the smaller tunneling gap over the nanotube than over the support. The width of the tunneling gap over the nanotube can decrease to zero, in this case one has point contact imaging [15]. In this imaging regime the tube is deformed in its topmost part due to the pressure exerted by the STM tip in contact with the tube, Figure 8a. However, after reducing slightly the value of the tunneling current from 1.2 nA to 1.0 nA, atomic resolution could be obtained on the same tube, Figure 8c. The periodicity of the triangular lattice along the line AB is 0.25 nm, while the corrugation amplitude is 0.05 nm, in good agreement with typical values for HOPG.

The point contact may have important effects in the STS measurements, too. Experimental STS curves of CNTs [16,17,18,19,20] frequently show some degree of asymmetry with respect to bias voltage polarity. The asymmetries for several scattering potentials, Figure 9, were calculated from the ratio of the wave packet tunneling probabilities for -/+ tip polarity. The simulation results [21] show that while the magnitude of the tunneling current is determined by the nanotube/tip tunnel gap, the asymmetry in the tunneling current is determined by the contact between the free nanotube and the support. Obviously, in practical cases the nanotube/support contact may be induced by the pressure of the STM tip on the topmost part of the nanotube laying on an atomically flat surface or it may be generated by mechanical deformation where the carbon nanotube crosses stepped surfaces, like the metallic electrodes over which carbon nanotubes are commonly placed for electric measurements. A second kind of asymmetry source is the particular tip geometry. The shape and aspect ratio of the

226

active microtip determines the probability current vortices[22] inside the tip and this effect introduces an asymmetry on the negative side of the STS spectrum.

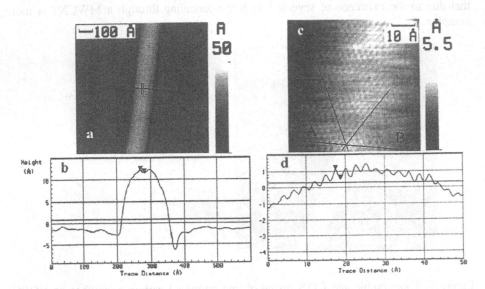

Figure 8. a) Large scale topographic image of a carbon nanotube imaged in point contact regime in the topmost part of the tube, $I_t = 1.2$ nA; $U_b = 100$ mV. b) Line cut along the line marked in a). c) Atomic resolution image on the same tube after reducing the tunneling current, $I_t = 1.0$ nA. d) Line cut along the line marked by AB, the two other lines in the image indicate the other two axes of the triangular lattice, note the corrugation amplitude of 0.05 nm.

There are two distinct classes of nanotubes on which atomic resolution was achieved: a) MWCNTs with diameters of several tens of nanometers [7,10], like in Figure 8c; and b) SWCNTs [18,19] with diameters typically in the 1 nm range. While the MWCNTs show a similar structure like HOPG, i.e., a triangular lattice composed of tunneling current maxima (light features), and sometimes Moiré like superstructures [7]; the SWCNTs show a triangular lattice of minima (dark features), corresponding to the empty centers of the hexagons building up the graphene sheet, Figure 10.

Computer simulation based on a theoretical model using a tight-binding π-electron Hamiltonian [23,24] was successfully used to calculate STM images of single-wall carbon nanotubes. The comparison of calculated STM images with experimental results allowed the identification of certain effects arising from the curvature of the measured object: i) only the topmost atoms of the nanotube will be "measured" by the STM tip (modelled by an s orbital) in their geometrically correct positions, the apparent distance of all other atoms from the topmost one will be inflated by a factor of $1 + s/R$ in the direction transversal to the tube axis, where s is the distance between the STM tip and the nanotube surface, and R is the radius of the nanotube; ii) the atoms all look the same on the nanotube, irrespective of its chirality, but the way in which the bonds between atoms appear in the STM image depends on chirality.

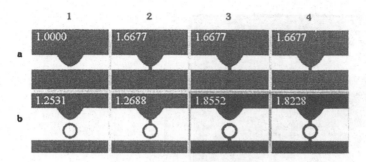

Figure 9. Normalized asymmetries [21] of wave packet transmission probability with respect to bias polarity. Panel *a1* is the reference STM tunnel junction without any point contact and without nanotube. In row *b*, a nanotube is inserted into the junction. In columns *2*, *3*, and *4*, a point contact is placed between the tip and the tube, between the tube and its support, and between both, respectively.

The HOPG-like atomic resolution images of large diameter MWCNTs may be attributed to the reduced curvature of the outer layers in these structures, so that an arrangement like the ABAB stacking of graphite can be achieved. While the SWCNTs behave like a single graphene sheet imaged by STM [25].

The first STS results on SWCNTs were reported simultaneously by two groups [18,19]. The experimental results are in good agreement with earlier theoretical predictions [1, 3, 26]. Metallic and semiconductor carbon nanotubes were found. The typical gap values found for semiconductor nanotubes with diameters in the 1.2 to 1.9 nm range were found to be of the order of 0.5 eV, while the metallic carbon nanotubes with diameters in the range of 1.1 to 2.0 nm had gaps of the order of 1.7 eV [18]. Both groups found that the measured gap values were proportional with $1/d_t$. The differences in the proportionality factors may arise from the differences in taking into account the systematic errors in the determination of the tube diameter due to convolution effects.

Spectroscopic measurements carried out on large diameter MWCNTs [20] showed in the central part of the nanotube a differential conductance close that of HOPG, but the carrier mobilities were found to be lower in the tubes as compared to graphite. Close to the tube ends capped by fullerene-like hemispheres a different electronic structure was found, which was attributed to the presence of pentagons.

A particularly interesting and relatively less investigated case is the case of nanotubes with a few walls. In Figure 11, the CITS image of bundles of single and double wall nanotubes grown by the decomposition of C_{60} in the presence of transition metals [27, 28] are shown. The typical tube diameters in the bundles are in the range of 1.0 – 2.8 nm. STS measurements indicate both metallic and semiconductor nanotubes, some tubes exhibit a DOS structure that may be interpreted as the combination of the individual DOS corresponding to two tubes. Earlier theoretical calculations showed that in the case of a MWCNT the total DOS will be the sum of the individual DOS of the tubes constituting the multishell structure [29].

228

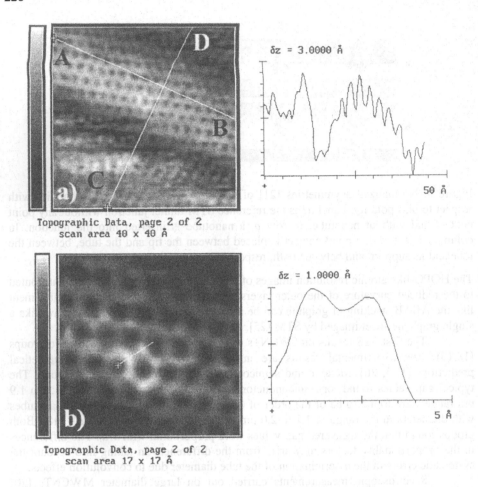

Figure 10. Atomic resolution STM image of single-wall carbon nanotube. a) Gray scale at left corresponds to 0.35 nm; AB is parallel with the tube axis, the line cut along CD is shown at right hand side. b) Detail of the tube seen in a), gray scale at left corresponds to 0.26 nm; as seen in the line cut (shown on right hand side of the image) taken along the marked line in b), the measured distance of the C atoms in the hexagon is 0.148 nm, while the amplitude of the corrugation from the empty center of the hexagon to the tunneling current maximum corresponding to the C atom is 0.8 nm.

Beyond its capabilities to characterize carbon nanotubes, STM is also suitable for being used as a "tool" to modify carbon nanotubes [30] and it is a powerful instrument in searching for new carbon nanostructures like the Y-branching of carbon nanotubes [31] and tightly wound, coiled single-wall carbon nanotubes [32].

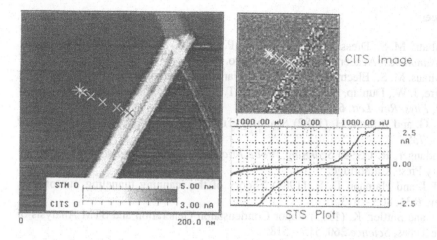

Figure 11. Topographic and CITS STM image of bundles of single and double wall carbon nanotubes. The CITS image is displayed at - 1.0 V. The spectra shown in the STS plot were taken in the points indicated by crosses. One may note the clear difference between the I-V curve of the metallic nanotube and the several, overlapping spectra corresponding to the HOPG.

3. Summary

Scanning tunneling microscopy to date is the only tool suitable for imaging individual carbon nanotubes, or related carbon nanostructures and in the same time to perform spectroscopic measurements on the imaged nano-object. This makes it extremely attractive. However, the way by which the scanning probe microscopy images are generated differs drastically from other imaging techniques like optical or electron microscopy. Therefore cautious image analysis and modeling are most helpful in separating useful information from eventual artifacts and factors that may influence the measurement. The imaging of supported three dimensional objects may strongly differ from the case of flat, homogenous, single crystalline surfaces, effects arising from the complexity of the tunneling interface, or from the Van der Waals, and mechanical interaction of tip and sample cannot be neglected.

STM is essentially a three dimensional imaging technique, therefore, it is strongly recommended that published STM images to be accompanied by gray scales and/or line cuts, to conserve the full amount of the experimentally acquired three dimensional information.

Acknowledgement

This work has been partly funded by OTKA grants T 030435, T025928, and TéT grant D-44/98 in Hungary, financial support from the Belgian OSTC and FNRS are gratefully acknowledge.

230

References

[1] Dresselhaus, M. S., Dresselhaus, G., Ecklund P. C. (1996) *Science of Fullerenes and Carbon Nanotubes*, Academic Press, San Diego.

[2] Dresselhaus, M. S., Electronic Properties of Nanotubes, *This Volume*.

[3] Mintmire, J. W., Dunlap, B. I., and White, C. T. (1992) Are Fullerene Tubules Metalic?, *Phys. Rev. Lett.* **68**, 631 - 634.

[4] Binnig, G. and Rohrer, H. (1982) Scanning Tunneling Microscopy, *Helv. Phys. Acta* **55**, 726 – 735.

[5] Wiesendanger, R. (1994) *Scanning Probe Microscopy and Spectroscopy*, Cambridge University Press, Cambridge.

[6] Tersoff J. and Hamann D. R. (1985) Theory of the scanning tunneling microscope, *Phys, Rev. B* **31**, 805 – 813.

[7] Ge, M. and Sattler, K. (1993) Vapor Condenstaion Generation and STM Analysis of Fullerene Tubes, *Science* **260**, 515 – 518.

[8] Jxhie, J., Sattler, K., Ge, M., Verkateswaran, N. Giant and supergiant lattices on graphite, *Phys Rev. B* **47**, 15835 – 15841.

[9] Olk, Ch. H., Heremans, J. P. (1994) Scanning tunneling spectroscopy of carbon nanotubes, *J. Mater. Res.* **9**, 259 – 262.

[10] Biró, L. P., Gyulai. J., Lambin, Ph., B.Nagy, J., Lazraescu, S., Márk, G. I., Fonseca, A., Surján, P., R., Szekeres, Zs., Thiry, P. A., Lucas, A. A. (1998) Scanning tunneling microscopy (STM) imaging of carbon nanotubes, *Carbon* **36**, 689 – 696.

[11] Márk, G. I., Biró, L. P., Gyulai, J. (1998) Simulation of STM images of three-dimensional surfaces and comparison with experimental data: Carbon nanotubes, *Phys. Rev. B* **58**, 12645 – 12648.

[12] Ref . [5], pp. 30 - 34.

[13] Biró, L. P., Lazarescu, S., Lambin, Ph., Thiry, P. A., Fonseca, A., B.Nagy, J., Lucas, A. A., (1997) Scanning tunneling microscope investigation of carbon nanotubes produced by catalytic decomposition of acetylene, *Phys. Rev. B* **56**, 12490 – 12498.

[14] Biró, L. P., B.Nagy, J., Lambin, Ph., Lazarescu, S., Fonseca, A., Thiry, P. A., Lucas, A. A., (1998) Scanning tunneling microscopy of carbon nanoyubes. Beyond the image, in H. Kuzmany, J. Fink, M. Mehring and S. Roth (eds.), *Molecular Nanostructures*, World Scientific, Singapore pp. 419 – 422.

[15] Agrait, N., Rodrigo, J. G., and Vieira, S., (1992) On the transition from tunneling regime to point contact: graphite, *Ultramicroscopy* **42 – 44**, 177 - 183.

[16] Hassanien, A., Tokumoto, M., Kumazawa, Y., Kataura, H., Maniwa, Y., Suzuki, S., and Achiba, Y., (1998) Atomic structure and electronic properties of single-wall carbon nanotubes probed by scanning tunneling microscope at room temperature, *Appl. Phys. Lett.* **81**, 3839 - 3841.

[17] Biró, L. P., Thiry, P. A., Lambin, Ph., Journet, C., Bernier, P., and A. A. Lucas, (1998) Influence of tunneling voltage on the imaging of carbon nanotube rafts by scanning tunneling microscopy, *Appl. Phys. Lett.* **73**, 3680 - 3682.

[18] Wildöer, J. W., Venema, L. C., Rinzler, G. R., Smalley, R. E., Dekker, C. (1998) Electronic structure of atomically resolved carbon nanotubes, *Nature* **391**, 59 - 62

[19] Odom, T. W., Huang, J-L., Kim. Ph., Lieber, Ch. M., (1998) Atomic structure and electronic properties of single-walled carbon nanotubes, *Nature* **391**, 62 - 64.

[20] Carroll, D. L., Redlich, P., Ajayan, P. M., Charlier, J. C., Blasé, X., De Vita, A., and Car, R, (1997) Electronic structure and localized states at carbon nanotube tips, *Phys. Rev. Lett.* **78**, 2811 - 2814.

[21] Márk, G. I., Biró, L. P., Gyulai, J., Thiry, P. A., Lambin, Ph. (1999) The use of computer simulation to investigate tip shape and point contact effects during scanning tunneling microscopy of supported nanostructures, in H. Kuzmany, J. Fink, M. Mehring and S. Roth (eds.), *Electronic Properties of Novel Materials – Science and Technology of Molecular Nanostructures*, American Institute of Physics, Melville, pp.323 – 327.

[22] Márk, G. I., Biró, L. P., Gyulai, J., Thiry, P. A., Lucas, A. A., and Lambin, Ph. (2000) Simulation of scanning tunneling spectroscopy of supported carbon nanotubes, *Phys. Rev. B* **62**, (in press).

[23] Meunier, V., and Lambin, Ph. (1998) Tight-binding computation of the STM image of carbon nanotubes, *Phys. Rev. Lett.* **81**, 5588- 5591.

[24] Lambin, Ph., Interpretation of the STM Images of Carbon Nanotubes, *This Volume*.

[25] Olk, C. H., Heremans, J., Dresselahaus, M. S., Speck, J. S., Nicholls, J. T. (1990) Scanning tunneling microscopy of a stage-1 $CuCl_{12}$ graphite intercalation compound, *Phys. Rev. B* **42**, 7524 - 7529.

[26] Charlier, J.-C., and Lambin, Ph. (1998) Electronic structure of carbon nanotubes with chiral symmetry, *Phys. Rev. B* **57**, R15037 – R15039.

[27] Biró, L. P., Ehlich, R., Tellgmann, R., Gromov, A., Krawez, N., Tschaplyguine, M., Pohl, M.-M., Véretsy, Z., Horváth, Z. E., Campbell, E. E. B. (1999) Growth of carbon nanotubes by fullerene decomposition in the presence of transition metals, *Chem. Phys. Lett.* **306**, 155 – 162.

[28] Biró, L. P., Ehlich, R., (submitted to Appl. Phys. Lett.) Room temperature growth of single and multi wall carbon nanotubes by [60]fullerene decomposition in the presence of transition metals.

[29] Lambin, Ph., Charlier, J.-C., Michenaud, J.-P., (1994) Electronic structure of coaxial carbon tubules in H. Kuzmany, J. Fink, M. Mehring, S. Roth (eds.), Progress in Fullerene Research, World Scientific, Singapore, pp. 131 – 134.

[30] Venema, L. C., Wildöer, J. W. G., Temminck Tuinstra, H. L. J., Dekker, C., Rinzler, A. G., Smalley, R., E., (1997) Length control of individual carbon nanotubes by nanostructuring with a scanning tunneling microscope, *Appl. Phys. Lett.* **71**, 2629 - 2631

[31] Nagy, P., Ehlich, R., Biró, L. P., Gyulai, J., (2000) Y-branching of single walled carbon nanotubes, *Appl. Phys. A* **70**. 481 - 483

[32] Biró. L. P., Lazarescu, S. D., Thiry, P. A., Fonseca, A., B.Nagy, J., Lucas, A. A., Lambin Ph. (2000) Scanning tunneling microscopy observation of tightly wound, single-wall coiled carbon nanotubes, *Europhys. Lett.* **50**, 494 - 500

INTERPRETATION OF THE STM IMAGES OF CARBON NANOTUBES

PH. LAMBIN

Département de Physique, Facultés Universitaires N.D.P.
61 Rue de Bruxelles, B 5000 Namur, Belgium

AND

V. MEUNIER

Department of Physics, North Carolina State University,
Raleigh NC 27695, USA.

Abstract. A theoretical modeling of scanning tunneling microscopy of the carbon nanotubes is presented. This theory is based on the standard perturbation description of elastic tunneling within a tight-binding description of the carbon π-electrons. The tip is treated as a single atom with an s wavefunction. Several simulations of topographic STM images of perfect single-wall nanotubes are illustrated with reference to available experimental data and other results from *ab-initio* calculations. The effects on the STM image of substituting a B atom for C in the (10,10) nanotube are also investigated.

1. Introduction

During the past few years, a number of exciting results have been obtained from scanning tunneling microscopy (STM) and spectroscopy (STS) on isolated single-wall carbon nanotubes [1–3]. STS has provided a beautiful confirmation of the theoretical prediction that a nanotube can be either metallic or semiconducting, depending on its two wrapping indices n and m. The precise determination of these two indices with STM remains a challenge, which demands the achievement of atomic resolution. Still, the carbon atoms cannot be resolved individually. In the topographical images, corrugation hollows appear at the centers of the hexagons of the honeycomb structure, defining a triangular lattice with parameter 0.246 nm. Recent

233

L.P. Biró et al. (eds.), Carbon Filaments and Nanotubes: Common Origins, Differing Applications, 233–244.
© 2001 *Kluwer Academic Publishers. Printed in the Netherlands.*

tight-binding [4] and *ab-initio* [5] calculations of the STM image support this interpretation. Around the hexagonal hollows, there is the network of protruding C-C σ bonds. Not all the bonds are equivalent, however. In the STM image, some bonds are enhanced with respect to the others, which break the honeycomb symmetry [6]. The angles between the σ bonds and the nanotube axis are related to the chiral angle of the nanotube. However, a geometric distortion of the lattice needs to be taken into account for a proper determination of the helicity of a chiral nanotube [7]. Even the measurement of the tube diameter with STM is not easy due to a tip shape convolution effect [8,9]. Some of these aspects, which are basic for a correct interpretation of the STM images of carbon nanotubes, are reviewed in the following sections.

2. STM theory

A formulation of the tunnel current in an STM is easy to derive when treating the coupling interaction v between tip (t) and sample (s) in first-order perturbation. The expression obtained for the elastic tunneling involves a sum over all states α and β of the tip and sample, respectively, located at the same energy E:

$$I = \frac{2\pi e}{\hbar} \int_{-\infty}^{+\infty} dE \left[f_t(E) - f_s(E) \right] \sum_{\alpha,\beta} |\langle \alpha | v | \beta \rangle|^2 \delta(E - E_\alpha) \delta(E - E_\beta) \quad (1)$$

where $f_t(E)$ and $f_s(E)$ are the occupation numbers of the two systems. In tight-binding, assuming one orbital per atom to simplify, the LCAO expansions of the electronic states write

$$|\alpha\rangle = \sum_{I \in t} \chi_I^\alpha |\eta_I\rangle \text{ and } |\beta\rangle = \sum_{J \in s} \psi_J^\beta |\theta_J\rangle \quad (2)$$

where η_I and θ_J are the atomic orbitals on the sites I and J of the tip and sample, respectively. In this basis set, $v_{IJ} = \langle \chi_I | v | \theta_J \rangle$ denotes the tight-binding coupling elements (see fig. 1(a)).

Let us introduce the quantity [10]

$$n_{JJ'}^s(E) = \sum_{\beta \in s} \psi_J^{\beta*} \delta(E - E_\beta) \psi_{J'}^\beta = \frac{-1}{\pi} \text{Im } G_{JJ'}^s(E + i0) . \quad (3)$$

In the last equality, which is valid with time reversal symmetry, $G_{JJ'}^s$ is an element of the Green's function of the sample. A diagonal element $n_{JJ}^s(E)$ of its imaginary part is the local density of states on the site J. A similar notation, $n_{II'}^t(E)$, is defined on the tip side. The current at 0 K deduced

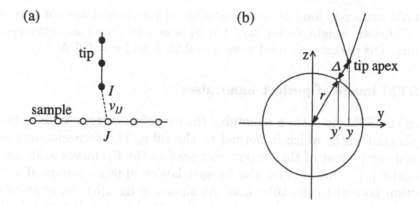

Figure 1. (a) Schematic representation of the tip–sample interface in STM. (b) When the STM current flows along the shortest path between the tip apex and a nanotube, the tip coordinate y is larger than that y' of the imaged atom.

from eq. 1 is

$$I = (2\pi)^2 \frac{e}{h} \int_{E_F^s - eV}^{E_F^s} dE \sum_{I, I' \in t} \sum_{J, J' \in s} v_{IJ} v_{I'J'}^* n_{II'}^t (E_F^t - E_F^s + eV_t + E) n_{JJ'}^s (E)$$

(4)

where V_t is the tip–sample bias potential, the E_F's are the Fermi levels of the unperturbed electrodes.

The STM images of carbon nanotubes presented in the next sections were computed from eq. 4 by considering just one atom I at the tip apex, with an s-wave orbital like in Tersoff-Hamann theory. The corresponding tip density of states, $n_{II}^t(E)$, was taken to be a Gaussian function of 6 eV full-width at half maximum. On the sample side, the carbon nanotubes were described by the usual π-electron tight-binding Hamiltonian [11], with first-neighbor hopping interaction $\gamma_0 = -2.75\,eV$, and the Green's function elements (eq. 3) were computed with the recursion technique. This technique, originally designed for the calculation of diagonal elements of the Green function [12], also gives access to non-diagonal elements [13].

The tip–sample coupling interactions were taken as sp hopping terms with the following form [4]

$$v_{IJ} = v_0\, w_{IJ}\, e^{-d_{IJ}/\lambda} \cos\theta_{IJ}$$

(5)

$$w_{IJ} = e^{-ad_{IJ}^2} / \sum_{J'} e^{-ad_{IJ'}^2}$$

(6)

where d_{IJ} is the distance between the tip atom I and the sample atom J, θ_{IJ} is the angle between the π orbital lobe on site J and the IJ direction. The coefficient v_0 is presumably small as compared to γ_0, but its actual value does not matter as long as absolute values of the current are not required. The Gaussian weight factor w_{IJ} (eq. 6) was introduced for convergence reasons. The parameters used were $\lambda = 0.85$ Å and $a = 0.6$ Å^{-2}.

3. STM image of perfect nanotubes

When the STM tip scans a nanotube, the tunneling current tends to follow the shortest path, which is normal to the tube. The corresponding non-vertical component of the current increases as the tip moves aside to the nanotube [14]. Due to this, the imaged lattice appears stretched in the direction normal to the tube axis. As shown in fig. 1(b), an atom of the nanotube that projects at location (x', y') on the horizontal plane is imaged when the tip has horizontal coordinates $x = x'$ and $y = y'(R + \Delta)/R$ when it is assumed that the tunneling current is radial. Here x is the coordinate parallel to the axis (perpendicular to the drawing), y is normal to the axis, R is the tube radius, and Δ is the tip–nanotube distance. The projected lattice is therefore stretched by the factor $(R + \Delta)/R$ in the y direction [4]. Due to this distortion, the angles between the three zig-zag chains of C atoms measured in an STM image deviate from $\pi/6$. The distortion can be corrected for by squeezing the y coordinates in such a way as to restore the correct angles [7].

Computed STM images of the topmost part of three metallic nanotubes are shown in fig. 2. In each case, the tip apex was positioned 0.5 nm above an atom. The local normal to the nanotube was taken as the z direction. The tip was moved along the two horizontal directions x and y, and its vertical position z was adjusted so as to reproduce the current computed at the initial position. Each image is a three-dimensional representation of the height $z(x, y)$ of the tip at constant current. The sharp corrugation hollows in the images correspond to the centers of the honeycomb hexagons. In general, there is a visible anisotropy of the C-C bonds, except in the armchair configuration. All these results are confirmed by calculations based on the Tersoff-Hamann theory with density functional theory [5].

Another representation of the same STM data as in fig. 2 is shown in fig. 3. Here, each image is a gray-scale two-dimensional map of the distance $\rho = R + \Delta$ of the tip apex to the nanotube axis, represented against the two horizontal coordinates x and y of the tip (see fig. 1(b)). In each map, the abscissa x is measured along the tube axis, and the ordinate y is perpendicular to it. We decided to represent the radial distance ρ of the tip rather its vertical position z in order to emphasize the atomic corrugation of the tube. Near the $y = 0$ median line, $\rho = \sqrt{y^2 + z^2}$ is close to the z coordinate as measured experimentally. The sharp corrugation hollows form dark spots at the centers of the hexagons. In the armchair (10,10)

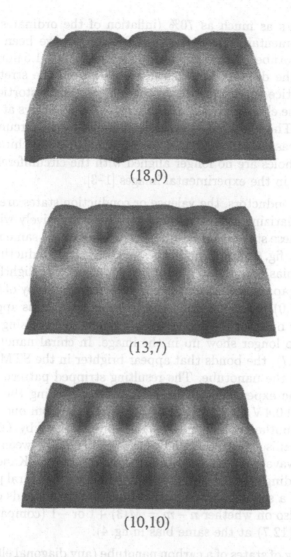

(18,0)

(13,7)

(10,10)

Figure 2. Three-dimensional constant-current images of the (18,0), (13,7), and (10,10) nanotubes. According to the simple rule $n-m = M(3)$, all the three nanotubes are metallic. The tip potential used in the calculations was 0.2 V. The image size is 1.2×0.8 nm^2.

nanotube, the largest protrusions are realized on the atoms, all the bonds look the same. In the zig-zag (18,0) nanotube, the largest protrusions are realized on the bonds parallel to the axis. There is a saddle point at the center of the inclined bonds. The image of the (13,7) chiral nanotube is intermediate between the two other ones: the individual atoms can still be discerned, like in the armchair nanotube, and one kind of bonds protrudes more than the others, like with the zig-zag geometry.

In the present calculations, the geometric image distortion discussed above represents as much as 70% (inflation of the ordinates by a factor of 1.7). Experimentally, distortions of 20 – 60 % have been observed on single-wall nanotubes with diameter in the range 1.2 – 1.5 nm [7]. In fig. 3, the effects of the distortion are clearly visible from the stretching of the honeycomb lattice superimposed on the images. The distortion is also responsible for the elongated shape of the corrugation hollows at the center of the hexagons. These elongated holes are parallel to the circumferential direction in the case of armchair and zig-zag nanotubes. In chiral nanotubes, the elongated holes are no longer aligned with the circumference, which is often observed in the experimental images [1–3].

With semiconductors, the valence or conduction states are explored selectively by polarizing the STM tip positively or negatively with respect to the sample. These states lead to different pictures of the same nanotube. As an illustration, fig. 4 shows the images of three semiconducting nanotubes computed for bias potentials equal to -0.4 and $+0.4$ V, slightly larger than half the band gap. In each case, there is a strong anisotropy of the bonds. In the zig-zag (17,0) nanotube, the bonds parallel to the axis appear stronger than the other ones for a positive tip potential. By reversing the polarity, these bonds no longer show up in the image. In chiral nanotubes such as (11,7) and (12,7), the bonds that appear brighter in the STM images form spirals around the nanotube. The resulting stripped pattern is commonly observed in the experimental images [15]. By switching the tip potential from -0.4 to $+0.4$ V, the protruding bonds change from one zig-zag chain of atoms to another, which rotates the spiral stripes by $60°$ in the image [16]. This anisotropy is related to the differences between bonding and anti-bonding wavefunctions, as demonstrated recently by Kane and Mele [6] from tight-binding calculations. The handedness of the spiral pattern in the STM image of a semiconducting nanotube not only depends on the sign of the bias but also on whether $n - m = M(3) +1$ or -1 (compare the images of (11,7) and (12,7) at the same bias in fig. 4).

The density of states of a carbon nanotube (any diagonal element $n_{JJ}(E)$, see eq. 3) is symmetric about the Fermi energy. Therefore, the asymmetry of the STM image of a semiconducting nanotube upon reversing bias must come from the non-diagonal elements of the Green function. In particular, the elements $n_{JJ'}$ between two first-neighbor sites are important because they condition the appearance of the C-C bonds in the STM image. These elements are plotted against energy in fig. 5 for three zig-zag nanotubes. They are odd functions of $E - E_F$. Consider first the (17,0) semiconductor. With a positive tip potential, the electrons tunnel from the sample to the tip, which means that the *occupied* states of the nanotube are probed. Near the top of the valence band, in particular, $n_{JJ'}$ is large and positive for the bonds parallel to the axis (solid curve fig. 5). That $n_{JJ'}$ is positive indicates

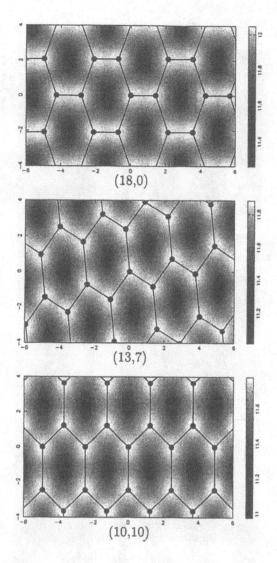

Figure 3. Gray-scale representation of the radial distance $\rho(x,y)$ at constant current of the STM tip apex above the topmost part of the (18,0), (13,7), and (10,10) nanotubes (same conditions as in fig. 2). All coordinates are in Å.

a bonding character [17]. As stated by eq. 4, a positive $n_{JJ'}$ element adds a positive contribution to the STM current when the tip is above a parallel JJ' bond. Still near the HOMO state, $n_{JJ'}$ is negative for the inclined bonds (dashed curve in fig. 5). These bonds have an antibonding character, which leads to a lowering of the current above them. All together, these results explain why the parallel bonds protrude more than the inclined ones in the STM image of (17,0) when $V_t = +0.4$ (fig. 4). At the bottom of the

240

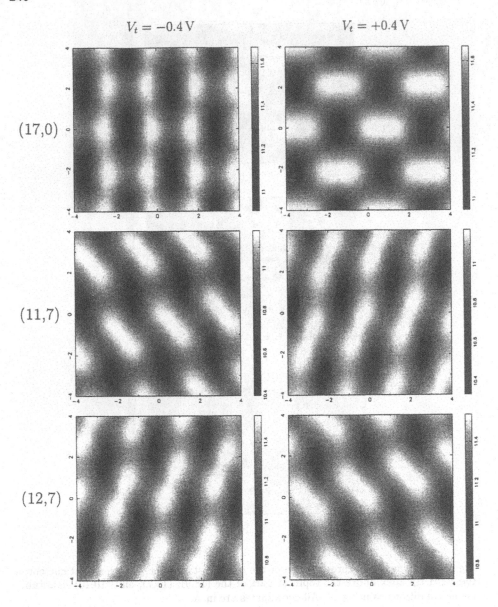

Figure 4. Same representation as in fig. 3 for the three semiconducting nanotubes (17,0), (11,7) and (12,7). The bias potential is 0.4 V, with opposite polarities. On the left-hand side figures, the tip is negative with respect to the sample. It is positive on the right-hand side.

conduction band, the signs of the $n_{JJ'}$ elements are inverted, and the contrast of the corresponding bonds changes in the STM image by reversing the bias. In the (16,0) nanotube, the sign of the $n_{JJ'}$ elements is opposite to that found in (17,0). The bonding bonds become antibonding and *vice-versa*, at least near the top of the valence band and near the bottom of the conduction band. One can therefore predict that the STM image of (16,0) for a given bias looks the same as that of (17,0) for the opposite bias. This effect, which differentiates the semiconducting nanotube according to $n - m$ = $M(3) \pm 1$, was illustrated for the pair of nanotubes (11,7) and (12,7) in fig. 4.

Metallic nanotubes are characterized by a plateau of low-level, constant density of states centered about the Fermi level. This plateau is built from two bands, with nearly linear and opposite slopes, that cross each other at the Fermi energy. Fig. 5 shows that the first-neighbor elements $n_{JJ'}$ are small all along the interval $-0.9 < E < +0.9$ eV corresponding to the metallic plateau of the (18,0) nanotube. The smallness of these elements compared to the density of states in the plateau (0.013 eV^{-1} per atom and per spin for (18,0)) explains why the bias potential influences the STM image of a metallic nanotube much less than that of a semiconductor.

4. STM images of nanotubes with a defect

A localized defect (impurity, vacancy, Stone-Wales defect) may have important effects on the STM image of a nanotube due to the scattering of the Bloch waves it produces [6]. A traveling Bloch wave is indeed partly reflected back by a defect, which, in one dimension, has important consequences on the electronic structure of the system. The interferences between the incident and the reflected waves lead to a stationary pattern that may be observable in STM with a small bias, or more directly in STS since this latter technique gives access to the density of sates. In a metallic sample, the reflection of a Fermi state by a defect gives rise to a so-called $2k_F$ oscillation [18], especially when the reflection coefficient is close to one. This oscillation of the density of states is noticeable in the case of an armchair nanotube, because the Fermi wavelength λ_F is three times the Bravais period a, which represents 0.75 nm. By comparison, other metallic nanotubes have a very long λ_F, more difficult to observe. Specifically, oscillations having the 0.75-nm period have been detected by STS in a short-length armchair nanotube [19]. Here the standing wave is due to the finite length of the system [5, 17]. Occasionally also, a $\sqrt{3} \times \sqrt{3}$ superstructure has been observed in the STM image, which could be the due to $2k_F$ oscillations produced by a defect in an armchair nanotube [16].

Calculations of the STM images of nanotubes containing topological defects such as an ending cap, a pentagon-heptagon pair, and a Stone-Wales rotation bond can be found in ref. [20]. In this Section, the case of a substitutional impurity is illustrated. Fig. 6 shows the computed STM

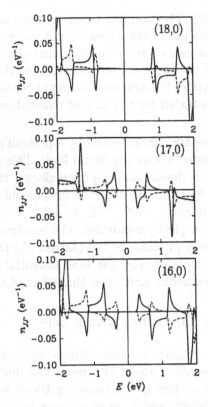

Figure 5. Plot of the $n_{JJ'}(E)$ elements coupling two first neighbor sites in zig-zag nanotubes. The solid curves correspond to the C-C bonds parallel to the axis, the dashed curves correspond the bonds inclined at 60°. The Fermi level is at zero energy.

image of a (10,10) nanotube containing a boron impurity at the center. This impurity was taken into account in the tight-binding Hamiltonian by shifting the on-site energy upwards by 3 eV. The B impurity and its three C neighboring sites form a Y-shaped, enhanced structure at the center of the image that locally breaks the honeycomb symmetry. There are several reasons for not observing density of states oscillations in these images. First, unlike with an ending cap for instance, Fermi waves can arrive from both sides, left and right, of the defect. Second, the reflection coefficient from a single impurity in (10,10) nanotube is very small [21]. And indeed, local density of states calculations indicate that $n_{JJ}(E_F)$ on the impurity site is very close to the one in the perfect tube. However, the density of states on the impurity is no longer constant around E_F but is a linear function of $E - E_F$ with negative slope. Below E_F, the states on the impurity have a larger density than in the perfect nanotube. This result explains why, in the STM image, the contrast between the impurity neighborhood and

$$V_t = -0.3\,\text{V}$$

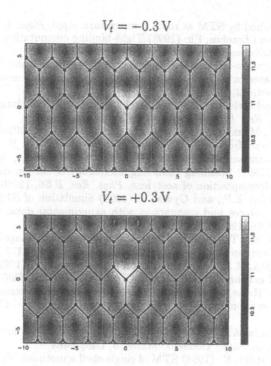

$$V_t = +0.3\,\text{V}$$

Figure 6. Computed constant-current $\rho(x,y)$ STM image of a (10,10) nanotube containing a B impurity at the central site for two bias polarities of the tip.

the rest of the nanotube is the strongest when the tip potential is positive. The reversed situation should arise with a N impurity. There is therefore a possibility to chemically identify B and N impurities from their STM signatures.

Acknowledgments

This work has been partly funded by the interuniversity research project on reduced dimensionality systems (PAI P4/10) of the Belgian Office for Scientific, Cultural, and Technical affairs. The authors acknowledge A. Rubio for helpful discussions.

References

1. Wildoer, J.W.G., Venema, L.C., Rinzler, A.G., Smalley, R.E., and Dekker, C. (1998) Electronic structure of atomically resolved carbon nanotubes, *Nature* **391**, 59-62.
2. Odom, T.W., Huang, J.L., Kim, Ph., and Lieber, Ch.M. (1998) Atomic structure and electronic properties of single-walled carbon nanotubes, *Nature* **391**, 62-64.
3. Hassanien, A., Tokumoto, M., Kumazawa, Y., Kataura, H., Maniwa, Y., Suzuki, S., and Achiba, Y. (1998) Atomic structure and electronic properties of single-wall

244

nanotubes probed by STM at room temperature, *Appl. Phys. Lett.* **73**, 3839-3841.

4. Meunier, V. and Lambin, Ph. (1998) Tight-binding computation of the STM image of carbon nanotubes, *Phys. Rev. Lett.* **81**, 5888-5891.

5. Rubio, A., Sanchez-Portal, D., Artacho, E., Ordejon, P., and Soler, J.M. (1999) Electronic states in a finite carbon nanotube: A one-dimensional quantum box, *Phys. Rev. Lett.* **82**, 3520-3523.

6. Kane, C.L. and Mele, E.J. (1999) Broken symmetry in STM images of carbon nanotubes, *Phys. Rev. B* **59**, R12759-R12762.

7. Venema, L.C., Meunier, V., Lambin, Ph., and Dekker, C. (2000) Atomic structure of carbon nanotubes from scanning tunneling microscopy, *Phys. Rev. B* **61**, 2991-2996.

8. Biró, L.P., Lazarescu, S., Lambin, Ph., Thiry, P.A., Fonseca, A., B.Nagy, J., and Lucas, A.A. (1997) Scanning tunneling microscopy of carbon nanotubes produced by catalytic decomposition of acetylene, *Phys. Rev. B* **56**, 12490-12498.

9. Mark, G.I., Biró, L.P., and Gyulai, J. (1998) Simulation of STM images of three-dimensional surfaces and comparison with experimental data: carbon nanotubes, *Phys. Rev. B* **58**, 12645-12648.

10. Kobayashi, K. and Tsukada, M. (1990) Simulation of STM image based on electronic states of surface/tip system, *J. Vac. Sci. Technol. A* **8**, 170-173.

11. Saito, R., Fujita, M., Dresselhaus, G. and Dresselhaus, M.S. (1992) Electronic structure of chiral graphene tubules, *Appl. Phys. Lett.* **60**, 2204-2206.

12. Haydock, R., Heine, V. and Kelly, M.J. (1975) Electronic structure based on the local atomic environment for tight-binding bands: II, *J. Phys. C: Solid St. Phys.* **8**, 2591-2605.

13. Inoue, J., Okada, A. and Ohta, Y. (1995) A block recursion method with complex wave vectors, *J. Phys.: Condens. Matter* **5**, L465-L468.

14. Ge, M. and Sattler, K. (1994) STM of single-shell nanotubes of carbon, *Appl. Phys. Lett.* **65**, 2284-2286.

15. Clauss, W. (1999) Scanning tunneling microscopy of carbon nanotubes, *Appl. Phys. A* **69**, 275-281.

16. Clauss, W., Bergeron, D.J., Freitag, M., Kane, C.L., Mele, E.J., and Johnson, A.T. (1999) Electron backscattering on single-wall carbon nanotubes observed by STM, *Europhys. Lett.* **47**, 601-607.

17. Meunier, V., Senet, P., and Lambin, Ph. (1999) Scanning tunneling spectroscopy signature of finite-size and connected nanotubes: A tight-binding study, *Phys. Rev. B* **60**, 7792-7795.

18. Briner, B.G. (1997) Looking at electronic wave functions on metal surfaces, *Europhys. News* **28**, 148-150.

19. Venema, L.C., Wildoer, J.W.G., Janssen, J.W., Tans, S.J., Temminck Tuinstra, H.L.J., Kouwenhoven, L.P., and Dekker, C. (1999) Imaging electron wave functions of quantized energy levels in carbon nanotubes, *Science* **283**, 52-55.

20. Lambin, Ph., Meunier, V., and Rubio, A. (2000) Simulation of STM images and STS spectra of carbon nanotubes, in D. Tománek and R.J. Enbody (eds.), *Science and Applications of Carbon Nanotubes*, Kluwer Academic Publisher, New York, pp. 17-34.

21. Kostyrko, T., Bartowiak, M., and Mahan, G.D. (1999) Localization in carbon nanotubes within a tight-binding model, *Phys. Rev. B* **60**, 10735-10738.

MECHANICAL PROPERTIES OF VAPOR GROWN CARBON FIBRES AND VGCF-THERMOPLASTIC COMPOSITES

F.W.J. Van HATTUM*, C.A. BERNARDO**, G.G. TIBBETTS[#]
*Center of Lightweight Structures, TUD-TNO, the Netherlands
**Department of Polymer Engineering, University of Minho, Portugal
[#]Materials and Processes Laboratory, General Motors R&D Center, USA

Abstract

A review of the mechanical properties of Vapor Grown Carbon Fibers and those of VGCF-thermoplastic composites is presented in this chapter. Research into the application of sub-micron VGCF in thermoplastic composites has demonstrated that, by combining advanced processing technologies and VGCF surface treatments, it is possible to obtain mechanical properties comparable to those attained when PAN-based carbon fibers are used as reinforcement.

1. Introduction

There has been a growing interest in Vapor Grown Carbon Fibers (VGCF), namely in their applications in carbon and polymer matrix composites. As a consequence, attention has been focused on the mechanical properties of these fibers, as they determine the overall properties of the composites. These properties, in turn, are related to the internal structure of the fibers.

In this chapter, the work done by a variety of researchers on the mechanical properties of VGCF is reviewed, and the different experimental methods used and the results obtained are compared and discussed. Next, the relations between the mechanical properties, structure, diameter and length of the fibers are examined. Finally, the results reported in the literature on the mechanical properties of VGCF-thermoplastic composites are described and critically evaluated.

As mentioned elsewhere in this book, there are two distinct types of vapor grown carbon fibers, produced by two different methods. The first method typically yields fibers centimeters in length, with diameters in the micrometer range. The mechanical properties reviewed herein were determined with fibers of this type.

For economic reasons, the composites whose properties are discussed below were processed with the second type of fibers. These fibers, which have diameters of about 0.1 μm and lengths from 10 up to 100 μm, are usually referred to as sub-micron Vapor Grown Carbon Fibers.

L.P. Biró et al. (eds.), Carbon Filaments and Nanotubes: Common Origins, Differing Applications, 245–254.

2. Fiber properties

2.1. STIFFNESS AND STRENGTH

Although the mechanical testing of fibers is a tedious task to perform, values of the modulus and tensile strength of VGCF can be found without difficulty in the literature [1-5]. The reported modulus values range from ca. 100 to above 1000 GPa. Tibbetts et al. [2] showed that the stiffness of VGCF depends on their diameters. In a given set of fibers, produced in the same experiment, when the diameters increased from 6 to 32 µm, the moduli decreased from about 300 to 120 GPa. The stiffness of pyrolitic carbon is related to the degree of preferred orientation of the graphitic basal planes. Accordingly, Tibbetts [1] successfully fitted the Ruland model, which relates the stiffness of a graphite fiber and the orientation angle of the graphitic planes, to VGCF stiffness-data, by measuring their orientation parameters. As the only fitting parameter used in the model is independent of the diameter, the author attributed the decrease in stiffness of thicker fibers to a decrease in their graphitic ordering, resulting from a longer deposition of pyrolitic carbon in the thickening period. This is in accordance with the finding that increasing the degree of graphitisation by heating the VGCF at high temperatures also increases their modulus significantly. In one example, the modulus of as-grown fibers more than doubled to about 500 GPa after a treatment at 2200°C [1]. Heat-treated fibers frequently fail in a sword-in-sheath mode. In this mode, a sequence of circumferential cracks along the inner rings allows consecutive cylinders to slide telescopically inside each other, gradually decreasing the load bearing capacity of the fiber [1]. It is clear from these results that the modulus and failure mode of VGCF are linked to their internal structure. Jacobsen et al. [4], using a low-strain vibrating-reed technique, reported Young's moduli of as-grown VGCF that are much higher than the highest ones previously reported [5]. However, it might be wise to reserve judgement on this complicated and indirect method.

Measurements of the tensile strength of VGCF are generally more consistent, varying from 2.5 to 3.5 GPa for fibers with diameters just below 10 µm [1,2,6] although Serp et al. [3] determined somewhat lower values. On the other hand, the dependence of strength on the diameter is in all cases very strong [1-3, 6]. Tibbetts et al. [2] explained this by noting that thicker fibers will have a larger number of flaws per unit length, and hence a greater probability of failure than thinner ones. This hypothesis is supported by the presence of cracks between consecutive cylindrical inner layers that may act as failure initiators. Accordingly, Tibbetts [1] was able to fit a Weibull distribution function to the strength versus diameter data.

2.2. EFFECT OF FIBRE MORPHOLOGY

Due to the production method, vapor grown carbon fibers can present a variety of morphologies. Several of these have been reported in the literature. To a certain extent, the morphology reflects the fibers' internal structure and has an effect on their properties. Van Hattum et al. [7,8] reported a systematic study of as-grown VGCF, covering their morphologies and mechanical properties. They identified seven distinct morphologies in which VGCF are commonly observed. These morphologies are shown in Figure 1.

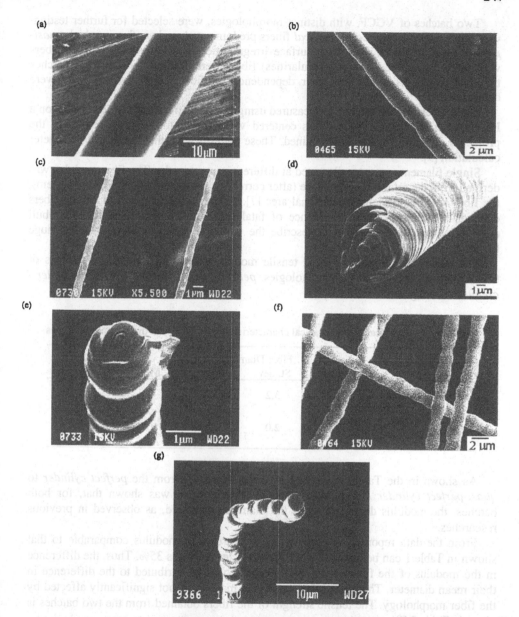

Figure 1. More common morphologies in which vapor grown carbon fibers can be observed: (a) *Perfect cylinder*; (b) *Quasi-perfect cylinder*; (c) *Cylinder with debris*; (d) *Finely-screwed thread* (e) *Palm tree trunk;* (f) *Lathe shaped;* (g) *Crenulated* (from reference [8], by courtesy of Elsevier)

Two batches of VGCF, with distinct morphologies, were selected for further testing. One batch, in which *perfect cylindrical* fibers predominated, and another in which *quasi-perfect* (straight fibers with some surface irregularities) and *crenulated* (curved fibers with more pronounced surface irregularities) fibers were dominant. From these batches the mechanical properties and their dependence on fiber length and diameter were determined.

The diameter of the fibers was measured using a laser diffraction technique based on a He-Ne laser source. Each fiber was centered with respect to the laser beam and the projected diffraction patterns determined. These patterns were then used in the diameter calculation [9].

Single filaments were tensile tested at different gauge lengths. The fiber modulus was derived from the total test compliance (after correction for the compliance of the system), the fiber length and the cross-sectional area [7]. As the tensile strength of carbon fibers depends significantly on the existence of fatal flaws in the tested length, a Weibull distribution function was used to describe the variation of tensile strength with gauge length.

The values of the diameters and tensile moduli determined for the two batches of fibers, with distinct dominant morphologies, *perfect cylinder* to *quasi-perfect cylinder / crenulated*, are given in Table 1 [8].

TABLE 1. Geometrical and mechanical characteristics of VGCF with distinct morphologies

Fiber morphology	Fiber Diameter (μm)				Modulus (GPa)
	Mean	St. dev.	Max.	Min.	
Perfect cylinder	11.7	3.2	20.7	4.4	140
Quasi-perfect cylinder/crenulated	18.3	2.0	24.8	10.7	110

As shown in the Table, the tensile modulus decreases from the *perfect cylinder* to *quasi-perfect cylinder / crenulated* fibers. Furthermore, it was shown that, for both batches, the modulus decreased when the diameter increased, as observed in previous researches.

From the data reported by Tibbetts [1], a decrease in modulus comparable to that shown in Table 1 can be obtained when the diameter increases 35%. Thus, the difference in the modulus of the fibers in the two batches could be attributed to the difference in their mean diameter. This means that the tensile modulus is not significantly affected by the fiber morphology. The tensile strength of the fibers obtained from the two batches is given in Table 2 [8].

The dependence of the strength on fiber length was in both cases well described using a Weibull distribution. Over the whole range of gauge lengths, a decrease in tensile strength from *perfect cylinder* to *quasi-perfect / crenulated* of approximately 45% can be observed. However, as shown in Figure 2, in this case, the effect cannot be attributed to the differences in diameter between the fibers of the two batches.

TABLE 2. Calculated tensile strength of VGCF with distinct morphologies

Fiber morphology	Perfect cylinder		Quasi-perfect cylinder/crenulated	
Fiber length (mm)	Tensile Strength (GPa)		Tensile Strength (GPa)	
	Mean ($\bar{\sigma}$)	St. dev.	Mean ($\bar{\sigma}$)	St. dev.
5	2.9	0.76	1.5	0.48
10	2.2	0.58	1.2	0.38
15	1.9	0.50	1.0	0.34
20	1.7	0.45	0.94	0.31
30	1.4	0.38	0.83	0.27
40	1.3	0.34	0.75	0.24

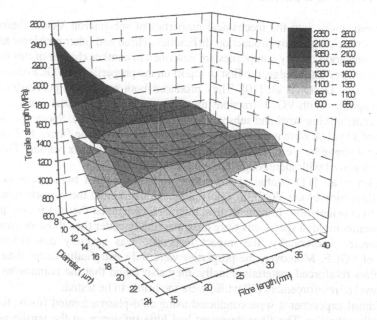

Figure 2. Three-dimensional plot of the fiber strength as a function of gauge length and average fiber diameter, for both *perfect cylinder* and *quasi-perfect cylinder/ crenulated* fibers (from reference [8], by courtesy of Elsevier)

The figure shows the combined effect of both diameter and gauge length on the strength, through a surface fit to the tensile data. The upper and lower surfaces correspond to the *perfect cylinder* and the *quasi-perfect cylinder / crenulated* data, respectively. Over the whole range of fiber diameters and lengths, the former tensile strength values are lower, indicating that the shape of the fibers affects the tensile strength.

Notice also that the strength decreases with increasing fiber diameter and/or fiber length, and that the influence of the diameter on strength decreases as the length increases. Similarly, the variation of tensile strength with the tested length becomes more pronounced for fibers with smaller diameters. It can be concluded that the overall dependence of tensile strength on both diameter and length is qualitatively similar for all fiber shapes, albeit quantitatively dependent on the morphology.

Finally, it should be noted that the values of fiber modulus and strength observed by van Hattum et al. [7,8] are in good agreement with those previously published by other researchers [1-3,6]. Although the strength values of the quasi-perfect/crenulated fibers are in the lower range of the values observed, they are in accordance with the values obtained by Serp et al [3], using similar fibers and test methods.

3. VGCF-thermoplastic composite properties

3.1. PC-VGCF COMPOSITES

Initial investigations into the mechanical properties of sub-micron VGCF-thermoplastic composites were plagued by production difficulties. These and other problems associated with obtaining good quality composites are described in another chapter of this book.

To the authors' knowledge, the first to report on the properties of VGCF-thermoplastic composites was Dasch et al. [10]. They produced both nylon and polycarbonate (PC) matrix composites with VGCF-fractions up to 30 vol. %. The material was compounded in a Brabender mixing bowl and subsequently compression molded. Specimens with fiber fractions of 30 vol.% were reported to be too brittle. By adding the fibers, both the tensile and flexural strengths were nominally increased above the matrix values. The flexural modulus increased with increasing fiber volume fraction.

Carneiro et al. [11] continued from this work, utilizing a twin-screw extruder to enhance dispersion of the VGCF in a polycarbonate matrix and then injection molding to enhance fiber orientation. In spite of this, they also concluded that the tensile properties of the injection-molded specimens were only marginally better than those of unreinforced polycarbonate. Additionally, the impact resistance was severely diminished by the addition of VGCF. Moreover, the properties were only marginally better than those of carbon black reinforced materials. Finally, they also found that the composites with the highest level of reinforcement (30 wt.%) were too brittle to be tested.

Additional experiments were conducted using cold-plasma treated fibers, to increase fiber-matrix adhesion. The fiber treatment had little influence on the tensile properties, whereas the impact resistance improved marginally. As a final attempt to obtain good properties, composites were also injection molded using a special process (SCORRIM), in which shear was applied during injection molding to enhance fiber orientation. Even with this process, the properties, including impact resistance, were only marginally improved. Further research showed the presence of polynuclear aromatic hydrocarbons (PAH) on the fiber surface. As these hydrocarbons can induce chemical stress cracking of the polycarbonate, the matrix can become severely brittle once the fibers are dispersed in it [11]. Removal of the PAH improved the properties of the composite, but not to a significant extent. Since in this research every effort was made to ensure good dispersion and wet-out of the fibers, the cause for the limited properties was thought to be poor adhesion between fibers and matrix.

3.2. PP-VGCF COMPOSITES

Tibbetts *et al.* [12] investigated the use of sub-micron VGCF in polypropylene (PP) and nylon (PA) matrices. The issue of fiber dispersion was dealt with in this research by controlled ball milling of the fibers prior to processing. By 'de-bulking' the VGCF in this way, smaller and easier to infiltrate fiber clumps were obtained, resulting in improved composite properties. The properties of the PP-VGCF composites were better than those of the PA-VGCF composites. This was attributed to the lower melt viscosity of polypropylene at the processing temperature, thus allowing for better infiltration of the fiber clumps.

In addition, the effect of the quality of the fibers' surface on the composite properties was investigated. VGCF were grown under different flow rates and subjected to different surface treatments, resulting in different amounts of polynuclear aromatics on the surface and different surface reactivities. Again the effect on composite properties was more pronounced for PP-VGCF. It was concluded that removing aromatics from the fibers' surface improves the composite properties. However, it does not seem necessary to eliminate them completely, as significant improvements were obtained without removing the final monolayer. Too strong a surface treatment could impair the fiber mechanical properties. The effect of fiber dispersion overshadows any influence of the fiber surface on the composite properties.

The results obtained in this research on PP-VGCF composites are shown in Figures 3 and 4 for tensile strength and modulus, respectively. In the figures, more recent data obtained by van Hattum *et al.* [13] for the same type of composites are also shown. In this research, high-shear equipment was used, thus bypassing the abovementioned infiltration problems. This resulted in composites with strength values similar to those of "conventional" PP-PAN composites, but stiffness values that were somewhat lower.

In any case, the property values shown in the figures are, in relative terms (that is referred to the corresponding ones of the unreinforced matrix), significantly higher than those of polycarbonate. This indicates that there is a better adhesion between polypropylene and vapor grown carbon fibers.

3.3. SURFACE REACTIVITY

From the results of Tibbetts et al. [12] it is evident that when VGCF are well dispersed, their surface characteristics can have a significant effect on the composite properties. Recently, interest in controlling those characteristics has grown, as they determine the interfacial properties of the fibers in a polymer matrix. Darmstadt et al. [14] showed that the surface reactivity of VGCF is lower than that of ex-pitch or ex-PAN fibers. However, it can be increased by oxidation to values approaching that of other carbon fibers.

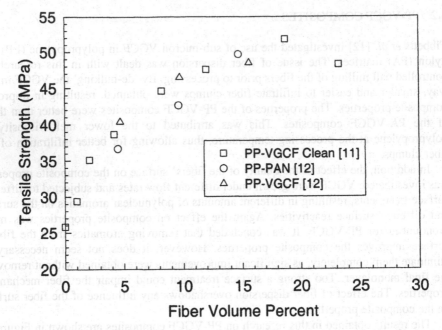

Figure 3. Tensile strength of polypropylene reinforced with various amounts of carbon fibers

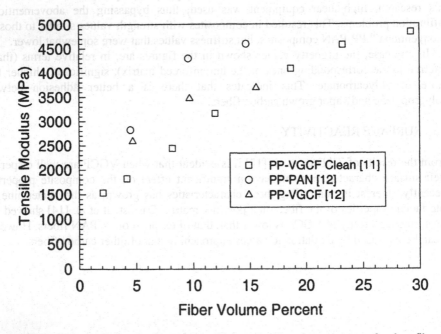

*Figure 4.*Tensile modulus of polypropylene reinforced with various amounts of carbon fibers

The effect of oxidation and graphitisation of the fibers on the properties of PP-VGCF composites were further investigated by Tibbetts et al. [15]. They concluded that less graphitised fibers adhere better to a polypropylene matrix than more graphitised ones. In addition, matrix-matrix interaction (and thus composite properties) may be improved by moderately oxidizing the fibers in either air or CO_2. The effect of different matrix surface treatments on the stiffness and strength of PP-VGCF composites is discussed elsewhere in this book.

Another detailed study of the properties of composites prepared with carefully oxidized VGCF has been recently performed [16]. This study clearly shows that the modification of the surface area and surface energy of the fibers has a profound effect on the tensile strength of PP-VGCF composites. It should be noted that the strength value of approximately 75 MPa mentioned therein is about 50% higher than the best ones reported so far (see Figure 3). Simultaneously, a production technique that allows the fabrication of VGCF with improved surface properties has been developed. This indicates that there is scope for further improvements, making VGCF really competitive with present commercial carbon fibers.

4. Conclusions

A review of the literature results indicates that the modulus of vapor grown carbon fibers depends mainly on the diameter of the fibers. On the other hand, their tensile strength depends not only in diameter and length but also on the fiber morphology.

Research into the application of sub-micron VGCF in thermoplastic composites has shown that it is feasible to process these composites using conventional (high-shear) equipment. The initial marginal improvements in composite properties over the unreinforced material were related to limited fiber-matrix adhesion.

Subsequent research using a polypropylene matrix has demonstrated that it is possible to obtain mechanical properties comparable to those attained when commercial PAN-based carbon fibers are used as reinforcement.

More advanced surface treatment of VGCF has led to major improvements in composite properties. Continuing this current line of research, a combination of advanced processing technologies with surface optimization, will lead to VGCF-thermoplastic composites with properties competitive with those of "conventional" carbon fiber thermoplastic composites.

5. Acknowledgements

This work was supported by the European Economic Community, through the Human Capital and Mobility Programme, under Grant Number CHCRXCT940457. The authors acknowledge the contributions of Drs. Antonio Madroñero (CENIM, Madrid, Spain) and José Luis Figueiredo (Faculdade de Engenharia, Porto, Portugal) to some of the results presented herein.

254

6. References

1. Tibbetts, G.G. (1990) Vapor-grown carbon fibers, in J.L. Figueiredo, C.A. Bernardo, R.T.K. Baker and K.J. Huttinger (eds.), *Carbon Fibers, Filaments and Composites*, Kluwer Academic Publishers, Dordrecht, pp. 73-94.
2. Tibbetts, G.G. and Beetz Jr, C.P. (1987) Mechanical properties of vapour-grown carbon fibres, *Journal of Physics D: Applied Physics* **20**, 292-297.
3. Serp, P., Figueiredo, J.L. and Bernardo, C.A. (1996) Influence of sulfur on the formation of vapor-Grown carbon fibers produced on a substrate using different iron catalyst precursors, in K.R. Palmer, D.T. Marx and M.A. Wright (eds.), *Carbon and Carbonaceous Composite Materials*, World Scientific Publishers, Singapore, pp. 134-147.
4. Jacobsen, R.L., Tritt, T.M., Guth, J.R., Ehrlich, A.C. and Gillespie, D.J. (1995) Mechanical properties of vapor-grown carbon fibers, *Carbon* **33**, 1217-1221.
5. Koyama, T., Endo, M. and Hishiyama, Y. (1974) Structure and properties of graphitized carbon fibers, *Japanese Journal of Applied Physics* **13**, 1933-1939.
6. Madroñero, A. (1994) Strength of short carbon fibres germinated and grown under hydrogen, *Materials: Science and Engineering* **A-185**, L1-4.
7. van Hattum, F.W.J., Benito-Romero, J.M., Madroñero, A. and Bernardo, C.A. (1997) Morphological, mechanical and interfacial analysis of vapour grown carbon fibres, *Carbon* **35**, 1175-1183.
8. van Hattum F.W.J., Serp, Ph., Figueiredo, J.L. and Bernardo, C.A. (1997) The effect of the morphology on the properties of vapour grown carbon fibres, *Carbon* **35**, 860-864.
9. Tzeng, S. S. (1990) Mechanical behavior of carbon fibers and accuracy evaluation of test data, Master Thesis, Rensselaer Polytechnic Institute, New York.
10. Dasch, C.J., Baxter, W.J. and Tibbetts, G.G., Thermoplastic composites using nanometer-size vapor-grown carbon fibres, 21st Biennial Conference on Carbon, Buffalo, USA, 1993, pp. 82-83.
11. Carneiro, O.S., Covas, J.A., Bernardo C.A., Caldeira, G., van Hattum, F.W.J., Ting, J.M., Alig, R.L. and Lake, M.L. (1998) Production and assessment of polycarbonate composites reinforced with vapour grown carbon fibres, *Composites Science and Technology* **58**, 401-407.
12. Tibbetts, G.G. and McHugh, J.J. (1999) Mechanical properties of vapor-grown carbon fiber composites with thermoplastic matrices, *Journal of Materials Research* **14**, 2871-2880.
13. van Hattum, F.W.J., Bernardo, C.A., Finegan, J.G. Tibbetts, G.G., Alig, R.L. and Lake, M.L. (1999) A study of the thermomechanical properties of carbon fiber- polypropylene composites, *Polymer Composites* **20**, 683-688.
14. Darmstadt, H., Roy, C., Kaliaguine, S., Ting, J.-M. and Alig, R.L. (1998) Surface spectroscopic analysis of vapour grown carbon fibres prepared under various conditions, *Carbon* **36**, 1183-1190.
15. Tibbetts, G.G., Finegan, J.C., Glasgow, D.G., Ting, J.-M. and Lake, M.L., Surface treatments for improving the properties of vapor-grown carbon fiber/polypropylene composites, 24th Biennial Conference on Carbon, Charleston, USA, 1999, pp. 58-59.
16. Glasgow, D.G., Lake, M.L., Tarasen, W.L., Tibbetts, G.G. and Finegan, J.C., Effect of surface treatments of carbon nanofibers on polypropylene composite properties, 6th Annual International Conference on Composites Engineering, Orlando, USA, 1999, pp. 837-838.

ATOMIC FORCE MICROSCOPY INVESTIGATION OF CARBON NANOTUBES

L. P. BIRÓ

Research Institute for Technical Physics and Materials Science
H-1525 Budapest, P. O. Box 49, Hungary, e-mail: biro@mfa.kfki.hu

Abstract. The basic principles of operation of atomic force microscopy (AFM), and the image formation mechanisms are discussed. In contact mode AFM the tip/sample interaction used to generate the image and to regulate the feedback loop is based on the deformation of the cantilever pressed against the sample. In tapping mode AFM this role is played by the flow of vibration energy from the piezoelectrically driven, vibrated cantilever into the sample. Superimposed on the significantly more pronounced tip/sample convolution effects than in the case of scanning tunneling microscopy, the two different kinds of interaction may generate different kinds of artifacts (compression of the tube, "snakeing", etc.)

The milestones of the AFM investigation of carbon nanotubes will be reviewed.

Contact, and tapping mode AFM measurements of carbon nanotubes grown in-situ by high energy, heavy ion irradiation will be used to discuss in more detail some particularities of the image formation.

1. Introduction

Beyond their remarkable electronic properties [1] the carbon nanotubes discovered a decade ago by Iijima [2], have exciting mechanical properties, too. The values of Young's modulus in the range of 1 TPa, found by measuring the thermal vibration amplitude of free standing carbon nanotubes inside the transmission electron microscope (TEM) [3], directed towards carbon nanotubes the attention of a new group of researchers, interested in mechanical properties. The range of likely applications of carbon nanotubes was extended to the manufacturing of new composites with superior stiffness/weight ratio than any presently known material.

To get better insight in the behavior of individual nanotubes under mechanical strain a tool was needed which is able to probe the mechanical behavior of one single carbon nanotube. The experimental tool most suitable to achieve this task is the atomic force microscope (AFM) [4], the second in the family of scanning probe microscopes, which followed very quickly the invention of the scanning tunneling microscope (STM).

On the other hand, carbon nanotubes proved to be most valuable in enhancing the capabilities of the AFM: by improving the resolution of the AFM in imaging deep trenches [5], by making it chemically sensitive [6], or allowing a significant increase in

255

L.P. Biró et al. (eds.), Carbon Filaments and Nanotubes: Common Origins, Differing Applications, 255–263.
© 2001 *Kluwer Academic Publishers. Printed in the Netherlands.*

the bit-density, and writing speed when an AFM tip with a carbon nanotube is used as "stylus" [7].

2. Atomic force microscopy

Atomic force microscopy was developed quickly after the invention of STM [4]. In principle, the concept of an AFM is very close to that of an STM [8, 9]. The major difference is given by the kind of the local probe (the tip), and its interaction with the sample. In order to be able to measure insulators, the tunneling had to be given up. It was replaced by another well know atomic interaction: the Van der Waals (VDW) interaction. Generally the VDW forces are usually of attractive character and increase rapidly as the distance between the atoms, or molecules is reduced. The force vs. distance dependence is described by:

$$F_{VDW}(s) \propto -\frac{1}{s^7} \qquad (1)$$

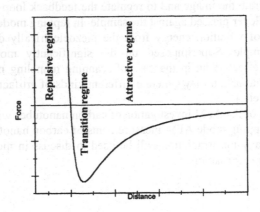

Figure 1. The regions dominated by attractive, or repulsive forces, and the region of transition.

This attractive force will increase untill the equilibrium distance between the two bodies is reached, after which the global force will decrease very abruptly, turning into a repulsive force arising from Coulomb repulsion of the ion cores [10]. The global force, Figure 1, will divide the region in the vicinity of the sample into three parts: a) the region dominated by the attractive force, b) the transition region, and c) the region dominated by the repulsive force. The third region is the useful one for contact mode AFM measurements. As one can see from Figure 1, in this region there is a linear correspondence between the displacement of the tip and the interaction force.

In practice a hard enough, sharp tip, mounted on a thin and elastic cantilever is brought within a distance to the surface to be imaged at which the character of the interaction force switches from attractive to repulsive. Conventionally this is regarded as the onset of the mechanical contact. The cantilever (or the sample) is mounted on a piezo actuator which can achieve the precise positioning (with a precision better than 0.1 nm) of the tip and of the sample with respect to each other. This piezo actuator is responsible for scanning in the plane of the sample, and for regulating the vertical distance between the tip and the sample via a feedback loop.

In contact mode AFM the detection of the forces of the order of 10^{-7} - 10^{-8} N, acting between a tip with a radius of curvature typically in the range of some tens of nm and the measured sample are possible due to the thin cantilever carrying the tip. The bending of the cantilever pressed against the sample is detected using a laser beam shone onto the back side of the cantilever in combination with optical interference, or in combination with a four-quadrant position sensitive detector. A similar feedback loop as in the case of the STM maintains a constant deformation of the cantilever, i.e., a constant force between the tip and sample. The topographic image is generated from the voltage applied to the tip piezo to maintain the interaction force constant.

For soft samples or for samples which may be deformed by the pressure of the AFM tip, and to reduce the frictional contribution to the contact mode AFM images, an operation mode based on vibrated cantilevers was introduced. In this "tapping mode" as it is frequently called a piezo driver vibrates tip with amplitude of the order of 100 nm, at a frequency of several hundreds of kHz. The interaction of the tip with the sample is "switched" from continuous to intermittent: it has a significant magnitude only in one of the maximum elongation positions of the vibrating cantilever. During each interaction period the tip will transfer a certain amount of vibration energy to the sample, therefore the amplitude of the tip in intermittent contact with the sample will be reduced as compared with the amplitude of the free tip. This reduction of the amplitude will be used in the feedback loop to maintain a "constant energy transfer" to the sample during the imaging, i.e., a constant difference between the free and the damped amplitude of the cantilever.

In conclusion, the contact mode AFM image is a constant force image, while the tapping mode AFM image is a constant energy transfer image. These differences may prove important when imaging inhomogeneous samples like carbon nanotubes on a support with different physical properties.

3. Atomic force microscopy of carbon nanotubes

The first AFM measurements on carbon nanotubes were carried out in order to compare TEM, STM and AFM images of carbon nanotubes [11]. The importance of tip/sample convolution effects in imaging carbon nanotubes by AFM was pointed out soon [12]. As already discussed in the case of the STM imaging of carbon nanotubes [9], when the radius of curvature of the sample and that of the tip become comparable, the convolution effects give significant contribution to the image formation.

If the imaged object has a smaller radius of curvature then the AFM tip the role of tip and sample will be interchanged. As the typical radius of curvature of the contact mode AFM tips - usually manufactured of silicon nitride to be hard enough as compared with the sample - is well over 10 nm, in a contact mode measurement, if regarded as infinitely stiff, a single wall carbon nanotube should image the tip. However, as showed by recent calculations, effects arising from the Van de Waals interaction with the surface [13] may produce deformation of SWCNTs and of MWCNTs, Figure 2. As shown by these calculations, the nanotubes may suffer radial and axial deformations due to the interaction with the substrate, or due to the reciprocal pressure exerted on each other by

258

crossing tubes, not to mention the pressure exerted by the tip during a contact mode AFM measurement. These deformation effects may significantly modify the measured height of the nanotube. The larger the diameter of the nanotune and smaller the number of the layers building up the tube, the larger deformations are to be expected. Therefore, tube height values measured in contact mode AFM cannot be automatically converted into diameter values, without careful image analysis. The study of measured height values versus the load applied by the AFM tip, and line cuts taken across the nanotube at different loads, may help in deciding to what extent the measured tube height was influenced by deformation.

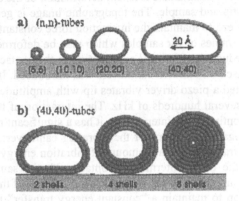

Figure 2. Radial deformations of adsorbed carbon nanotubes calculated using molecular mechanics. The deformations shown are true representation of the results. a) Radial compressions of adsorbed single-walled nanotubes; b) reduction of the deformation with the increase of the number of inner shells [13].

AFM was successfully used to produce the deformation of nanotubes on purpose, for example, measuring in this way the Young's modulus of the nanotubes. Multi wall carbon nanotubes (MWCNT) were fixed on a cleaved MoSe$_2$ surface by depositing a grid of SiO pads [14]. The Young's modulus was measured by bending the nanotubes in the plane of the support applying a point load by the AFM tip. It was found that the Young's modulus has a value of 1.28 ± 0.59 TPa with no dependence on tube diameter.

Large angle (> 90°), repeated bending of MWCNTs on a mica surface was carried out using the AFM tip as a tool [15]. It was found that carbon nanotubes withstand several cycles of bending without undergoing catastrophic failure.

The Young's modulus of MWCNTs, and single wall carbon nanotubes (SWCNT) ropes was measured on a polished alumina ultrafiltration membrane [16]. On such a substrate, nanotubes occasionally lay over the pores suspended over the hollow of the pore. This geometry allows the deformation of the nanotube in a direction perpendicular to the substrate, like for a clamped beam. For arc grown nanotubes the Young modulus was found to be 810 ± 410 GPa.

As measured in this geometry, opposite to graphite, the irradiation of the nanotubes with 2 MeV electrons did not caused the reduction of the Young's modulus [16].

The along-axis electrical transport through carbon nanotubes was measured using nanotubes fixed under an Au contact and using a conductive AFM tip as mobile contact sliding along the tube [17]. Resistivities in the range of 8 – 117 (ohm •m) were found; the nanotube diameters used to calculate the resistivities were taken from transversal line cuts taken through the nanotubes.

A recent overview of various other applications of the AFM in nanotube science is given by Muster and coworkers [18].

A recently developed non-conventional way of producing carbon nanotubes is based on the bombardment of a graphite target by high energy (E > 100 MeV), heavy ions [19]. Where dense, nuclear cascades reach the surface of the target, sputtering craters are produced with diameters in the 1 μm range, and a depth of tens of nm. Frequently, carbon nanotubes emerge from these craters, or are found on the flat bottom of the craters, Figure 3.

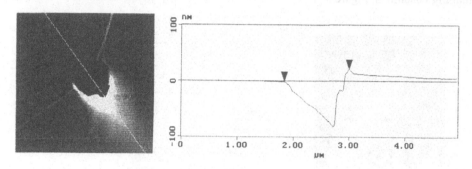

Figure 3. Tapping mode AFM image of a crater produced by sputtering on HOPG irradiated with high energy ions. The line cut shows the depth and width of the crater. Note the two carbon nanotubes emerging towards the upper right and lower left corner.

A major difference of the samples prepared by high energy irradiation, as compared with the usual nanotube samples for AFM is as follows: when produced by ion irradiation, the nanotubes grow in-situ and they do not undergo ultrasonication in some organic solvent. During the deposition process of nanotubes onto highly oriented pirolytic graphite (HOPG) from suspension, before the organic solvent is completely evaporated, the carbon nanotubes have the possibility to orient themselves with respect to the HOPG surface [20, 21]. As showed by recent calculations [22] a carbon nanotube placed on HOPG and rotated around an axis normal to the graphene sheet, will have six positions oriented at 60°, in which the HOPG lattice will be in registry with the hexagonal network of the outer wall of the nanotube. In these positions the absolute value of the interaction energy between the nanotube and the HOPG will have a sharp maximum, producing the "orientational locking" of the nanotube [22]. The interaction energy will be significantly smaller in the intermediate orientations.

When measured with contact mode AFM, except the distortion arising from convolution, and compression of tubes, the only other effect observed was the slight overshoot of the nanotube on the upper terrace of the step crossed by the MWCNT. The nanotube raises above its normal distance from the plateau as shown by the difference between the thin line (measured line cut) and the thick line (plateau), Figure 4. A similar effect was evidenced earlier during STM measurements of single wall carbon nanotubes

260

crossing steps, and confirmed by theoretical calculations [23]. Recent measurements performed using a combined AFM-STM instrument, operated using the AFM feedback while acquiring both the AFM and STM images of a carbon nanotube crossing an Au electrode pad, showed that in the region where the overshooting is expected, the STM current flowing through the nanotube to the Au contact has a minimum [24]. This suggests that electronic properties of the nanotube are modified under the mechanical stress which produces the overshooting. This effect may influence the transport measurements carried out by laying carbon nanotubes over pre-deposited Au contact pads.

When measuring the same samples using a tapping mode AFM, it was frequently found that the nanotubes apparently vibrate in a plane perpendicular to the plane of the substrate. As the other topographic features present in the image were imaged regularly, it can be excluded that the apparent vibration arises due to improper imaging conditions, Figure 5.

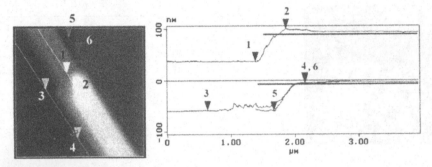

Figure 4. Contact mode AFM image of a carbon nanotube crossing a step. The line cuts taken along the lines marked in the image are shown at the right hand side. The vertical distance of the markers 1-2 is 64 nm, while the vertical distances between markers 3-4, and 5-6 are 53 nm, and 56 nm, respectively.

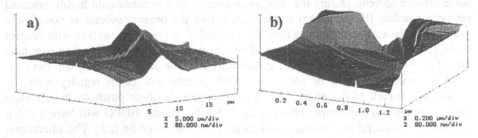

Figure 5. Carbon nanotubes produced by high energy irradiation of graphite. a) Carbon nanotube emerging from a crater and crossing a surface fold produced during the cleavage of HOPG. b) Two carbon nanotubes crossing each other on the flat bottom of the same crater from which emerges the long tube in a). Note the wave-like profile of the tube in a) and of the tube going from left to right in b).

Using a simple computer model to get more insight in the way in which the tapping mode AFM image is generated, and comparing the simulation with the experimental results, it was found that the wave-like profile of the nanotube arises due to the combination of three periodic motions: i) the vertical vibration of the tapping mode AFM tip; ii) the scanning of the AFM tip in the horizontal plane; iii) the oscillation of the nanotube in the (horizontal) plane of the support. The horizontal oscillation is made possible by the lack of orientational locking [22] which minimizes the strength of interaction between the nanotube and its support. The horizontal oscillation is transformed into an apparent vertical vibration, due to the discreteness of the sampling performed by the AFM tip, due to the contribution of convolution effects, and due to the fact that in consecutive scan lines the AFM tip will encounter the nanotube in positions corresponding to different elongations in its horizontal plane of vibration [25].

Another particularity of the image formation in tapping mode AFM is the generation of the image at "constant energy transfer" as already mentioned earlier. The capability of a nanotube floating on the VDW potential to dissipate the vibration energy fed into it by the vibrated AFM tip is smaller than the capability of the bulk HOPG to dissipate the vibration energy. This makes that the piezo actuator will have to bring the AFM tip closer to the nanotube then to the HOPG to achieve the same damping of the free vibration amplitude. This is illustrated in Figure 6. One may note that where the nanotube crosses the elongated surface defect, its measured height is smaller than in the region where the nanotube has a good mechanical contact to the HOPG, i.e., in the regions where it can transfer the vibration energy more effectively to the substrate. This decrease in the apparent height is similar to the decrease, which may alter the real height of a nanotube in tapping mode AFM images. Due to the less effective energy dissipation when the tip taps on the nanotube as compared with the bulk HOPG, the height values of carbon nanotubes measured by tapping mode AFM may be underestimated.

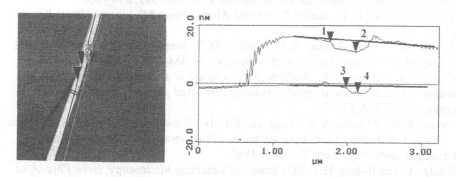

Figure 6. Tapping mode AFM image of a multi wall carbon nanotube crossing a surface defect. The vertical distance between markers 1-2 is 0.34 nm, while between markers 3-4 is 0.18 nm.

4. Summary

Atomic force microscopy to date is the most suitable tool for imaging individual carbon nanotubes, or related carbon nanostructures and in the same time to

262

measure mechanical, and other physical characteristics of a single nano-object. However the way the AFM images are generated differs drastically from other imaging techniques like optical, electron microscopy, or STM. Therefore cautious image analysis and modeling are most helpful in separating useful information from eventual artifacts and factors that may influence the measurement. The imaging of supported three dimensional objects may strongly differ from the case of flat, homogenous surfaces, effects arising from the complexity of the Van der Waals, and mechanical interaction of tip and sample cannot be neglected.

Like STM, AFM is also a three dimensional imaging technique, therefore, it is strongly recommended that in publications the AFM images to be accompanied by gray scales and/or line cuts, to conserve the full amount of the experimentally acquired three dimensional information.

Acknowledgement

This work has been partly funded by OTKA grants T 030435, and T025928 in Hungary. Prof. J. Gyulai, Prof. H. Ryssel, and Dr. L. Frey are gratefully acknowledged for helpful discussions.

References

[1] Dresselhaus, M. S., Electronic Structure and Applications of Carbon Nanotubes, *This volume*

[2] Iijima, S (1991) Helical microtubules of graphitic carbon, *Nature* **354**, 56–58.

[3] Treacy, M. M. J., Ebbesen, T. W., Gibson, J. M. (1996) Exceptionally high Young's modulus observed for individual carbon nanotubes, *Nature* **381**, 678 - 680

[4] Binnig G., Quate, C. F., Gerber Ch. (1986) Atomic Force Microscope, *Phys. Rev. Lett.* **56**, 930 – 933.

[5] Dai, H. Hafner, J. H., Rinzler, A. G., Colbert, D. T., Smalley, R. E. (1996) Nanotubes as nanoprobes in scanning probe microscopy, *Nature* **384**, 147 – 150.

[6] Wong, S. S., Joselevich, E., Woolley, A. T., Cheung C. L., Lieber, Ch. M. (1998) Covalently functionalized nanotubes as nanometre-sized probes in chemistry and biology, *Nature* **394**, 52 – 55.

[7] Cooper. E. B., Manalis, S. R. , Fang, H., Dai, H., Matsumoto, K., Minne, S. C., Hunt, T., Quate, C. F. (1999) Terabit-per-square-inch data storage with the atomic force microscope, *Appl. Phys. Lett.* **75**, 3566 -3568

[8] Binnig, G. and Rohrer, H. (1982) Scanning Tunneling Microscopy, *Helv. Phys. Acta* **55**, 726 – 735.

[9] Biró, L. P., Márk, G. I., Scanning tunneling microscopy investigation of carbon nanotubes, *This volume*

[10] Ciraci, S., Baratoff, A. and Batra, I. P. (1990) Tip-sample interaction effects in scanning-tunneling and atomic-force microscopy, *Phys. Rev. B* **41**, 2763 – 2775.

[11] Gallagher. M. J., Chen, Dong., Jacobsen, B. P., Saris, D., Lamb, L. D., Tinker, F. A., Jiao, J. Huffman, D. R., Seraphin, S., and Zhou, D. (1993) Characterization of carbon nanotubes by scanning probe microscopy *Surf. Sci. Lett.* **281**, L335 – L340

[12] Höper, R., Workman, R. K., Chen, D. Sarid, D., Ydav, T., Withers, J. C., Loufty, R. O. (1994) Single shell carbon nanotubes imaged by atomic force microscopy, *Surface Science* **311**, L731 – L736

[13] Hertel, T., Walkup, R. E., Avouris, Ph., (1998) Deformation of carbon nanotubes by surface van der Waals forces, *Phys. Rev. B* **58**, 13870 - 13873

[14] Wong, E. W., Sheehan, P. E., Lieber, Ch. M. (1997) Nanobeam mechanics: elasticity, strength, and toughness of nanorods and nanotubes, *Nature* **227**, 1971 - 1975

[15] Falvo, M. R., Clary, G. J., Taylor II, R. M., Chi, V., Brooks Jr., F. P., Washburn S., Superfine, R. (1997) Bending and buckling of carbon nanotubes under large strain, *Nature* **389**, 582 – 584.

[16] Salvetat, J.-P., Bonard, J.-M., Thomson, N. H., Kulik, A. J., Forró, L., Benoit, W., Zuppiroli, L. (1999) Mechanical properties of carbon nanotubes, *Appl. Phys. A* **69**, 255 – 260.

[17] Dai, H., Wong, E. W., Lieber, Ch. M., (1996) Probing Electrical Transport in Nanomaterials: Conductivity of individual carbon nanotubes, *Science* **271**, 523 - 526

[18] Muster, J., Duesberg, G. S., Roth, S., Burghard, M. (1999) Application of scanning force microscopy in nanotube science, *Appl. Phys. A* **69**, 261 – 167.

[19] Biró, L. P., Szabó, B., Márk, G. I., Gyulai. J., Havancsák, K., Kürti, J., Dunlop, A., Frey, L., Ryssel, H. (1999) Carbon nanotubes produced by high energy (E > 100 MeV), heavy ion irradiation of graphite, *Nucl. Instr. and Meth. B.* **148**, 1102 - 1105

[20] Biró, L. P., Gyulai. J., Lambin, Ph., B.Nagy, J., Lazraescu, S., Márk, G. I., Fonseca, A., Surján, P., R., Szekeres, Zs., Thiry, P. A., Lucas, A. A. (1998) Scanning tunneling microscopy (STM) imaging of carbon nanotubes, *Carbon* **36**, 689 – 696.

[21] Liu, J., Rinzler, A. G., Dai, H., Hafner, J. H., Bradley, R. K., Boul, P. J., Lu, A., Iverson, T., Shelimov, K., Huffman, C. B., Rodriguez-Macias, F., Shon, Y-S., Lee, T. R., Colbert D. T., Smalley, R. E. (1998) Fullerene Pipes, *Science* **280**, 1253 –1256.

[22] Buldum, A., Lu, J. P. (1999) Atomic Scale Sliding and Rolling of Carbon Nanotubes, *Phys. Rev. Lett.* **83**, 5050 – 5053.

[23] Biró, L. P., Lazarescu, S., Lambin, Ph., Thiry, P. A., Fonseca, A., B.Nagy, J., Lucas, A. A., (1997) Scanning tunneling microscope investigation of carbon nanotubes produced by catalytic decomposition of acetylene, *Phys. Rev. B* **56**, 12490 – 12498.

[24] Clauss, W., Freitag, M., Bergereon, D. J., Johnosn, A. T. (1999) Characterization of Single Wall Carbon Nanotubes by Scanning Tunneling and Scanning Force Microscopy, in H. Kuzmany, J. Fink, M. Mehring and S. Roth (eds.), *Electronic Properties of Novel Materials – Science and Technology of Molecular Nanostructures*, American Institute of Physics, Melville, pp.308 – 312.

[25] Biró, L. P., Márk, G. I., Gyulai, J., Rozlosnik, N., Kürti, J., Szabó, B., Frey, L., Ryssel, H. (1999) Scanning probe method investigation of carbon nanotubes produced by high energy ion irradiation of graphite, *Carbon* **37**, 739 – 744.

STRUCTURAL AND ELECTRONIC PROPERTIES OF CARBON NANOTUBE JUNCTIONS

PH. LAMBIN
Département de Physique, Facultés Universitaires N.D.P.
61 Rue de Bruxelles, B 5000 Namur, Belgium

AND

V. MEUNIER
Department of Physics, North Carolina State University,
Raleigh NC 27695, USA.

Abstract. A short review is given on the theoretical and experimental works devoted to nanotube molecular connections. For the present study, several such junctions were generated on the computer and their structures were optimized with the help of empirical atomic potentials. Local electronic densities of states were evaluated in the interfacial regions from a tight-binding Hamiltonian. In some metal-semiconductor hybrids, interesting oscillations of the local density of states take place in front of the semiconducting section. The oscillations are due to interferences between incident and reflected Bloch waves when these cannot propagate in the semiconductor.

1. Introduction

The research of potential applications of carbon nanotubes has become a field of growing interest. In this respect, realizing a connection between two nanotubes possessing different electronic properties offers the possibility to construct elements of logical devices working at the nanoscopic scale [1, 2]. That possibility exists since a few such connections have recently been identified with a scanning force microscope and their transport properties were studied experimentally [3].

In the simplest case, a junction between two different nanotubes can be realized by introducing a pentagon-heptagon pair (or several pairs) that

L.P. Biró et al. (eds.), Carbon Filaments and Nanotubes: Common Origins, Differing Applications, 265–274.
© *2001 Kluwer Academic Publishers. Printed in the Netherlands.*

preserves the triple coordination of each C atom in the interfacial region [4]. This insertion bends the structure to an angle that depends on the relative positions of the pentagon and heptagon [5] and on the number of pentagon-heptagon pairs [6]. Occasionally, carbon nanotubes presenting one or more sharp bends have been observed experimentally by TEM [6,7] and, as mentioned above, by AFM [3].

2. Structure

A nanotube junction consists of two cylinders joined by a conical section. A two-dimensional representation of a junction is shown in fig. 1. The chiral vectors $\vec{C}_{h1} = n_1 \vec{a}_1 + m_1 \vec{a}_2$ and $\vec{C}_{h2} = n_2 \vec{a}_1 + m_2 \vec{a}_2$ of the two nanotubes are represented by the segments AH and DE, respectively. With no loss of generality, it can be assumed that the nanotube radii satisfy $R_1 > R_2$. Also, the wrapping indices n and m can always be chosen in such a way that the angle Φ between $\vec{C}_{h1}$ and $\vec{C}_{h2}$ be smaller or equal to $\pi/6$. This is because the chiral vectors can be rotated by $\pm\pi/3$ without changing the atomic structure of the nanotubes. The three-dimensional structure of the connection is obtained by superposing the edge AB onto HG, BC onto GF, and CD onto FE. In so doing, BCFG becomes the conical section. The continuity of the graphene layer across the BC $\equiv$ GF segments will be realized if BC and GF have the same length and have equivalent directions. This latter condition implies that the angle between GF and BC is $\pi/3$. Important consequences follow from that result. First, $\angle ABC + \angle HGF = 5\pi/3$, which means removing a $\pi/3$ wedge that corresponds to the existence of a pentagon at B ($\equiv G$). Second, $\angle BCD + \angle GFE = 7\pi/3$, and this corresponds to a heptagon at C ($\equiv F$).

By equating the length of the segments BC and GF, knowing that the angle between them is $\pi/3$, one arrives to the expression the so-called junction vector $\vec{j}$, which connects the pentagon to the heptagon along the BC segment [5,6]

$$\vec{j} = (m_2 - m_1)\vec{a}_1 + (n_1 + m_1 - n_2 - m_2)\vec{a}_2 \qquad (1)$$

and its length is $j = |\vec{C}_{h1} - \vec{C}_{h2}|$.

As a particular important case, the pentagon is adjacent to the heptagon whenever $|\vec{j}| = |\vec{a}_1| = |\vec{a}_2| = a$. Eq. 1 leads to six possibilities: $n_1 - n_2 = \pm 1$ and $m_1 - m_2 = 0$, $n_1 - n_2 = 0$ and $m_1 - m_2 = \pm 1$, $n_1 - n_2 = \pm 1$ and $m_1 - m_2 = \mp 1$. The defect is then equivalent to an edge dislocation [8]. An example of such a defect is shown in fig. 2 with the connection between the two enantiomer (9,10) and (10,9) nanotubes. This junction has a mirror plane that bisects the pentagon-heptagon pair. Adjacent pentagon-heptagon pairs may appear spontaneously in armchair nanotubes under a

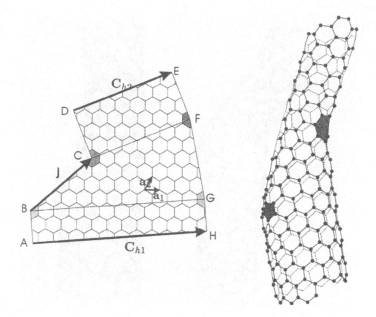

Figure 1. Left: two-dimensional development of a conical connection between the nanotubes (11,1) (chiral vector $\vec{C}_{h1}$) and (5,3) (chiral vector $\vec{C}_{h2}$). Right: resulting three dimensional structure of the junction (11,1)/(5,3). The pentagon and heptagon rings have been shaded.

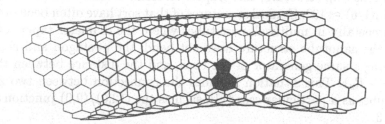

Figure 2. (9,10)/(10,9) junction realized with a fused pentagon-heptagon pair (shaded surface). The defect is an edge dislocation: it corresponds to the insertion of a zig-zag chain of atoms (•) ending at a vertex shared by the two odd-membered rings. This extra row of atoms changes the chirality from (9,10) to (10,9) and bends the structure.

tensile load, with the formation of Stone-Wales defects. These consists in two fused pentagon-heptagon pairs, which form two dislocations with opposite Burgers vectors. Further relaxation of the strain energy can be released by separating the two pairs [9]. The separation of the two dislocations leads to two junctions in series.

When the two connected nanotubes have the same chirality, the polygon BCFG in fig. 1 becomes an isosceles trapezium with $\angle GBC = \angle BGF = \pi/3$.

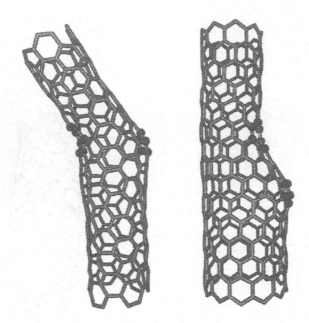

Figure 3. Computer-generated models of the $(6,4)/(8,-1)$ (left) and $(12,0)/(9,0)$ (right) junctions. The pentagon and heptagon are indicated by dark balls.

In the rolled-up structure, this trapezium forms a cone with opening angle $2\arcsin(1/6) \approx 19°$. Conical structures of that sort have often been observed experimentally in multiwall nanotubes where the diameter tapers off [10]. Here, the nanotube structure does not bend, the pentagon and heptagon are along a same generator of the cone, and the distance between them is $2\pi(R_1 - R_2)$. This situation includes the connection between two zig-zag nanotubes as a particular case [11], such as the $(12,0)/(9,0)$ junction shown in fig. 3.

In the two-dimensional representation of the connection (fig. 1), the angle Φ between the nanotube axes is the angle between their chiral vectors, $\vec{C}_{h1}$ and $\vec{C}_{h2}$. This angle may not exceed $30°$ as mentioned above. That limit corresponds to bent connections (knees) proposed by Dunlap [4], an example of which is the $(6,4)/(8,-1)$ junction between two chiral nanotubes shown in fig. 3. Here, the pentagon and heptagon are at opposite sides. All intermediate cases exist between this situation and the one discussed above where the pentagon and heptagon are along the same line. In the $(11,1)/(5,3)$ junction shown in the right-hand side of fig. 1 for instance, the odd-membered rings are neither along a same generator neither on opposite sides of the cone.

The two-dimensional construction described in fig. 1 is useful to define the table of connection between the atoms in the junction and to obtain

TABLE 1. Nanotube junction, the junction vector $\vec{j}$ (eq. 1), the angle between the chiral vectors Φ, and the bend angle of the three-dimensional structure.

junction	$\vec{j}$	Φ (°)	bend angle (°)
(6,4)/(8,-1)	(-5,3)	30	34
(5,5)/(9,0)	(-5,1)	30	36
(6,6)/(10,0)	(-6,2)	30	36
(10,10)/(18,0)	(-10,2)	30	36
(8,0)/(7,1)	(1,0)	6.6	12
(12,0)/(9,0)	(0,3)	0	0

a rough approximation of the structure in three dimensions. The atomic structure can then be optimized by minimizing the total energy of the system. We performed this optimization for several junctions with an empirical C-C potential [12] and found the results listed in table 1. In the three-dimensional structure, the axes of the two terminal nanotubes are not located in a same plane unless there is a mirror plane in the junction. The angle between the axes was always found larger than the two-dimensional angle Φ. Han *et al* [6] arrived at the same conclusion with the Tersoff potential. The largest bends are realized in Dunlap knees, were the angle can reach 36°. Single-wall nanotube kinks presenting such large angles were observed recently [3].

3. Electronic properties

The ground-state electronic properties of nanotube junctions have been the subject of a great deal of theoretical work [5, 6, 11–15] The states close to the Fermi level of the nanotubes and their junctions are dominated by the π electrons, which are well-described by the tight-binding Hamiltonian

$$H = \sum_I |I\rangle \epsilon_I \langle I| + \sum_{I,J} |I\rangle \gamma_{I,J} \langle J| \tag{2}$$

In a homopolar material such as graphite and in the absence of charge transfer, the diagonal elements ϵ_I all take the same value that can be chosen as the zero of energy. The hopping terms $\gamma_{I,J}$ can be restricted to the first-neighbor pairs. Since all bonds in the structure have the same length, which remains approximately true in any junction between nanotubes, one is left with a single hopping parameter, γ_0, between first-neighbor atoms. We used $\gamma_0 = -3$ eV. The Hamiltonian is then proportional to the adjacency matrix of the atomic structure, which can be defined from two-dimensional representation of the junction, fig. 1.

The local density of states (DOS) on a given atomic site, I, is

$$
\begin{aligned}
n_I(E) &= \sum_k \langle I|\psi_k\rangle \delta(E - E_k)\langle\psi_k|I\rangle \\
&= (-1/\pi)\sum_k \langle I|\psi_k\rangle \mathrm{Im}(E - E_k + i0)^{-1}\langle\psi_k|I\rangle \\
&= (-1/\pi)\mathrm{Im}\langle I|(E - H + i0)^{-1}|I\rangle
\end{aligned}
\tag{3}
$$

where the ψ_k are the eigenstates of H with energies E_k. The last equality follows from the the completeness of the ψ_k's. Eq. 3 relates the local density of states to the imaginary part of a diagonal element of the Green's function,

$$
G_{II}(z) = \langle I|(z - H)^{-1}|I\rangle = \cfrac{1}{z - a_1 - \cfrac{b_1^2}{z - a_2 - \cdots}}
\tag{4}
$$

where z is the energy E plus a positive vanishingly-small imaginary part. The recursion technique was used to calculate $G_{II}(z)$ in the form of a continued fraction [16]. This algorithm provides us with the set of continued-fraction coefficients $a_1, b_1 \ldots$ up to a large number of levels.

The electronic properties of two typical junctions are illustrated in fig. 4. The first junction is the metal/metal (12,0)/(9,0) connection shown in fig. 3. Here, the pentagon and heptagon are seen face on, near the center of the drawing of the structure. They are separated by two hexagons. The curves on the right-hand side are local densities of states per C atom which were averaged over the sites located around successives sections labelled a, b ... o. The Fermi energy E_F is at zero energy. The DOS of the (9,0) and (12,0) tubes are similar: there is a plateau of constant density around the Fermi level. Both the extension and the height of this metallic plateau are inversely proportional to the nanotube diameter. The plateau is bounded on both sides by a pair of nearly-degenerate van Hove singularities. As seen in fig. 4, the transition between the DOS levels and shapes operates fastly across the interface in the (12,0)/(9,0) junction. There is a little electron-hole asymmetry in the curves f–g due to the presence of the odd-membered rings. Below and above the conical connection, the van Hove singularities readily put in their places and get sharper and sharper (the small imaginary part added to the energy to force the convergence of the continued fractions broadens the peaks).

The other system illustrated in fig. 4 is a connection between the semi-conducting (10,0) and metallic (6,6) nanotubes. This is a Dunlap knee similar to the bent structure shown in fig. 3. The local DOS's are displayed

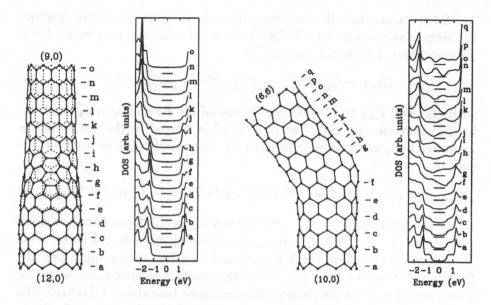

Figure 4. Structure and variations of the local DOS in the (12,0)/(9,0) (left) and (10,0)/(6,6) junctions. The densities of states have been computed with 250 levels of continued fraction (eq. 4), the horizontal bars indicate zero DOS.

on the right-hand side of the drawing of the structure. The curves again correspond to annular sections of the structure indicated with the letters a, b ... q. The band gap of the (10,0) nanotube has already developed at the section a, that is to say less then 1 nm below the interface. However, the van Hove singularities that bound the gap at $\pm 0.5\,\mathrm{eV}$ are not yet well defined there. In the (6,6) metallic nanotube, the plateau of density of states does not form right above the interface. Instead, quasi-periodic oscillations of the DOS develop, with a period equal to the distance between annular sections, 1.5 times the Bravais translation a of the nanotube. These oscillations are the consequence of a standing wave pattern formed by the interferences of Bloch states that travel down the metallic nanotube and are reflected backwards by the junction when their energies fall within the forbidden band gap of the semiconductor.

The existence of quantum interferences in this part of the junction can be substantiated as follows. Let ψ_{k_1} be a Bloch state of the metallic nanotube with wave vector $k_1 = k_F + \delta k$, where k_F is the Fermi wave vector. Apart from a normalization factor, the total wave function in the metallic section, and at a large distance from the interface, is $\psi = \psi_{k_1} + \rho_1 \psi_{k_1}^* + \rho_2 \psi_{k_2}^*$ with $k_2 = k_F - \delta k$. Indeed, there are two electron bands crossing at k_F with equal and opposite slopes, ψ_{k_1} and ψ_{k_2} have the same energy, and both contribute to the reflected wave with coefficients ρ_1 and ρ_2. If one labels $I = 1$,

2 ... 24 the atoms in a Bravais unit cell of the (6,6) nanotube (for instance the sites in sections p and q in fig. 4), the coefficient of ψ on the site $I + T$ obtained from I by a translation T is

$$C_{I+T} = C_{1,I}e^{ik_1T} + \rho_1 C_{1,I}^* e^{-ik_1T} + \rho_2 C_{2,I}^* e^{-ik_2T} \qquad (5)$$

with $C_{1,I}$ and $C_{2,I}$ the LCAO coefficients of the Bloch states ψ_{k_1} and ψ_{k_2} in the reference unit cell. The local DOS is related to the square of the wave function, as expressed above (eq. 3). One readily obtains from the last equation

$$|C_{I+T}|^2 = a_I + b_I \cos(2k_1T + \beta_I) + c_I \cos(2k_2T + \gamma_I) + f_I \cos(2\delta k\, T + \phi_I) \quad (6)$$

where the amplitudes (a_I, b_I ...) and phases (β_I, γ_I ...) depend on the moduli and phases of $C_{1,I}$, $C_{2,I}$, ρ_1 and ρ_2. For an energy close to the Fermi level, δk is small. The square coefficient given by eq. 6 is then a quasi periodic function of T with period π/k_F. For an armchair nanotube, $k_F = 2\pi/3a$, which corresponds to the period 1.5a mentioned here above. Of course, 1.5a is not a Bravais translation of the nanotube. The DOS curves of fig. 4 emphasize this period because the DOS's were taken from successive sections separated by 0.5a. It is worth mentioning that oscillations having the Bravais period $3a \approx 0.75$ nm have been observed by STS in a line scan along a short a piece of an armchair nanotube [17]. In that case, the standing wave pattern is due to the finite length of the nanotube. Coming back to the metal-semiconductor junctions, $2k_F$ oscillations of the electron density of states taking place in the metallic part are typical of the armchair configuration. When the local densities of states are averaged around successive tube sections as presented here, there are no oscillations in a metallic zigzag nanotube, because $k_F = 0$, and not either in a chiral nanotube because k_F is small [12].

4. Transport

Electron transport in a single-wall nanotube with a sharp bend of approximately 40° has been studied experimentally [3]. Both metal–metal and metal–semiconductor junctions have been identified from transport measurments. The first kind of junction exhibited a strong reduction of conductance with respect to perfect nanotube, and anomalous temperature dependencies. Non-linear and asymmetric current–voltage characteristics were observed in the metal–semiconductor junction, with I–V curves that resembled the ones of a rectifying diode.

On the theory side, the Landauer formula can be used to calculate the tunnel conductance between two leads, here the perfect nanotubes on both sides of the junction [18]. For the case of a connection between two metallic nanotubes, Tamura and Tsukada found a universal law of the conductance at small bias [19] which, with the effective-mass approximation, can be represented analytically as [20]

$$g = \frac{2e^2}{h} \frac{8}{(R_1/R_2)^3 + (R_1/R_2)^{-3} + 2} \tag{7}$$

where R_1 and R_2 are the radii of the connected tubes. The interesting property of this formula is that is does not depend on the bend angle of the junction. When $R_1 = R_2$, the conductance is that of a perfect nanotube, equal to two quantum units because there are two chanels corresponding to the two subbands crossing at the Fermi level. That the conductivity decreases with increasing R_1/R_2 is due to the radius discontinuity at the interface which partly reflects the Bloch functions backwards.

It is interesting to note that a metal–metal junction may present a large conductance gap if the Fermi states on both sides belong to different representations of the rotational symmetry group of the connected system. This peculiarity has been demonstrated for a (12,0)/(6,6) matched junction based on six pentagon-heptagon pairs symmetrically arranged around a circumference [18].

The case of metal–semiconductor is more complicate. Due to the band gap, the junction must be polarized to drive the current across. The profile of the potential drop at the interface is unknown and must be determined by a self-consistent calculation [21]. In the absence of doping, an all-carbon metal–semiconductor junction should have a symmetric I–V curve, with no rectifying properties. The experimental observation of rectification implies therefore some doping, possibly through a charge transfer from the electrodes on which the nanotube is deposited [3]. Due to the charge transfer, there is a bending of the valence and conduction bands of the semiconductor, and this bending extends over a long distance [22]. Model calculations then show that the Schottky barrier that forms at the junction is reduced by application of a bias voltage V. When the Fermi level is pinned right at the top of the valence band of the semiconductor, the onset of (non-tunneling) conductance across the junction is at $eV = E_g/2$ for forward bias (positive V on the semiconducting terminal) and at $-2E_g$ for reverse bias, with E_g the semiconducting gap [22]. This behavior explains the observed strong asymmetry of the I–V characteristics of the metal–semiconductor junction [3].

Acknowledgments

This work has been partly funded by the interuniversity research project on reduced dimensionality systems (PAI P4/10) of the Belgian Office for Scientific, Cultural, and Technical affairs.

274

References

1. Service, R.F. (1996) Mixing nanotube structures to make a tiny switch, *Science* **271**, 1232.
2. Dresselhaus, M. (1996) Carbon connections promise nanoelectronics, *Phys. World* **9**(May), 18-19.
3. Yao, Z., Postma, H.W.Ch., Balents, L., and Dekker, C. (1999) Carbon nanotube intramolecular junctions, *Nature* **402**, 273-276.
4. Dunlap, B.I. (1992) Connecting carbon tubules, *Phys. Rev. B* **46**, 1933-1936; Dunlap, B.I. (1994) Relating carbon tubules, *Phys. Rev. B* **49**, 5643-5649.
5. Saito, R., Dresselhaus, G., and Dresselhaus, M.S. (1996) Tunneling conductance of connected carbon nanotubes, *Phys. Rev. B* **53**, 2044-2050.
6. Han, J., Anantram, M.P., Jaffe, R.L., Kong, J., and Dai, H. (1998) Observation and modeling of single-wall carbon nanotube bend junctions, *Phys. Rev. B* **57**, 14983-14989.
7. Terrones, M., Hsu, W.K., Hare, J.P., Kroto, H.W., Terrones, H., and Walton, D.R.M. (1996) Graphite structures: from planar to spheres, toroids and helices, *Phil. Trans. R. Soc. London A* **354**, 2025-2054.
8. Lauginie, P. and Conard, J. (1997) New growing modes for carbon: Modelization of lattice defects, structures of tubules and onions, *J. Phys. Chem. Solids* **58**, 1949-1963.
9. Buongiorno Nardelli, M., Yakobson, B.I., and Bernholc, J. (1998) Brittle and ductile behavior in carbon nanotubes, *Phys. Rev. Lett.* **81**, 4656-4659.
10. Iijima, S., Ichihashi, T., and Ando, Y. (1992) Pentagons, heptagons and negative curvature in graphite microtubule growth, *Nature* **356**, 776-778.
11. Charlier, J.C., Ebbesen, T.W., and Lambin, Ph. (1996) Structural and electronic properties of pentagon-heptagon pair defects in carbon nanotubes, *Phys. Rev. B* **53**, 11108-11113.
12. Meunier, V., Henrard, L., and Lambin, Ph. (1998) Energetics of bent carbon nanotubes, *Phys. Rev. B* **57**, 2591-2596.
13. Lambin, Ph., Fonseca, A., Vigneron, J.P., B.Nagy, J., and Lucas, A.A. (1995) Structural and electronic properties of bent carbon nanotubes, *Chem. Phys. Lett.* **245**, 85-89.
14. Chico, L., Crespi, V.H., Benedict, L.X., Louie, S.G., and Cohen, M.L. (1996) Pure carbon nanoscale devices: Nanotube heterojunctions, *Phys. Rev. Lett.* **76**, 971-974.
15. Blase, X., Charlier, J.C., De Vita, A., and Car, R. (1997) Theory of composite $B_x C_y N_z$ nanotube heterojunctions, *Appl. Phys. Lett.* **70**, 197-199.
16. Haydock, R., Heine, V. and Kelly, M.J. (1972) Electronic structure based on the local atomic environment for tight-binding bands, *J. Phys. C: Solid St. Phys.* **5**, 2845-2858.
17. Venema, L.C., Wildoer, J.W.G., Janssen, J.W., Tans, S.J., Temminck Tuinstra, H.L.J., Kouwenhoven, L.P., and Dekker, C. (1999) Imaging electron wave functions of quantized energy levels in carbon nanotubes, *Science* **283**, 52-55.
18. Chico, L., Benedict, L.X., Louie, S.G., and Cohen, M.L. (1996) Quantum conductance of carbon nanotubes with defects, *Phys. Rev. B* **54**, 2600-2606.
19. Tamura, R. and Tsukada, M. (1997) Conductance of nanotube junctions and its scaling law, *Phys. Rev. B* **55**, 4991-4998.
20. Tamura, R. and Tsukada, M. (1998) Analysis of quantum conductance of carbon nanotube junctions by the effective-mass approximation, *Phys. Rev. B* **58**, 8120-8124.
21. Farajian, A.A., Esfarjani, K., and Kawazoe, Y. (1999) Nonlinear coherent transport through doped nanotube junctions, *Phys. Rev. Lett.* **82**, 5084-5087.
22. Odintsov, A.O. (1999) Schottky barrier in carbon nanotube heterojunction, preprint TU Delft.

APPLICATIONS OF SUBMICRON DIAMETER CARBON FILAMENTS

D.D.L. CHUNG
Composite Materials Research Laboratory
State University of New York at Buffalo
Buffalo, NY 14260-4400, U.S.A.

Abstract

The applications of submicron diameter carbon filaments grown catalytically from carbonaceous gases are reviewed. These relate to structural applications, electromagnetic interference shielding, electromagnetic reflection, surface electrical conduction, DC electrical conduction, field emission, electrochemical applications, thermal conduction, strain sensors, porous carbons and catalyst supports.

1. Introduction

Submicron diameter carbon filaments are mainly grown catalytically from carbonaceous gases at 500-700°C [1,2], although they include carbon nanotubes [3], which typically have diameter in the nanometer range. Due to the higher yield in production, the former is more abundant than the latter and applications involving the former are better developed than those involving the latter. Submicron carbon filaments are to be distinguished from conventional carbon fibers, which are made by pyrolysis of pitch or polymer [4-6]. They are also to be distinguished from vapor grown carbon fibers (VGCF), which are prepared by pyrolysis of carbonaceous gases to non-catalytically deposit carbon on catalytically grown submicron diameter carbon filaments at 950-1100°C [7-14]. Carbon filaments differ from both conventional and vapor grown carbon fibers in their small diameter. Conventional carbon fibers typically have diameter around 10 μm and VGCF have diameters up to 10 μm. Both carbon filaments and VGCF are not continuous, though the latter can be longer than the former. In contrast, conventional carbon fibers can be continuous. In spite of the discontinuous nature of carbon filaments, the aspect ratio can be quite high, due to the small diameter. Carbon filaments tend to be not straight. Thus, they typically have a morphology that resembles cotton wool. In this paper, the term "filaments" refers to filaments of submicron diameter, whereas the term "fibers" refers to fibers of diameter greater than 1 μm.

Carbon filaments are commercially available, though not in large volumes. As the amount of usage increases, the price of carbon filaments will fall. According to Applied Sciences Inc. (Cedarville, Ohio), which manufactures catalytically grown carbon filaments, the price will fall to U.S. $4-6 per kilogram. This price is even lower than that of short isotropic pitch based carbon fibers. In order for the usage to increase, applications must be developed.

Much research has been conducted to understand the process of catalytic growth of carbon filaments [1,2]. However, relatively little attention has been given to the applications of submicron diameter carbon filaments, although considerable

275

L.P. Bíró et al. (eds.), Carbon Filaments and Nanotubes: Common Origins, Differing Applications, 275–288.
© 2001 *Kluwer Academic Publishers. Printed in the Netherlands.*

progress has recently been made. This paper provides a review of this rapidly growing area, including structural applications, electromagnetic reflection, electromagnetic interference shielding, surface electrical conduction, DC electrical conduction, field emission, electrochemical applications, thermal conduction, strain sensors, porous carbons and catalyst supports.

2. Structural applications

Structural applications are most important for conventional continuous carbon fibers. Due to their texture with the carbon layers at an angle from the filament axis, discontinuous nature, small diameter and the consequent large interfacial area, carbon filaments (catalytically grown, of diameter $0.1 \mu m$) are far less effective than continuous carbon fibers as reinforcements in composites. Carbon filaments are even not as effective as short carbon fibers (even those based on isotropic pitch) at the same volume fraction, for both thermoplastic [15,16] and cement matrices [17]. For example, in a thermoplastic matrix, carbon filaments at 19 vol.% give a tensile strength of 27 MPa [16], whereas carbon fibers (isotropic pitch based, 3000 μm long) at 20 vol.% give a tensile strength of 64 MPa [15]. In a cement paste matrix, carbon filaments at 0.51 vol.% give a tensile strength of 1.2 MPa, whereas carbon fibers (isotropic pitch based, 5 mm long) give a tensile strength of 1.7 MPa [17]. Although the filaments do not reinforce as well as the fibers, they still reinforce under tension. For instance, the tensile strength, modulus and ductility and the compressive modulus are all increased by the filaments (0.51 vol. %), but the compressive strength and ductility decrease [17].

More success has been found in hybrid composites involving both carbon filaments (catalytically grown, of diameter $0.1 \mu m$) and conventional carbon fibers [18] than in composites involving carbon filaments only [16,17]. The use of carbon filaments in the interlaminar region between adjacent layers of conventional continuous carbon fibers in an epoxy-matrix composite enhances both transverse and longitudinal vibration damping ability (due to the large area of the interface between filaments and matrix). It increases also the storage modulus in the transverse direction (due to the presence of filaments that are oriented near the direction perpendicular to the fiber layers). Even though the filament volume fraction (0.6%) is negligibly low compared to the fiber volume fraction (56.5%) in the hybrid composite, the filaments are effective. The longitudinal storage modulus is only slightly decreased by the filament addition [18]. Another form of hybrid composite involves the catalytic growth of carbon filaments on conventional carbon fibers for the purpose of providing mechanical interlocking between adjacent fibers in a composite [19,20]. The interfacial shear strength between fiber and a polymer matrix is increased by over 4.75 times after the growth of filaments on the fiber [20]. Furthermore, the specific surface area is increased from about 1.0 m^2/g to 250-300 m^2/g [20]. However, the practicality of composite fabrication using such hybrid fibers remains an issue.

Due to the clingy morphology and small diameter of the carbon filaments, their dispersion in a composite requires more care than that of conventional short carbon fibers. A method of dispersing the carbon filaments in a thermoplastic matrix involves (i) dispersing the filaments in an isopropyl alcohol aqueous solution with the help of a trace amount of a dispersant, such as Triton X102 of Rohm and Haas Co. (Philadelphia, Pennsylvania); (ii) mixing the slurry with thermoplastic powder at room temperature by using a kitchen blender, such that the concentration of isopropyl alcohol in the aqueous solution is adjusted so that the thermoplastic particles are suspended in the solution, (iii) draining the solution, (iv) drying at 120°C, and (v) hot pressing uniaxially above the glass transition temperature of the thermoplastic and at 1000 psi (6.9 MPa) for 0.5 hours.

The mixing in step (ii) causes very little filament breakage, so the aspect ratio of the filaments after mixing remains high (typically >1000) [15,20]. In this method, the thermoplastic must be in the form of fine particles (the finer, the better), due to the small diameter of the carbon filaments. As thermoplastics are mostly in the form of pellets rather than particles, the choice of thermoplastics is limited. In the case of a thermosetting resin such as epoxy, dispersion of the carbon filaments requires dilution of the resin with a solvent so as to lower the viscosity, and subsequent mixing of the filament-resin slurry by using a vigorous means, such as a blender. Due to the strong effect of the form of the matrix raw material on the dispersion of the carbon filaments, the properties (both mechanical and electromagnetic) of the composites significantly depend on the matrix material.

Catalytically grown carbon filaments tend to have a layer of polyaromatic hydrocarbons on their surface, due to the process in which the filaments are grown [22]. The hydrocarbon layer can be removed by cleansing with a solvent, such as acetone and methylene chloride [22]. The total or partial removal of the hydrocarbon layer improves the bonding between the filaments and a thermoplastic matrix. This is suggested by the fact that the volume electrical resistivity of the composite is much lower [21,23] and the mechanical properties improve [24] when the hydrocarbon layer on the filaments has been removed prior to incorporation in the matrix.

Surface treatment of the filaments by oxidation helps the mechanical properties of cement-matrix and polymer-matrix composites [25,26]. The bond between carbon and a cement matrix is weak compared to that between carbon and a polymer matrix. Therefore, surface treatment of carbon for improving that bond is particularly important. The treatment of catalytically grown carbon filaments with ozone gas (0.3 vol.% in air, 160°C, 10 min) increases the tensile strength, modulus and ductility, and the compressive strength, modulus and ductility of cement pastes, relative to the values for pastes with the same volume fraction of untreated filaments [25]. Similar effects apply to the ozone treatment of carbon fibers [26,27], for which it has been shown that the treatment improves the wetting by water, the degree of fiber dispersion, and bond strength with cement, in addition to increasing the surface oxygen concentration [27].

Ceramic-matrix (Al_2O_3 or $MgAl_2O_4$ as matrix) composite powders that contain in-situ formed carbon nanotubes may be useful for structural applications [28].

3. Electromagnetic interference shielding, electromagnetic reflection and surface electrical conduction

Electromagnetic interference (EMI) shielding [29-32] is in critical demand due to the interference of wireless (particularly radio frequency) devices with digital devices and the increasing sensitivity and importance of electronic devices. EMI shielding is one of the main applications of conventional short carbon fibers [33]. Due to the small diameter, carbon filaments (catalytically grown, of diameter 0.1 μm) are more effective at the same volume fraction in a composite than conventional short carbon fibers for EMI shielding, as shown for both thermoplastic [15,16] and cement [17,34] matrices. For example, in a thermoplastic matrix, carbon filaments at 19 vol. % give an EMI shielding effectiveness of 74 dB at 1

GHz [16], whereas carbon fibers (isotropic pitch based, 3000 μm long) at 20 vol.% give a shielding effectiveness of 46 dB at 1 GHz [15]. In a cement-matrix composite, fiber volume fractions are typically less than 1%. Carbon filaments at 0.54 vol.% in a cement paste give an effectiveness of 26 dB at 1.5 GHz [17], whereas carbon fibers (isotropic pitch based, 3 mm long) at 0.84 vol. % in mortar give an effectiveness of 15 dB at 1.5 GHz [34]. These effectiveness measurements were made with the same fixture and about the same sample thickness. A low volume fraction of the filler is attractive for maintaining ductility or resilience in the polymer-matrix composite, as both ductility and resilience decrease with increasing filler volume fraction. Resilience is particularly important for EMI shielding gaskets and electric cable jackets. In addition, a low volume fraction of the filler reduces the material cost and improves the processability of the composites, whether polymer-matrix or cement-matrix composites.

The greater shielding effectiveness of the filaments compared to the fibers is due to the skin effect, i.e., the fact that high frequency electromagnetic radiation interacts only with the near surface region of an electrical conductor. However, carbon filaments are still not as effective as nickel fibers of diameter 2 μm at the same volume fraction, as shown for a thermoplastic matrix [23]. On the other hand, by coating a carbon filament with nickel by electroplating, a nickel filament (0.4 μm diameter) with a carbon core (0.1 μm diameter) is obtained [23,35]. The nickel filaments (0.4 μm diameter) are more effective than the nickel fibers (2 μm diameter) for shielding, due to their small diameter. At 1 GHz, a shielding effectiveness of 87 dB was attained by using only 7 vol.% nickel filaments in a thermoplastic matrix [23]. The shielding is almost all by reflection rather than absorption.

The high radio wave reflectivity of carbon filament (0.1 μm diameter) reinforced cement paste (at 1 GHz, 10 dB higher than plain cement paste) makes carbon filament concrete attractive for use in lateral guidance in automatic highways [36]. Automatic highways refer to highways, which provide fully automated control of vehicles, so that safety and mobility are enhanced. In other words, a driver does not need to drive on an automatic highway, as the vehicle goes automatically, with both lateral control (steering to control position relative to the center of the traffic lane) and longitudinal control (speed and headway). Current technology uses magnetic sensors together with magnetic highway markings to provide lateral guidance, and uses radar to monitor the vehicle position relative to other vehicles in its lane for the purpose of longitudinal guidance. Cement paste containing 0.5 vol. % carbon filaments exhibits reflectivity at 1 GHz that is 29 dB higher than the transmissivity. Without the filaments, the reflectivity is 3-11 dB lower than the transmissivity.

The electromagnetic technology has various advantages compared to the magnetic technology. Amongst these, low material cost (reflecting concrete is estimated to be 30% more expensive than conventional concrete, thus much less expensive than concrete with embedded magnets or magnetic strips), low labor cost (same as conventional concrete, thus much less than concrete with embedded magnets or magnetic strips), and low peripheral electronic cost (off-the-shelf oscillator and detector). Also noteworthy are good mechanical properties (reflecting concrete exhibiting better mechanical properties and lower drying shrinkage than conventional concrete, whereas embedded magnets weaken concrete), good reliability (less affected by weather, as frequency, impedance and power selectivity provides tuning capability), and high durability (demagnetization and marking detachment not being issues). Moreover, the magnetic field from a magnetic marking can be shielded by electrical conductors (such as steel) between the marking and the vehicle, whereas the electromagnetic field cannot be easily shielded.

The surface impedance of carbon filament composites, nickel filament composites and nickel fiber composites are low. In particular, at 1 GHz, the surface impedance is comparable to that of copper for a thermoplastic matrix composite with 7 vol. % nickel filaments and a thermoplastic matrix composite with 13 vol. % nickel fibers (2 μm diameter) [16]. The surface impedance is higher for carbon filament composites than nickel filament composites or nickel fiber (2 μm diameter) composites at similar filler volume fraction [16]. Although carbon filaments have a lower density than nickel filaments, a thermoplastic matrix composite with 7 vol. % nickel filaments has the same specific surface conductance (surface conductance divided by the density, where conductance is the reciprocal of impedance) as one with 19 vol. % carbon filaments [16]. The low surface impedance is valuable for applications related to electrostatic discharge protection and microwave waveguides.

4. DC electrical conduction

The submicron diameter carbon filaments are useful as an electrically conducting additive for enhancing the DC electrical conductivity of a polymer-matrix composite [37-41], provided that they are properly dispersed [16,21,39]. An application that benefits from the enhanced conductivity pertains to solid rocket propellants, for which enhanced conductivity decreases the incidence of dangerous electric discharge events [38]. Another application pertains to molecular optoelectronics and involves the use of carbon nanotubes [40,41].

5. Field emission

The high aspect ratio, small radius of curvature at the tip, high chemical stability and high mechanical strength of carbon nanotubes are advantageous for field emission, i.e., the emission of electrons under an applied electric field. Field emission is relevant to various electronic devices, including high current electron sources, flat-panel displays and light source bulbs [42-53]. The nanotube as emitters provide significantly brighter displays than either cathode ray tubes or Spindt tip based displays [3]. In addition, field emission electron sources are energy saving compared with thermionic ones, because no heating is necessary to emit electrons from the cathode surface [45]. Moreover, carbon nanotubes are free of any precious or hazardous element [45]. The alignment of the nanotubes is desired for this application. In addition, the nanotubes should have closed, well-ordered tips.

6. Electrochemical applications

Due to its electrical conductivity and chemical resistance, carbon is an important material for electrochemical applications, particularly electrodes for electrochemical cells [54-62], double-layer capacitors [61,63-67] and energy storage [68,69]. The small diameter of carbon nanotubes is advantageous for microelectrode arrays [70-73]. Capacitors exhibiting fast response (100 Hz) and high specific capacitance (100 F/g) have been attained by using carbon filaments [68]. Porous tablets of carbon nanotubes have been fabricated by using a polymeric binder for use as polarizable electrodes in capacitors [67].

Electrolyte absorptivity, specific surface area, surface chemistry and crystallographic structure are important for electrodes. Carbon black [59] is the most common type of carbon for these applications, though the use of conventional carbon fibers [54-60,63-66], VGCF [61,62] and carbon filaments [68,74-77] has been investigated. The removal of the hydrocarbon layer on the carbon filaments improves the electrochemical behavior, as indicated by the electron transfer rate across the electrode-electrolyte interface [22].

Catalytically grown carbon filaments of diameter 0.1 μm have been shown to be superior to carbon black in lithium primary cells, which use carbon as a porous electrode (current collector) [75] and as an electrically conductive additive in a non-conducting electrode [76]. The current collector of the lithium/thionyl chloride (Li/SOCl$_2$) cell conventionally uses carbon black, which needs a teflon binder. Due to the cleansing ability of thionyl chloride (the catholyte), carbon filaments used in place of carbon black do not require solvent cleansing prior to use. As the filaments tend to cling together, a binder is not necessary, in contrast to carbon black. Using the same paper-making process, the carbon filaments can be made into a thinner sheet than carbon black. The thinness is valuable for enhancing the energy density (particularly the specific gravimetric energy density) of the cell, as the area over which the lithium anode faces the carbon current collector is increased. In addition, the packing density is lower for the filament sheet than the carbon black sheet, so that the catholyte absorptivity is higher for the filament sheet than the carbon black and consequently the energy density is further increased [75].

The MnO$_2$ cathode of a Li/MnO$_2$ primary cell is itself electrically non-conducting, so a conductive additive, typically carbon black, is mixed with the MnO$_2$ particles. The use of catalytically grown carbon filaments of diameter 0.1 μm in place of carbon black as the conductive additive causes the running voltage near cell end-of-life to decline gradually, in contrast to the abrupt end-of-life when carbon black is used. The gradualness toward end-of-life is due to a high electron transfer rate and a high rate of electrolyte absorption. In order for the filaments to be effective, they need to undergo solvent cleansing prior to use [76].

By using catalytically grown carbon filaments of diameter around 80 Å, a double-layer capacitor of specific capacitance 102 F/g at 1 Hz has been achieved [77].

The catalytic growth of carbon filaments on carbons provides a way of modifying the surface for the purpose of improving the electrochemical behavior [78]. The resulting carbons are called hairy carbons. Particularly abundant hair growth occurs when the carbon is carbon black, due to the confinement of the catalyst size by the pores. Hair growth followed by an oxidation heat treatment, results in further improvement in the electrochemical behavior. The particulate nature of hairy carbon black is in contrast to the fibrous nature of carbon filaments or carbon fibers. The particulate nature facilitates dispersion, while the hairiness makes the use of a binder not necessary for electrode forming.

7. Thermal conduction

Due to the low temperature (500-700°C) during the catalytic growth of carbon filaments, carbon filaments may be only slightly crystalline (i.e., almost totally amorphous) after fabrication. However, subsequent heat treatment at 2500-3000°C causes graphitization [22], which is expected to result in a large increase in the thermal conductivity.

A high thermal conductivity is valuable for use of the filaments in composites for thermal management, which is critically needed for heat dissipation from electronic packages, space radiators and plasma facing. Due to the small diameter of the carbon filaments, single filament thermal conductivity measurement is difficult. The thermal conductivity of conventional pitch-based carbon fiber is 603 W/m.K for P-120 [79], 750 W/m.K for P-100-4 [80], 1000 W/m.K for P-X-5 [80] and 1055 W/m.K for K1100 [79]; that of VGCF is 2540-2680 W/m.K [80].

The thermal conductivity of carbon filament composites has not been reported. However, polymer matrix [81], aluminum matrix [82,83] and carbon matrix [82,84-86] composites containing VGCF exhibit thermal conductivities up to 466, 642 and 910 W/m.K respectively. For comparison, a polymer matrix composite containing conventional pitch-based continuous carbon fibers (P-120) exhibits a thermal conductivity of 245 W/m.K [80].

The electrical resistivity of a carbon filament (diameter 0.1 μm) polymer-matrix composite is higher than that of a nickel filament (0.4 μm diameter, with a 0.1 μm diameter carbon core) composite or a nickel fiber (2 μm diameter) composite at the same filler volume fraction and with the same matrix polymer [23]. At 13 vol. % carbon filaments, the DC resistivity of the composite is 0.37 Ω.cm; at 13 vol.% nickel filaments, the resistivity is 0.0035 Ω.cm [23]. The high resistivity of the carbon filament composite (in which the filaments were cleansed of the hydrocarbon surface layer prior to incorporation in the matrix) is attributed to the high resistivity of the filaments, which were not graphitized, and to the large filament-matrix interface area per unit volume.

8. Strain sensors

Strain sensors refer to sensors of strain, which relates to stress. The strain sensed includes reversible and irreversible strains. Due to the advent of smart structures, strain sensors are increasingly needed for structural vibration control and in situ structural health monitoring. Composites containing conventional short carbon fibers have their volume electrical resistivity change reversibly upon reversible strain, thus allowing the composites to serve as strain sensors. In the case of a composite with a ductile matrix (such as a polymer matrix), this phenomenon is due to the change in the distance between adjacent fibers in the composite and is referred to as piezoresistivity [87]. Tension causes this distance to increase, thereby increasing the resistivity; compression causes this distance to decrease, thereby decreasing the resistivity. In the case of a composite with an elastomer matrix (such as a silicone), the phenomenon is different in both direction and origin; the resistivity decreases upon tension, as observed for a silicone-matrix composite with 0.4 μm diameter nickel filaments [88]. This reverse piezoresistivity effect is probably due to the increase in filament alignment upon tension. In the case of a composite with a brittle matrix (such as a cement), the phenomenon is not reverse but is yet different in origin. It is due to the slight (< 1 μm) pull-out of the fiber (short) bridging a crack as the crack opens and the consequent increase in the contact electrical resistivity of the fiber-matrix interface [89-91]. Tension causes a crack to open, thereby increasing the resistivity; compression causes a crack to close, thereby decreasing the resistivity.

The use of carbon filaments in place of conventional short carbon fibers (based on isotropic pitch) in a polymer-matrix composite improves the reproducibility and linearity of the piezoresistivity effect (not reverse) [92]. This is because of the small diameter of the filaments, which results in (i) a large number of filaments per unit volume of the composite, (ii) reduced tendency for the filaments to buckle upon compression of the composite, and (iii) reduced tendency for the matrix at the junction of adjacent filaments to be damaged. Furthermore, the use of the filaments enhances the tendency for the reverse piezoresistivity effect [93].

The use of carbon filaments in place of conventional short carbon fibers (based on isotropic pitch) in a cement-matrix composite results in increased noise in the electromechanical effect [17]. This is because of the bent morphology and large aspect ratio of the filaments, which hinder the pull-out of filaments. Thus, carbon filaments are not attractive for cement-matrix composite strain sensors.

9. Porous carbons

Porous carbons with high porosity (above 50 vol. %) and/or high specific surface area are useful for numerous non-structural applications, such as electrodes, catalysts, catalyst support, filters, chemical absorbers, molecular sieves, membranes, dental and surgical prosthetic devices and thermal insulators [94-102]. Carbons with high porosity can be made from carbon fibers, which may be bound by a binder such as a polymer, pitch or carbon. Alternatively they can be made by carbonizing organic fibers that are bound by a binder. In either case, a pore forming agent may be used, although it is not essential. Porous carbons can also be made from a polymer that is not in the form of fibers, through foaming and carbonization. Due to the large diameter (typically 10 μm or more) of the carbon or organic fibers for the fiber-based porous carbons, and due to the foaming process for the polymer-based porous carbons, the pores in the resulting carbons are large (> 40 μm in mean size). As a result, the porous carbons are low in strength (< 7 MPa under compression) and in the specific geometric surface area (SGSA, < 1,100 cm^2/cm^3). This problem is more serious for the polymer-based porous carbons than the fiber-based porous carbons. By using carbon filaments (catalytically grown, of diameter 0.1 μm) in place of carbon fibers (typically 10 μm in diameter), a porous carbon of 4 μm mean pore size, SGSA 35,000 cm^2/cm^3 and compressive strength 30-35 MPa was obtained [103].

Carbons with high specific surface include the conventional activated carbon bulk [104-106], activated carbon fibers [107], fine carbon particles [108], carbon aerogels [109] and carbon nanotubes [110]. Other than being used as adsorbents for purification and chemical processing [104-106], these carbons are used as catalytic materials, battery electrode materials, capacitor materials, gas (e.g., hydrogen) storage materials and biomedical engineering materials. A hydrogen storage capacity of approximately 1.5 wt. % has been reached at ambient temperature and a hydrogen pressure of 12.5 MPa [110].

A problem concerning porous carbon materials relates to the need for high-surface-area porous carbon materials with mesopores and/or macropores for some applications. In fact, many macromolecules and ions encountered in purification, catalysis and batteries cannot penetrate the surface of the carbon without such pores. According to IUPAC, pores are classified into four types, namely macropores (diameter> 500 Å), mesopores (20 Å <diameter <500 Å), micropores (8 Å <diameter <20 Å) and micro-micropores (diameter <8 Å). Most pores are micropores in conventional activated carbons. The pore volume in activated carbon fibers (including pitch-based, PAN-based and rayon-based carbon fibers) is occupied by micropores (mainly) and micro-micropores. The pore volume in carbon aerogels is occupied by mesopores (mainly) and micropores. On the other hand, the specific surface area of carbon aerogels are low (e.g., 650 m^2/g [106]) compared to activated carbons (as high as 3000 m^2/g).

By surface oxidation of carbon filaments in ozone at 150°C and then activation in CO_2 + N_2 (1:1) at 970°C a mesoporous carbon (83% of total pore volume being > 30 Å pore size, 17% of total pore volume being <30 Å pore size), with a total pore volume of 1.55 cm^3/g and high specific surface area (1310 m^2/g) can be obtained [111]. Without activation (whether ozone treated or not), the filaments have only 44-57% of the total pore volume being > 30 Å pore size and the specific surface area is low (41-54 m^2/g). Activation by CO_2 greatly increased the specific surface area. This is in contrast to conventional carbon fibers, which have essentially no pores. The porous nature of the filaments is attributed to the fact that the filaments are made from carbonaceous gases. The separation between adjacent filaments in a filament compact is of the order of 0.1 μm, thus providing macropores within the compact. These elongated macropores serve as channels that facilitate fluid flow. The combination of mesopores within each filament and macropores between the filaments is in contrast to carbon aerogels, which have micropores within each particle and between particles and mesopores between chains of interconnected particles. The mesoporous activated carbon filaments have mean mesopore size (BJH) 54 Å.

10. Catalyst support

Catalytically grown carbon filaments, even without activation, have been shown to be an effective catalyst (e.g., Ni and Fe based particles) support material [112-117]. The catalytic activity for the conversion of hydrocarbons is higher than that when the catalyst particles are supported on either active carbon or γ-alumina [113,115]. The dispersion of a catalyst on carbon filaments is improved by prior surface treatment of the filaments by oxidation (e.g., using nitric acid) [118]. A platinum catalyst supported by carbon filaments is more active and filters more easily than that supported by activated carbon, as shown for the hydrogenation of nitrobenzene [118].

11. Conclusion

Applications of submicron diameter carbon filaments include structural applications, EMI shielding, electromagnetic reflection, surface electrical conduction, DC electrical conduction, electrochemical applications, thermal conduction, strain sensors and catalyst support. Most applications involve the incorporation of the carbon filaments in composites, most commonly polymer-matrix composites. Particularly promising applications include (i) the use of carbon filaments between layers of conventional continuous carbon fibers for improving the vibration damping ability and the storage modulus in the transverse direction, (ii) the coating of the carbon filaments with nickel to form nickel filaments for use as a filler in polymer-matrix composites for EMI shielding, electrostatic discharge protection and microwave waveguides, (iii) the use of carbon filaments as a filler in concrete for lateral guidance in automatic highways, (iv) the use of carbon filaments as a porous electrode (current collector) and as electrically conducting additives in a non-conducting electrode for lithium primary cells, (v) the use of carbon filaments for double-layer capacitors, (vi) the use of carbon filaments as a filler in a polymer-matrix composite strain sensor, (vii) the use of carbon nanotubes and activated carbon filaments for adsorption, hydrogen storage and catalyst support, and (viii) the use of carbon nanotubes for field emission.

Acknowledgement

This work was supported by the Defense Advanced Research Projects Agency of the U.S. Department of Defense.

References

1. Rodriguez, N.M. (1993) A review of catalytically grown carbon nanofibers, *J. Mater. Res.* **8**(12), 3233-3250.
2. Chitrapu, P., Lund, C.R.F. and Tsamopoulos, J.A. (1992) A model for the catalytic growth of carbon filaments, *Carbon* **30**(2), 285-293.
3. Subramoney, S. (1999) Carbon nanotubes – a status report, *The Electrochemical Society Interface* **18**(4), 34-37.
4. Chung, D.D.L. (1994) *Carbon Fiber Composites*, Butterworth-Heinemann, Boston.
5. Peebles, L.H. (1994) *Carbon Fibers*, CRC Press, Boca Raton.
6. Dresselhaus, M.S., Dresselhaus, G., Sugihara, K., Spain, I.L. and Goldberg, H.A. (1988) Springer Series in Materials Science, *Graphite Fibers and Filaments*, Vol. 1, Springer-Verlag, Berlin.
7. Endo, M. (1988) Grow carbon fibers in the vapor phase, *CHEMTECH* **18**(9), 568-576.
8. Tibbetts, G.G. (1989) Vapor-grown carbon fibers: status and prospects, *Carbon* **27**(5), 745-747.
9. Tibbetts, G.G., Gorkiewicz, D.W. and Alig, R.L. (1993) A new reactor for growing carbon fibers from liquid- and vapor-phase hydrocarbons, *Carbon* **31**(5), 809-814.
10. Kato, T., Matsumoto, T., Saito, T., Hayashi, J.-H., Kusakabe, K. and Morooka, S. (1993) Effect of carbon source on formation of vapor-grown carbon fiber, *Carbon* **31**(6), 937-940.
11. Mukai, S.R., Masuda, T., Harada, T. and Hashimoto, K. (1996) Dominant hydrocarbon which contributes to the growth of vapor grown carbon fibers, *Carbon* **34**(5), 645-648.
12. Masuda, T., Mukai, S.R., Fujikawa, H., Fujikata, Y. and Hashimoto, K. (1994) Rapid vapor growth carbon fiber production using the intermittent liquid pulse injection technique, *Mater. Manufacturing Proc.* **9**(2), 237-247.
13. Ishioka, M., Okada, T. and Matsubara, K. (1992) Mechanical properties of vapor-grown carbon fibers prepared from benzene in Linz-Donawitz converter gas by floating catalyst method, *J. Mater. Res.* **7**(11), 3019-3022.
14. Tibbetts, G.G. (1990) Carbon fibers from vapor phase hydrocarbons, *SAE Transactions 99 (Sect. 1)*, Soc. of Automotive Engineers, Warrendale, PA, 246-249.
15. Li, L. and Chung, D.D.L. (1994) Electrical and mechanical properties of electrically conductive polyethersulfone composite, *Composites* **25**(3), 215-224.
16. Shui, X. and Chung, D.D.L. (1997) Nickel filament polymer-matrix composites with low surface impedance and high electromagnetic interference shielding effectiveness, *J. Electron. Mater.* **26**(8), 928-934.
17. Fu, X. and Chung, D.D.L. (1996, 1997) Submicron carbon filament cement-matrix composites for electromagnetic interference shielding, *Cem. Concr. Res.* **26**(10), 1467-1472; **27**(2), 314.
18. Hudnut, S.W. and Chung, D.D.L. (1995) Use of submicron diameter carbon filaments for reinforcement between continuous carbon fiber layers in a polymer-matrix composite, *Carbon* **33**(11), 1627-1631.
19. Downs, W.B. and Baker, R.T.K. (1991) Novel carbon fiber-carbon filament structures, *Carbon* **29**(8), 1173-1179.
20. Downs, W.B. and Baker, R.T.K. (1995) Modification of the surface properties of carbon fibers via the catalytic growth of carbon nanofibers, *J. Mater. Res.* **10**(3), 625-633.
21. Shui, X. and Chung, D.D.L. (1993) Conducting polymer-matrix composites containing carbon filaments of submicron diameter, *38th Int. SAMPE Symp. Exhib.*, Advanced Materials: Performance Through Technology Insertion, Vince Bailey, Gerald C. Janicki and Thomas Haulik (eds.), Book 2, 1869-1875.
22. Shui, X, Frysz, C.A. and Chung, D.D.L. (1995) Solvent cleansing of the surface of carbon filaments and its benefit to the electrochemical behavior, *Carbon* **33**(12), 1681-1698.

23. Shui, X. and Chung, D.D.L. (1995) Submicron nickel filaments made by electroplating carbon filaments as a new filler material for electromagnetic interference shielding, *J. Electron. Mater.* **24**(2), 107-113.

24. Tibbetts, G.G. and McHugh, J.J. (1999) Mechanical properties of vapor-grown carbon fiber composites with thermoplastic matrices, *J. Mater. Res.* **14**(7), 2871-2880.

25. Fu, X. and Chung, D.D.L. (1998) Submicron-diameter-carbon-filament cement-matrix composites, *Carbon* **36**(4), 459-462.

26. Caldeira, G., Maia, J.M., Carneiro, O.S., Covas, J.A. and Bernardo, C.A. (1997) Production and characterization of innovative carbon fibers-polycarbonate composites, *ANTEC '97: Plastics – Saving the Planet, Conf. Proc.*, SPE, Brookfield, Conn., **2**, 2352-2356.

27. Fu, X., Lu, W. and Chung, D.D.L. (1998) Ozone treatment of carbon fiber for reinforcing cement, *Carbon* **36**(9), 1337-1345.

28. Laurent, Ch., Peigney, A. Quenard, O. and Rousset, A. (1997) Novel ceramic matrix nanocomposite powders containing carbon nanotubes, *Key Eng. Mater.* **132-136**(Pt 1), 157-160.

29. Mottahed, B.D. and Manoocheheri, S. (1995) Review of research in materials, modeling and simulation, design factors, testing, and measurements related to electromagnetic interference shielding, *Polym.-Plast. Technol. Eng.* **34**(2), 271-346.

30. Neelakanta, P.S. and Subramaniam, K. (1992) Controlling the properties of electromagnetic composites, *Adv. Mater. Proc.* **141**(3), 20-25.

31. Lu, G., Li, X. and Jiang, H. (1996) Electrical and shielding properties of ABS resin filled with nickel-coated carbon fibers, *Composites Sci. Tech.* **56**, 193-200.

32. Kaynak, A., Polat, A. and Yilmazer, U. (1996) Some microwave and mechanical properties of carbon fiber-polypropylene and carbon black-polypropylene composites, *Mater. Res. Bull.* **31**(10), 1195-1206.

33. Jana, P.B. and Mallick, A.K. (1994) Studies on effectiveness of electromagnetic interference shielding in carbon fiber filled polychloroprene composites, *J. Elastomers and Plastics* **26**(1), 58-73.

34. Chiou, J.-M., Zheng, Q. and Chung, D.D.L. (1989) Electromagnetic interference shielding by carbon fiber reinforced cement, *Composites* **20**(4), 379-381.

35. Shui, X. and Chung, D.D.L. (2000) Submicron diameter nickel filaments and their polymer-matrix composites, *J. Mater. Sci.* **35**, 1773-1785.

36. Fu, X. and Chung, D.D.L. (1998) Radio wave reflecting concrete for lateral guidance in automatic highways, *Cem. Concr. Res.* **28**(6), 795-801.

37. Chellappa, V. and Jang, B.Z. (1995) Electrical conduction in thermoplastic elastomer matrix composites containing catalytic chemical vapor deposited carbon whisker, *J. Mater. Sci.* **30**(19), 4879-4883.

38. Farriss, C.W., II, Kelley, F.N. and Von Meerwall, E. (1995) Use of carbon fibril additives to reduce the DC resistivity of elastomer-based composites, *J. Appl. Polymer Sci.* **55**(6), 935-943.

39. Sandler, J., Shaffer, M.S.P., Prasse, T., Bauhofer, W., Schulte, K. and Windle, A.H. (1999) *Polymer* **40**, 5967-5971.

40. Curran, S.A., Ajayan, P.M., Blau, W.J., Carroll, D.L., Coleman, J.N., Dalton, A.B., et. al., (1998) Composite from poly(m-phenylenevinylene-co-2,5-dioctoxy-p-phenylenevinylene) and carbon nanotubes: a novel material for molecular optoelectronics, *Adv. Mater.* **10**(14), 1091-1093.

41. Dai, L. (1999) Advanced synthesis and microfabrications of conjugated polymers, C60-containing polymers and carbon nanotubes for optoelectronic applications, *Polymers for Adv. Tech.*, **10**(7), 357-420.

42. Ma, R.Z., Xu, C.L., Wei, B.Q., Liang, J., Wu, D.H. and Li, D.J. (1999) Electrical conductivity and field emission characteristics of hot-pressed sintered carbon nanotubes, *Mater. Res. Bull.* **34**(5), 741-747.

43. Bonard, J.-M., Salvetat, J.-P., Stockli, T., Forro, L. and Chatelain, A. (1999) Field emission from carbon nanotubes: perspectives for applications and clues to the emission mechanism, *Appl. Phys. A*: Materials Science & Processing **69**(3), 245-254.

44. Huang, S., Dai, L. and Mau, A.W.H. (1999) Patterned growth and contact transfer of well-aligned carbon nanotube films, *J. Phys. Chem. B.* **103**(21), 4223-4227.

45. Saito, Y. and Uemura, S. (2000) Field emission from carbon nanotubes and its application to electron sources, *Carbon* **38**(2), 169-182.

46. Saito, Y., Hamaguchi, K., Mizushima, R., Uemura, S., Nagasako, T., Yotani, J., et al., (1999) Field emission from carbon nanotubes and its application to cathode ray tube lighting elements, *Appl. Surface Sci.* **146**(1), 305-311.

47. Saito, Y. (1998) New forms of carbon – fullerenes and nanotubes, *Nihon Enerugi Gakkaishi/Journal of the Japan Institute of Energy* **77**(9), 867-875 (Japanese).

48. Habermann, T., Goehl, A., Janischowsky, K., Nau, D., Stammler, M., Ley, L. (1998) Field emission characterization of carbon nanostructures for cold cathode applications, *Proc. of the IEEE International Vacuum Microelectronics Conference*, IEEE, Piscataway, NJ, 200-201.

286

49. Dean, K.A., VonAllmen, P. and Chalamala, B.R. (1998) Thermal field emission behavior of single walled carbon nanotubes, *Proc. of the IEEE International Vacuum Microelectronics Conference*, IEEE, Piscataway, NJ, 196-197.

50. Saito, Y., Hamaguchi, K., Uemura, S., Uchida, K., Tasaka, Y., Ikazaki, F., et al., (1998) Field emission from multi-walled carbon nanotubes and its application to electron tubes, *Appl. Phys. A: Materials Science & Processing* **67**(1), 95-100.

51. Sinitsyn, N.I., Gulyaev, Y.V., Torgashov, G.V., Chernozatonskii, L.A., Kosakovskaya, Z.Y., Zakharchenko, Y.F., et al., (1997) Thin films consisting of carbon nanotubes as a new material for emission electronics, *Appl. Surface Sci.* **111**, 145-150.

52. Gulyaev, Y.V., Chernozatonskii, L.A., Kosakovskaya, Z.Y., Musatov, A.L., Sinitsin, N.I. and Torgashov, G.V. (1996) Carbon nanotube structures – a new material of vacuum microelectronics, *Proc. of the IEEE International Vacuum Microelectronics Conference*, IEEE, Piscataway, NJ, 5-9.

53. Sinitsyn, N.I., Gulyaev, Y.V., Devjatkov, N.D., Golant, M.B., Alekseyenko, A.M., Zakharchenko, Y.F., (1999) Potentials of vacuum microelectornics on the way to constructing microwave vacuum integrated circuits, *Radiotekhnika* (4), 8-17 (Russian).

54. Verbrugge, M.W. and Koch, D.J. (1996) Lithium intercalation of carbon-fiber microelectrodes, *J. Electrochem. Soc.* **143**(1), 24-31.

55. Endo, M., Nakamura, J.-I., Emori, A., Sasabe, Y., Takeuchi, K. and Inagaki, M. (1994) Lithium secondary battery based on intercalation in carbon fibers as negative electrode, *Molecular Crystals and Liquid Crystals Science and Technology*, Section A: Molecular Crystals and Liquid Crystals, Proc. 7th Int. Symp. on Intercalation Compounds **244**, 171-176.

56. Takamura, T., Kikuchi, M., Awano, H., Tatsuya, U. and Ikezawa, Y. (1995) Carbon surface conditioning produces and anode suitable for heavy-duty discharge in Li secondary batteries, *Materials Research Society Symp. Proc.*, Materials for Electrochemical Energy Storage and Conversion – Batteries, Capacitors and Fuel Cells, 1995 Spring Meeting, **393**, 345-355.

57. Tamaki, T. Characteristics of mesophase pitch-based carbon fibers as anode materials for lithium secondary cells, *Ibid* 357-365.

58. Yazami, R., Zaghib, K. and Deschamps, M. (1994) Carbon fibers and natural graphite as negative electrodes for lithium ion-type batteries, *J. Power Sources* **52**, 55-59.

59. Chusid, O., Ein Ely, Y., Aurbach, D., Babai, M. and Carmeli, Y. (1993) Electrochemical and spectroscopic studies of carbon electrodes in lithium battery electrolyte systems, *J. Power Sources* **43-44**, 47-64.

60. Endo, M., Nishimura, Y., Takahashi, T., Takeuchi, K. and Dresselhaus, M.S. (1996) Lithium storage behavior for various kinds of carbon anodes in Li ion secondary battery, *J. Phys. Chem. Solids* **57**(6-8), 725-728.

61. Endo, M., Okada, Y. and Nakamura, H. (1989) Lithium secondary battery and electric double layer capacitor using carbon fibers electrode, *Synth. Met.* **34**(1-3), 739-744.

62. Zaghib, K., Tatsumi, K., Abe, H., Ohsaki, T., Sawad, Y. and Higuchi, S. (1998) Optimization of the dimensions of vapor-grown carbon fiber for use as negative electrodes in lithium-ion rechargeable cells, *J. Electrochem. Soc.* **145**(1), 210-215.

63. Biniak, S., Dzielendziak, B. and Siedlewski, J. (1995) The electrochemical behaviour of carbon fibre electrodes in various electrolytes – double-layer capacitance, *Carbon* **33**(9), 1255-1263.

64. Tanahashi, I., Yoshida, A. and Nishino, A. (1990) Activated carbon fiber sheets as polarizable electrodes of electric double layer capacitors, *Carbon* **28**(4), 477-482.

65. Ishikawa, M., Morita, M., Ihara, M. and Matsuda, Y. (1994) Electric double-layer capacitor composed of activated carbon fiber cloth electrodes and solid polymer electrolytes containing alkylammonium salts, *J. Electrochem. Soc.* **141**(7), 1730-1734.

66. Matsuda, Y., Morita, M., Ishikawa, M. and Ihara, M. (1993) New electric double-layer capacitors using polymer solid electrolytes containing tetraalkylammonium salts, *J. Electrochem. Soc.* **140**(7), L109-L110.

67. Ma, R.Z., Liang, J., Wei, B.Q., Zhang, B., Xu, C.L. and Wu, D.H. (1999) Study of electrochemical capacitors utilizing carbon nanotube electrodes, *J. Power Sources* **84**(1), 126-129.

68. Lipka, S.M. (1998) Carbon nanofibers and their applications for energy storage, *Proc. 13th Annual Battery Conference on Applications and Advances*, IEEE, Piscataway, NJ, 373-374.

69. Che, G., Lakshmi, B.B., Martin, C.R. and Fisher, E.R. (1999) Metal-nanocluster-filled carbon nanotubes: catalytic properties and possible applications in electrochemical energy storage and production, *Langmuir* **15**(3), 750-758.

70. Burghard, M., Duesberg, G., Philipp, G., Muster, J. and Roth, S. (1998) Controlled adsorption of carbon nanotubes on chemically modified electrode arrays, *Adv. Mater.* **10**(8), 584-588.

71. Davis, J.J., Coles, R.J. and Hill, H.A.O. (1997) Protein electrochemistry at carbon nanotube electrodes, *J. Electroanalytical Chem.* **440**(1-2), 279-282.

72. Grobert, N., Terrones, M., Osborne, A.J., Terrones, H., Hsu, W.K., Trasobares, S., (1998) Thermolysis of C60 thin films yields Ni-filled tapered nanotubes, *Appl. Phys. A: Materials Science & Processing* **67**(5), 595-598.

73. Britto, P.J., Santhanam, K.S.V., Rubio, A., Alonso, J.A. and Ajayan, P.M. (1999) Improved charge transfer at carbon nanotube electrodes, *Adv. Mater.* **11**(2), 154-157.

74. Shui, X., Chung, D.D.L. and Frysz, C.A. (1994) Hairy carbon electrodes studied by cyclic voltammetry and battery discharge testing, *J. Power Sources* **47**(3), 313-320.

75. Frysz, C.A., Shui, X. and Chung, D.D.L. (1996) Use of carbon filaments in place of carbon black as the current collector of a lithium cell with a thionyl chloride-bromine chloride catholyte, *J. Power Sources* **58**(1), 55-66.

76. Frysz, C.A., Shui, X. and Chung, D.D.L. (1996) Carbon filaments and carbon black as a conductive additive to the manganese dioxide cathode of a lithium electrolytic cell, *J. Power Sources* **58**(1), 41-54.

77. Niu, C., Sichel, E.K., Hoch, R., Moy, D. and Tennent, H. (1997) High power electrochemical capacitors based on carbon nanotube electrodes, *Appl. Phys. Lett.* **70**(11), 1480-1482.

78. Shui, X., Frysz, C.A. and Chung, D.D.L. (1997) Electrochemical behavior of hairy carbons, *Carbon* **35**(10-11), 1439-1455.

79. Lundblad, W.E., Starrett, H.S. and Wanstrall, C.W. (1994) Technique for the measurement of the thermal conductivity of graphite and carbon fibers, *Proc. 26th Int. SAMPE Technical Conference, 50 Years of Progress in Materials and Process Science Technology* 759-764.

80. Nysten, B. and Issi, J.-P. (1990) Composites based on thermally hyperconductive carbon fibers, *Composites* **21**(4), 339-343.

81. Ting, J.-M., Guth J.R., and Lake, M.L. (1995) Light weight, highly thermally conductive composites for space radiators, *Proc. 19th Annual Conf. Composites*, Advanced Ceramics, Materials, and Structures, Cocoa Beach, Fl, Jan. 1995, American Ceramic Soc., Ceramic Engineering and Science Proc. **16**(4), 279-288.

82. Ting, J.-M., Lake, M.L. and Duffy, D.R. (1995) Composites based on thermally hyper-conductive vapor grown carbon fiber, *J. Mater. Res.* **10**(6), 1478-1484.

83. Ting, J.-M. and Lake, M.L. (1995) Vapor grown carbon fiber reinforced aluminum composites with very high thermal conductivity, *J. Mater. Res.* **10**(2), 247-250.

84. Ting, J.-M. and Lake, M.L. (1994) High heat flux composites for plasma-facing materials, *J. Nuclear Mater.* **212**(1), pt. B, 1141-1145.

85. Ting, J.-M. and Lake, M.L. (1995) Vapor-grown carbon-fiber reinforced carbon composites, *Carbon* **33**(5), 663-667.

86. Ting, J.-M. and Lake, M.L. (1994) An innovative semiconductor base – diamond/(carbon-carbon) composite, *Diamond and Related Materials* **3**(10), 1243-1248.

87. Wang, X. and Chung, D.D.L. (1995) Short carbon fiber reinforced epoxy as a piezoresistive strain sensor, *Smart Mater. Struct.* **4**, 363-367.

88. Shui, X. and Chung, D.D.L. (1997) A new electromechanical effect in discontinuous filament elastomer-matrix composites, *Smart Mater. Struct.* **6**, 102-105.

89. Chen, P.-W. and Chung, D.D.L. (1996) Carbon fiber reinforced concrete as an intrinsically smart concrete for damage assessment during static and dynamic loading, *ACI Mater. J.* **93**(4), 341-350.

90. Chen, P.-W. and Chung, D.D.L. (1996) Concrete as a new strain/stress sensor, *Composites, Part B* **27B**, 11-23.

91. Chen, P.-W. and Chung, D.D.L. (1993) Carbon fiber reinforced concrete as a smart material capable of non-destructive flaw detection, *Smart Mater. Struct.* **2**, 22-30.

92. Shui, X. and Chung, D.D.L. (1996) Piezoresistive carbon filament polymer-matrix composite strain sensor, *Smart Mater. Struct.* **5**, 243-246.

93. Chellappa, V., Chiou, Z.W. and Jang, B.Z. (1995) Electromechanical and electrothermal behaviours of carbon whisker reinforced elastomer composites, *J. Mater. Sci.* **30**(17), 4263-4272.

94. Hucke, E.E. (1975) Methods of producing carbonaceous bodies and the products thereof, U.S. Patent 3,859,421.

95. Wang, J. (1981) Reticulated vitreous carbon – a new versatile electrode material, *Electrochim. Acta* **26**, 1721-1726.

96. Strohl, A.N. and Curran, D.J. (1979) Controlled potential coulometry with the flow-through reticulated vitreous carbon electrode, *Anal. Chem.* **51**, 1050-1053.

97. Blaedel, W.J. and Wang, J. (1980) Characteristics of a rotated porous flow-through electrode, *Anal. Chem.* **52**(11), 1697-1700.

98. Agarwal, I.C., Rochon, A.M., Gesser, H.D. and Sparling, A.B. (1984) Electrodeposition of six heavy metals on reticulated vitreous carbon electrode, *Water Res.* **18**, 227-232.

99. Sylwester, A.P., Aubert, J.H., Rand, P.B., Arnold, C., Jr. and Clough, L.R. (1987) Low-density microcellular carbonized polyacrylonitrile (PAN) foams, *Polymer. Mater. Sci. Eng.* **57**, 113-117.

100. Sylwester, A.P. and Clough, R.L. (1989) Electrically conductive reticulated carbon composites, *Synth Met.*, **29**(2-3), 253-258.

101. Oren, Y. and Soffer, A. (1983) Graphite felt as an efficient porous electrode for impurity removal and recovery of metals, *Electrochim. Acta* **28**, 1649-1654.

102. Lestrade, C., Guyomar, P.Y and Astruc, M. (1981) Electrochemical removal of dilute heavy metals with carbon felt porous electrodes, *Environ. Technol. Lett.* **2**, 409.

103. Shui, X. and Chung, D.D.L. (1996) High-strength high-surface-area porous carbon made from submicron-diameter carbon filaments, *Carbon* **34**(6), 811-814; **34**(9), 1162.
104. Dubinin, M.M., Polyakov, N.S. and Petukhova, G.A. (1993) Porous structure and surface chemistry of active carbons, *Adsorption Science Tech.* **10**(1-4), 17-26.
105. Chiang, H.-L., Chiang, P.C. and You, J.H. (1995) The influences of O₃ reaction on physico-chemical characteristics of activated carbon for benzene adsorption, *Toxicological Environmental Chem.* **47**(1-2), 97-108.
106. Takeuchi, Y. and Itoh, T. (1993) Removal of ozone from air by activated carbon treatment, *Sep. Technol.* **3**(3), 168-175.
107. Alcaniz-Monge, J., Cazorla-Amoros, D., Linares-Solano, A., Yoshida, S. and Oya, A. (1994) Effect of the activating gas on tensile strength and pore structure of pitch-based carbon fibres, *Carbon* **32**(7), 1277-1283.
108. Ghosal, R., Kaul, D.J., Boes, U., Sanders, D., Smith, D.M. and Maskara, A. (1995) Specialty carbon adsorbents with a tailored pore structure and their properties, advances in porous materials, *Materials Research Society Symp. Proc.* **371**, 413-423.
109. Fung, A.W.P., Wang, Z.H., Lu, K., Dresslehaus, M.S. and Pekala, R.W. (1993) Characterization of carbon aerogels by transport measurements, *J. Mater. Res.* **8**(8), 1875-1885.
110. Ströbel, R., Jörissen, L., Schliermann, T., Trapp, V., Schütz, W., Bohmhammel, K., Wolf, G. and Garche, J. (1999) Hydrogen adsorption on carbon materials, *J. Power Sources* **84**(2), 221-224.
111. Lu, W. and Chung, D.D.L. (1997) Mesoporous activated carbon filaments, *Carbon* **35**(3), 427-430.
112. Kim, M., Rodriguez, N.M. and Baker, R.T.K. (1995) Carbon nanofibers as a novel catalyst support, *Mater. Res. Soc. Symp. Proc.*, Synthesis and Properties of Advanced Catalytic Materials **368**, 99-104.
113. Rodriguez, N.M., Kim, M.-S. and Baker, R.T.K. (1994) Carbon nanofibers. A unique catalyst support medium, *J. Phys. Chem.* **98**(50), 13108-13111.
114. Park C. and Baker, R.T.K. (1998) Impact of the graphite nanofiber structure on the behavior of supported nickel, *Mater. Res. Soc. Symp. Proc.*, Recent Advances in Catalytic Materials **497**, 145-150.
115. Chambers, A., Nemes, T., Rodriguez, N.M. and Baker, R.T.K. (1998) Catalytic behavior of graphite nanofiber supported nickel particles. 1. Comparison with other support media, *J. Phys. Chem.* **102**, 2251-2258.
116. Park, C., Rodriguez, N.M. and Baker, R.T.K. (1997) Use of graphite nanofibers as a novel catalyst support medium for hydrogenation reactions, *Mater. Res. Soc. Symp. Proc.*, Advanced Catalytic Materials **454**, 21-26.
117. Park, C. Baker, R.T.K. and Rodriguez, N.M. (1995) Modification of the catalytic activity of iron by the addition of nickel, *Coke Formation and Mitigation Preprints* – American Chemical Society, Div. Petroleum Chem., ACS, Washington, D.C. **40**(4), 646-648.
118. Geus, J.W., Van Dillen, A.J. and Hoogenraad, M.S. (1995) Carbon fibrils mechanism of growth and utilization as a catalyst support, *Mater. Res. Soc. Symp. Proc.*, Synthesis and Properties of Advanced Catalytic Materials **368**, 87-98.

THE ROLE OF RHEOLOGY IN THE PROCESSING OF VAPOR GROWN CARBON FIBER / THERMOPLASTIC COMPOSITES

C.A. BERNARDO*, F.W.J. van HATTUM**, O.S. CARNEIRO*, J.M. MAIA*

*Department of Polymer Engineering, Universidade do Minho, Portugal
**Centre of Lightweight Structures, TUD-TNO, the Netherlands

Abstract.

The work done on the processing of thermoplastics reinforced with sub-micron VGCF is reviewed in this chapter. It is feasible, and indeed easy, to produce these composites by means of conventional processing technologies. The main issues affecting the production of good-quality composites are fiber alignment, fiber-matrix adhesion and fiber dispersion. Rheological experiments can give valuable quantitative information on these issues, by illuminating the relationship between melt structure, processing and properties.

1. Introduction

Carbon fibers are increasingly used in high-technology applications due to their good mechanical, thermal and electrical properties. Incorporating these fibers in a polymer matrix yields composites with properties that can be tailored to requirement. Depending on fiber length, content and orientation, the composites will have properties that are somewhere between the constituents' properties. Carbon fibers are traditionally used in polymer composites with thermoset matrices, like epoxy and polyester. Typical manufacturing methods are filament winding and hand lay-up. In this way, it is possible to manufacture composite materials having outstanding properties that can be applied in high technology areas, such as aerospace engineering. However, the nature of the manufacturing processes and the relatively high price of the composites have limited their applicability in a large number of areas. If carbon fibers are employed in combination with thermoplastic polymer matrices, different manufacturing methods, such as compression and injection molding, can be used. These methods allow the production of parts with a greater freedom of design at high rates, thus opening new areas of application. However, traditional carbon fibers are relatively expensive when compared, for instance, to glass fibers. This remains a major limitation to their use in composites for high consumption markets.

Vapor Grown Carbon Fibers (VGCF) are distinctively different from traditional carbon fibers in their way of production, properties and prospect of low-cost fabrication.

L.P. Biró et al. (eds.), Carbon Filaments and Nanotubes: Common Origins, Differing Applications, 289–300.
© 2001 *Kluwer Academic Publishers. Printed in the Netherlands.*

Hence, they offer an opportunity for the production of thermoplastic matrix composites at low cost, if conventional manufacturing techniques can be used in the process.

One of the objectives of the present chapter is to review the work done on the processing of thermoplastics filled with VGCF. For comparison, thermoplastics filled with a commercial carbon black, processed under the same conditions, were also measured. As the rheological characteristics of VGCF-thermoplastic composites determined in the melt are good indicators of their processability, this data was also included in the review.

2. Vapor Grown Carbon Fibers

As explained in another chapter of this book, there are two distinct types of vapor grown carbon fibers, corresponding to two different production methods. In the first method, the fibers are grown on catalyst seeded substrates in a batch process, typically yielding fibers centimeters in length. In the second method, the fibers are grown inside a continuous flow reactor. In this way, fibers with a diameter of about 0.1 μm and lengths of 10 to 100 μm can be obtained. This latter method allows production at much lower costs and higher rates, thus making VGCF commercially more attractive. The research on the fabrication of thermoplastic composites by injection or compression molding has therefore focused on this type of fiber, usually referred to as sub-micron (diameter) VGCF. However, the distinctive nature of sub-micron VGCF compared to other carbon fibers engenders specific processing problems, as the fibers emerge from the flow reactor in a highly entangled state, with a low apparent density. This makes them quite difficult to manipulate and/or incorporate in highly viscous polymer melts.

3. Composite Manufacturing

As VGCF are relatively new, most of the research has been focused on their growth and properties. As a consequence, the investigation on the application of these fibers in thermoplastic matrices has only recently commenced, and the amount of work done in this area is relatively limited.

To the authors' knowledge, the first work on the application of VGCF in thermoplastic polymer matrix composites was published in 1993 by Dasch et al. [1]. The VGCF were grown by a continuous process developed by Tibbetts et al. [2], and not subjected to any subsequent surface treatment. Fibers prepared in this way are called as-grown VGCF in this text. Their average diameter and length were reported as 80 nm and 2.5 μm, respectively, giving an average aspect ratio of about 30. Thermoplastic composites were prepared with both polycarbonate and nylon. The fibers and the polymer were mixed in a Brabender mixing bowl at a temperature slightly above the melting point of the resin. Material from the mixing bowl was compression molded to form the test specimens. Although the manufacturing method was far from optimum, the results demonstrate the potential of using conventional, easy-to-use, equipment for the fabrication of VGCF-thermoplastic composites. SEM-micrographs evidenced the wet-out of the as-grown VGCF by the polymer. The properties of the fibers, inferred from the composite's properties, were similar to those of PAN-based carbon fibers. One of the observations of this early research was the brittle behavior of the composites at fiber volume fractions above 30%. The authors report specimens cracking or even breaking in the mould due to residual stresses after cooling. Moreover, the equipment did not orient the fibers during processing, thus limiting the final composite's properties.

Dutta *et al.* [3] reported the fabrication of VGCF composites with a polycarbonate matrix. The VGCF were produced by Applied Sciences, Inc. Dry-mixed blends of VGCF and polycarbonate were prepared in a single screw extruder, in an attempt to obtain a better dispersion than that reported by Dasch *et al.* In spite of that, SEM-micrographs of the extrudates show a poor dispersion of the VGCF. The fibers showed a clear tendency to agglomerate, even at fractions as low as 2% by weight. At a higher concentration (5 wt. %) the dispersion was even poorer. The authors suggested that it was necessary to obtain a more homogeneous blend, and pointed out that a single screw extruder is not a very efficient mixer.

Following the above research, Carneiro *et al.* [4] reported an assessment of the practical feasibility of producing VGCF/ polycarbonate composites via conventional polymer processing techniques. The VGCF used in this research were also supplied by Applied Sciences, Inc. and were therefore similar to those mentioned above. Prior to processing, the fibers were de-bulked with a surfactant to facilitate handling and incorporation into the polymer melt. The final composites were processed in three steps. First, the fibers and the polycarbonate were dry mixed to obtain a homogeneous blend. Then, the blend was fed to an extruder and composite extrudates with 5, 10, 20 and 30 wt.% fibers were produced. To assure a good dispersion of the fibers, a co-rotative twin-screw extruder was used in this step. A schematic cross-section of the machine is shown in Figure 1.

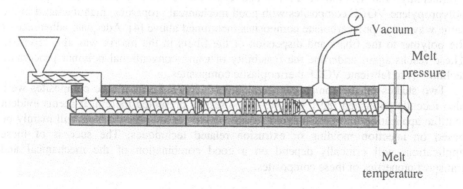

Figure 1. Schematic of the extruder used in the production of the composites

The operating conditions were optimized in prior runs, using the same polycarbonate with carbon black as filler. The extrudate was then pelletized and fed to a normal injection molding machine where the final test specimens were obtained. The machine parameters were set to maximize melt homogeneity and the degree of fiber orientation. No special difficulties were observed in this procedure. The authors therefore concluded that the production of VGCF-thermoplastic composites is feasible using conventional technologies. However, the composite with 30 wt. % fibers was too brittle to be injection molded, as previously observed by Dasch et al. [1]. Furthermore, in SEM observations of the composites' fracture surfaces, the polymer did not seem to wet the VGCF, indicating poor adhesion between fibers and matrix.

Tibbetts et al. [5] investigated the production of composites with sub-micron VGCF in both polypropylene and nylon-6,6. They used a fairly simple bench top injection molder, in which the fibers were gradually added to the polymer melt. The rotational and vertical movement of a rotor in a cylindrical heated cup performed a limited mixing of fibers and resin. Finally, by pulling down a lever, the mixture was injected into a mould to obtain the final composite. From SEM observations the authors concluded that both the polycarbonate and the nylon wetted the fibers well. The composites, however, had extensive uninfiltrated fiber areas. The low shear generated by the equipment was apparently not able to break up the large entangled fiber structures.

The infiltration problem was investigated by ball-milling the as-grown fibers before processing. SEM micrographs showed that ball milling can effectively break up the fiber clumps, thus allowing better melt infiltration. However, the resulting composite properties were very sensitive to the duration of the process, as long ball-mixing times tended to reduce the length of the fibers. This effect was revealed by an initial increase in the composite properties, followed by a sharp decrease as the ball-milling time increased. Hence, ball milling can only facilitate processing in a limited way. The authors concluded that, in order to obtain better performance composites, an optimum dispersion/distribution of the fibers is essential. High shear processes, such as extrusion can be used to attain these effects. Additionally, infiltration can be improved by previous dispersion and mixing, use of higher pressures or less viscous polymers and longer infiltration times.

Recently, van Hattum et al. [6] and Carneiro et al. [7], successfully produced polypropylene-VGCF composites with good mechanical properties, manufactured in the same way as the polycarbonate composites mentioned above [4]. Adequate adherence of the polymer to the fibers and dispersion of the fibers in the matrix was also reported. These results again underline the feasibility of using conventional polymer processing technology to fabricate VGCF thermoplastic composites.

Two studies on the compression molding of VGCF-thermoplastic composites were also recently published [8,9]. In spite of the results presented therein, it seems evident that the application of VGCF-thermoplastic composites in wider markets will mainly be based on injection molding or extrusion related techniques. The success of these applications will critically depend on a good combination of the mechanical and transport properties of these composites.

4. Rheology

From the above results, it can be concluded that the smooth processing of composites with good mechanical properties depends on three main factors: i) fiber alignment; ii) fiber-matrix adhesion; and iii) fiber dispersion in the polymer melt. Each of these factors can be studied by means of rheological experiments and theory for different types of vapor grown carbon fibers and distinct polymer matrices, as will be shown in the subsequent paragraphs.

4.1. FIBRE ALIGNMENT

The effect of the shear generated during the processing of polymer matrix composites on the alignment of the fibers can be studied by rotational and capillary rheometries. In rotational rheometry experiments the material is placed between two parallel rotating disks. The shear rates reached in these experiments are low, typically equal to, or below $10\ s^{-1}$. In capillary rheometry the material is forced to flow from a large barrel into a small die. With this set-up, shear rates in the range of those occurring during processing, 1 to $1000\ s^{-1}$, can be obtained.

One of these studies was reported by Carneiro et al. [4], who investigated the rheological behavior of polycarbonate-VGCF composites. The shear viscosity of the polymer and the composites, determined both by rotational and capillary rheometries in the low and high shear rate ranges, respectively, are represented jointly in Figure 2.

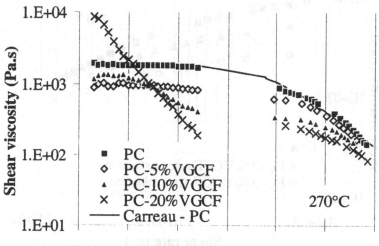

Figure 2. Shear viscosity of polycarbonate-and polycarbonate-VGCF composites determined by rotational (low shear) and capillary (high shear) rheometries (from ref. [4], by courtesy of Elsevier Science Ltd.)

The behavior depicted in the figure is not that commonly observed in this type of representations. The viscosity of the composites at low shear rates can be higher than that of the unreinforced matrix material. In the shear-thinning region, however, the viscosity decreases with increasing fiber content, and continues to do so in the high shear rates region, *i.e.*, above $100\ s^{-1}$. This behavior, already observed in other polycarbonate systems, is consistent with the decrease in the consumption of power that occurred during the extrusion of the composites. The decrease was attributed to the effect of shear-induced fiber alignment [10], resulting in a layered stream. Very high shear rates can occur locally, leading to a lower viscosity than otherwise anticipated. Increasing the fiber content will induce an even higher fiber alignment, decreasing further the viscosity [11]. In addition, poor fiber-matrix adhesion, as observed by Carneiro *et al.* on the same system [4], can cause inter-layer slip, enhancing the effect. Furthermore, it can be observed that there is a discontinuity between the viscosities measured by parallel-plate and capillary rheometries, the former being lower than the latter. This effect can be explained by the differences in the deformation history between the two systems. Although the shear rates are higher in the initial stages of the capillary experiments, there is no previous shear-induced matrix orientation, as is the case in the parallel-plate rheometry. Consequently the viscosities are higher, but this effect is reversed as the shear rates increase.

Another effect related to matrix alignment that shows the viscosity of unreinforced polycarbonate and 20 wt.% VGCF polycarbonate composites is observed in Figure 3.

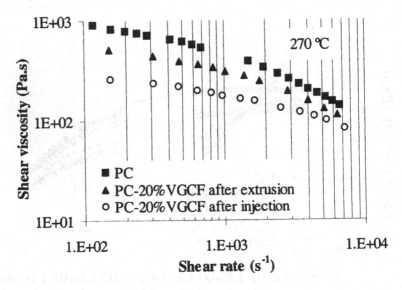

Figure 3. Shear viscosity of polycarbonate and 20 wt.% VGCF polycarbonate composites after extrusion and after injection (from ref. [10], by courtesy of SPE, the Society of Plastics Engineers)

In the case of the composites, the viscosity was determined after extrusion and after extrusion and subsequent injection molding, respectively. The viscosity of the latter is consistently lower. Furthermore, the shear viscosities of both composites are lower than that of the polycarbonate. This effect is not related the degradation of the polymer, as the molecular weight of samples of extruded and injected molded unreinforced polycarbonate did not differ significantly. The difference can be explained by induced matrix alignment and possibly matrix breakage that was larger in the injection molded samples which were subjected to higher shear rates. This explanation is supported by the 'time-sweep' data depicted in Figure 4, determined by rotational rheometry on the same material, processed by one or both the above techniques sequentially [10].

In a first run, performed at a constant shear rate of 0.2 s⁻¹, the viscosity decreases progressively with time, stabilizing after approximately 33 minutes at around 700 Pa.s. At the beginning of a second run, 30 minutes later, the viscosity is still the same, remaining constant throughout the entire run. The matrix orientation induced in the first run is thus irreversible. This indicates that the orientation generated by extrusion is maintained and augmented during injection molding, thus decreasing the viscosity.

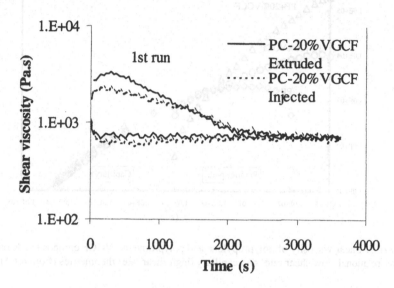

Figure 4. Time sweep viscosity data of composites with different levels of matrix alignment (from ref. [10], by courtesy of SPE, the Society of Plastics Engineers)

4.2. FIBER-MATRIX ADHESION

In rheological experiments, fiber-matrix adhesion will generally be observed through changes in the viscosity. Furthermore, good fiber-matrix adhesion would ideally rule out the effect of inter-layer slip described above. Thus, the comparison of rheological data on different systems will provide valuable information on the level of fiber-matrix adhesion, which will be reflected in the composites' mechanical performance.

The rheological curves presented in Figure 5 were obtained by Maia *et al.* for polypropylene and polypropylene-VGCF composites [12], prepared in the same manner as the polycarbonate-VGCF composites [4].

By comparing figures 5 and 2, it is clear that the reinforcing capability of the fibers is much more evident in this case. In fact, the viscosity of the composites in capillary flow is always higher than that of the unreinforced polymer. In rotational flow, at shear rates above 1 s⁻¹, only the viscosity of the 20 wt.% composite is lower than that of the unreinforced material. Since the surface energies of polypropylene and VGCF are similar, the results can be explained by increased adhesion between fiber and matrix. Increased adhesion will also decrease inter-layer slip, which will result in a further increase of the viscosity. However, shear-induced fiber alignment still plays a role in the occurrence of layered flow, since the viscosity of the 20 wt.% composite at high shear rates becomes lower than that of the unreinforced polymer.

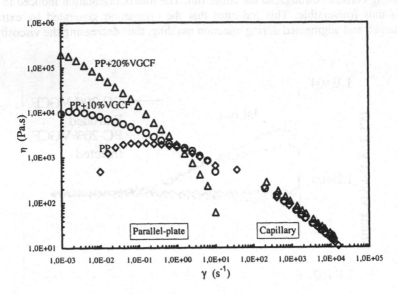

Figure 5. Shear viscosity of polypropylene and polypropylene-VGCF composites determined by rotational (low shear rate) and capillary (high shear rate) rheometries (from ref. [12])

Carneiro *et al.* [4] submitted vapor grown carbon fibers to a cold oxygen plasma treatment to increase their surface reactivity and thus improve the interaction with the polycarbonate matrix. Composites were then prepared with the plasma treated fibers. Figure 6 depicts the shear viscosities of the unreinforced polymer and that of the 20 wt.% VGCF (plasma treated)-polycarbonate composite, after extrusion and extrusion and subsequent injection molding.

In contrast with the results obtained with the as-grown fibers (see Figure 3), the viscosity of the composites is now greater than that of unreinforced polycarbonate, irrespective of the type of processing. This is obviously a result of improved fiber-matrix interaction. Again, extruded composites have a higher viscosity than those extruded and injection molded.

4.3. FIBRE DISPERSION

Rheological experiments can also be used to obtain qualitative information about the level of fiber dispersion in a composite. One such experiment is depicted in Figure 2. In that figure it can be observed that, at low shear rates, the viscosity of the 20 wt.% composite is significantly greater than that of the other systems. The difference is about one order of magnitude.

This effect may be explained by the occurrence of fiber-fiber interactions and/or the development of weak structures of non-aligned fibers above a critical concentration [13].

As the shear rate increases, the fiber alignment also increases while these interactions/structures weaken, thereby decreasing the viscosity.

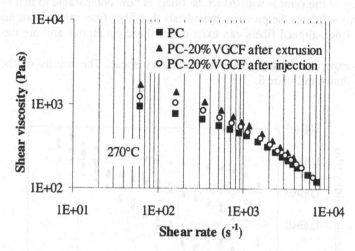

Figure 6. Shear viscosity of polycarbonate and 20 wt.% VGCF (plasma treated) - polycarbonate composites, after extrusion and after injection (from ref. [4], by courtesy of Elsevier Science, Ltd.)

Testing pre-sheared samples, and comparing with normally processed ones can verify the above effect. Pre-shearing consists in exposing the samples to a shear rate of 0.5 s^{-1} for 15 minutes prior to testing. Figure 7 shows rotational shear flow curves for pre-sheared polycarbonate-VGCF composites. For comparison, results taken from Figure 2, for non pre-sheared samples, are also shown.

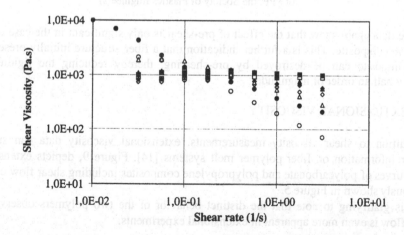

Figure 7. Effect of pre-shear on the viscosity of polycarbonate-VGCF composites. Fiber concentrations, in weight %: (●) 0% □ PC, (◊) 5%, (Δ) 10%, (O) 20%; open symbols: experiments done with pre-shear; full symbols: no pre-shear (from ref. [10], by courtesy of SPE, the Society of Plastics Engineers)

It can be observed that when pre-shear is applied there is a marked decrease in viscosity for higher fiber concentrations at higher shear rates. However, the low shear rate viscosity of the sample with 20 wt. % fibers is now comparable to that of the other systems. This validates the previous hypothesis that fiber-fiber interactions and/or weak structures of non-aligned fibers can exist under those conditions and are destroyed by pre-shear.

Similar experiments were done in oscillatory shear. The results of the dynamic rigidity are shown in Figure 8.

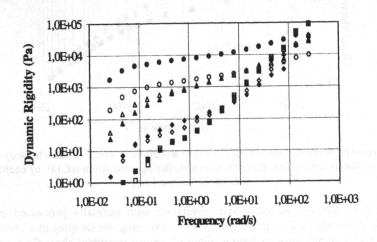

Figure 8. Effect of pre-shear on the dynamic rigidity of polycarbonate and polypropylene VGCF composites. Symbols as in Figure 7 (from ref. [10], by courtesy of SPE, the Society of Plastics Engineers)

The data again show that the effect of pre-shear is only significant in the case of the 20 wt.% composite. This is a further indication that a fiber structure initially present in that composite can be destroyed by pre-shearing, thereby reducing the rigidity by roughly half an order of magnitude.

4.4. EXTENSIONAL VISCOSITY

In addition to shear viscosity measurements, extensional viscosity data can supply further information on fiber-polymer melt systems [14]. Figure 9, depicts extensional flow curves of polycarbonate and polypropylene composites including shear flow curves previously shown in Figure 5.

It is gratifying to note that the distinct behavior of the two polymers observed in shear flow is even more apparent in extensional experiments.

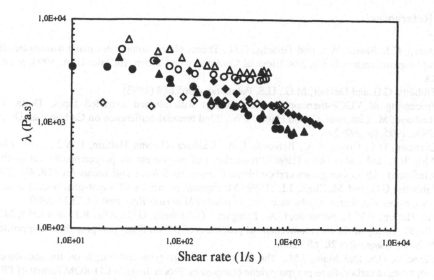

Figure 9. Extensional viscosity of polypropylene and polycarbonate
composites. Fiber concentrations, in weight %: (◊) 0%, (O) 10%, (Δ) 20%;
open symbols: polypropylene; full symbols: polycarbonate (from ref. [12])

In the case of polypropylene, the extensional viscosity increases with increasing fiber content, as would be expected. On the other hand, the viscosity of the VGCF-polycarbonate composites is lower than that of the unreinforced material. Note, however, that the extensional viscosity of the two polycarbonate composites is similar, as opposed to their shear viscosity (Figure 2). This is a further indication that the inter-layer slip effect referred to above is indeed occurring.

5. Conclusions

The results presented in this chapter demonstrate that it is feasible, and indeed easy, to produce thermoplastic matrix composites, reinforced with sub-micron VGCF, by means of conventional processing technologies.

The main issues affecting the production of good-quality composites are fiber alignment, fiber-matrix adhesion and fiber dispersion. Rheological experiments can give valuable quantitative information on these issues, by helping to understand the relationship between melt structure, processing and properties.

6. Acknowledgements

This work was supported by the European Economic Community, through the Human Capital and Mobility Programme, under Grant Number CHCRXCT940457. The authors are indebted to Applied Sciences, Inc. (Cedarville, Ohio, USA) for providing the Vapor Grown Carbon Fibers.

7. References

1. Dasch, C.J., Baxter, W.J. and Tibbetts, G.G., Thermoplastic composites using nanometer-size vapor-grown carbon fibers, 21st Biennial Conference on Carbon, Buffalo, USA, 1993, pp. 82-83.
2. Tibbetts, G.G. and Devour, M.G., U.S. Patent no. 4,565,684 (1993).
3. Processing of VGCF-thermoplastic composites with aligned extruded tapes, Dutta, D., Husband, M., Ciminelli, C. and Hager, J.W., 22nd biennial conference on Carbon, San Diego, USA, 1995, pp. 292-293.
4. Carneiro, O.S., Covas, J.A., Bernardo C.A., Caldeira, G., van Hattum, F.W.J., Ting, J.M., Alig, R.L. and Lake, M.L. (1998) Production and assessment of polycarbonate composites reinforced with vapour grown carbon fibres, *Composites Science and Technology* **58**, 401-407.
5. Tibbetts, G.G. and McHugh, J.J. (1999) Mechanical properties of vapor-grown carbon fiber composites with thermoplastic matrices, *Journal of Materials Research* **14**, 2871-2880.
6. van Hattum, F.W.J., Bernardo, C.A., Finegan, J.G. Tibbetts, G.G., Alig, R.L. and Lake, M.L. (1999) A study of the thermomechanical properties of carbon fiber- polypropylene composites, *Polymer Composites* **20**, 683-688.
7. Carneiro, O.S. and Maia, J.M., The influence of fibre type and length on the rheological properties of carbon-fibre/polypropylene composites, Proceedings in CD-ROM format of PPS-15, Hertogenbosch, the Netherlands, 1999.
8. Zhang, C., Yi, X., Yui, H., Asai, Sh., Sumita, M. (1998) Morphology and electrical properties of short carbon fibre-filled polymer blends: high density polyethylene/ poly(methyl methacrylate), *J. Appl. Polym. Sci.* **69**, 1813-1819.
9. Patton, R.D., Pittman, C. U. Jr., Wang, L., Hill, J. R. (1999) Vapor grown carbon fiber composites with epoxy and poly(phenylene sulfide) matrices, *Composites, Part A* **30**, 1081-1091.
10. Caldeira, G., Maia, J.M., Carneiro, O.S., Covas, J.A. and Bernardo, C.A. (1998) Production and characterization of vapor grown carbon fiber-polycarbonate composites, *Polymer Composites* **19**, 147-151.
11. Gupta, R.K. (1994) Particulate suspensions, in S.G. Advani (ed.), *Flow and Rheology of Polymer Composites Manufacturing*, Elsevier, Amsterdam.
12. Maia, J.M. and Carneiro, O.S. (1999) The influence of matrix-fibre interactions on the rheological properties of carbon-fibre composites of thermoplastic matrix, 2nd Conference on Material Forming, ESAFORM 99, Guimarães, Portugal, 1999, pp.455-458.
13. Mills, N.J. (1971) Rheology of filled polymers, *Journal of Applied Polymer Science* **15**, 2791-2805.
14. Yanovski, Y.G. (1993) *Polymer Rheology: Theory and Practice*, Chapman & Hall, London

TENSILE, ELECTRICAL AND THERMAL PROPERTIES OF VAPOR GROWN CARBON FIBERS COMPOSITES

C.A. BERNARDO*, S. A. GORDEYEV**, J.A. FERREIRA**
*Department of Polymer Engineering, Universidade do Minho, Portugal
**Department of Physics, Universidade do Minho, Portugal

Abstract

This chapter presents a study of the DC electrical resistivity and thermal conductivity of VGCF filled polypropylene (PP). The electrical resistivity exhibits a characteristic percolating behavior. Because of the low degree of graphite perfection, the fraction of VGCF required to achieve percolation was higher than expected. Also non-linear I-V characteristics and time dependent electrical resistivity effects were observed. The thermal conductivity of the composites agrees with the predictions of the effective medium theory. To enhance both the mechanical and transport properties an alternative technique must be used. It is well known that most crystalline polymers can be self-reinforced by forming a fibrilar structure, if stretched at high temperature. However, the production of carbon fiber reinforced composites in fibrilar form is not easy. In fact, the dimensional characteristics of the fibrilar structure are much smaller than the average fiber length in a composite. Due to their small dimensions, sub-micron VGCF could be suitable fillers to incorporate in such a structure. Hence, the feasibility of processing highly oriented thermoplastic composites reinforced with VGCF is also studied herein. The mechanical behavior of these composites, as well as their DC response over a wide range of applied electrical fields, are reported, showing that this technique offers good prospects for property enhancement.

1. Introduction

Vapor grown carbon fibers (VGCF) are a new class of carbon fibers whose structure and properties have been described in considerable detail in a number of publications [1,2]. When heat-treated, VGCF are known to exhibit some of the best transport characteristics among carbon fibers [3]. VGCF grown by a continuous process [4] in a commercial scale have diameters of about 0.1-0.2 micrometers and lengths ranging from 10 to 100 μm.

It was shown recently that it is possible to reinforce thermoplastics with VGCF in a straightforward manner via standard injection [5,6] and compression [7,8] molding techniques. However, the mechanical properties of those composites were not significantly better than those of other composites reinforced with different carbon fibers. On the other hand, the addition of VGCF to a thermoplastic matrix can increase

L.P. Biró et al. (eds.), Carbon Filaments and Nanotubes: Common Origins, Differing Applications, 301–314.

electrical and thermal conductivities, which are important for many applications. In a recent study [9], the thermal conductivity of PP filled with VGCF was shown to be similar to that of PP/PAN composites, except at the highest fiber concentrations. Although in both systems the conductivity increased substantially with the addition of fibers, a threshold behavior for heat transfer was not observed. This was explained by the fact that heat conduction involves phonon propagation and scattering, so that there are no abrupt changes in this property when the fiber concentration increases. Therefore, the concentration dependence of the thermal conductivity was in good agreement with the predictions of the effective medium theory.

On the other hand, to provide a continuous path for electrical charge carriers through a polymer a three-dimensional network of conductive filler particles is needed. The formation of such a network is possible only at filler concentrations above some critical (threshold) value, characterized by a sharp drop in the electrical resistance. This phenomenon is known as percolation [10]. Theoretical models predict that the threshold for percolation of spherical non-permeable particles is about 16 vol. % [11]. Therefore, direct current (DC) electrical conductivity of polymer composites filled with isotropic particles depends primarily on the filler volume fraction.

Conductive carbon black (CB) is the filler more often used to provide electrical conductivity in polymers [12, 13]. Due to the tendency of carbon black to aggregate, the conduction threshold of CB-filled polymers may be significantly lower than that predicted by the percolation theory. For instance, the resistivity of injection or compression molded composites with 8 to 10 vol. % carbon black varies from 10^{-2} to 10 Ωm [12]. In some novel grades of CB, this level of resistivity can be attained with the addition of 3 to 4 vol. % [14].

Short carbon fibbers (CF) are considered the most advanced fillers for the purpose mentioned above, due to extremely high axial strength and stiffness, as well as high electrical conductivity [15]. As short carbon fibbers can act as extra nucleating agents, their effect on the mechanical performance is most pronounced when crystallisable polymers are used as matrices. The effect on the electrical properties can also be quite significant. Percolation theory predicts that the percolation threshold depends not only on the fibers' concentration, but also on their aspect ratio and geometrical arrangement in the matrix [16]. In particular, in polymers filled with isotropically oriented CF with high aspect ratio (>20), the onset of percolation can occur at volume fractions as low as 1% [17], increasing rapidly upon alignment of the fibers.

On the other hand, it is known that the greatest advantage of most semi-crystalline polymers is their ability to self-reinforce via the formation of a fibrillar structure when stretched at high temperature [18]. The nucleating ability of the particles is less important in this case, and their adhesion to the matrix and mechanical properties become critical. However, there is a major problem in the use of carbon fibers to reinforce polymers in fibrillar form. Conventional carbon fibers have diameters of several microns, and their length in a composite after fabrication varies from 100 to 300 μm. The typical dimensional characteristics of the fibrillar structure are much smaller [19]. Since it is extremely difficult to embed carbon fibers in such a structure without serious distortions, no successful attempts to self-reinforce thermoplastics filled with carbon fibers have been reported in the literature.

On the other hand, due to their dimensions, sub-micron VGCF can be used to manufacture highly oriented thermoplastic composites [20]. Samples of oriented PP filled with VGCF were fabricated using traditional fiber technology, including spinning of a polymer melt and subsequent stretching of the spun material to high draw ratios. The effect of fiber content and processing conditions on the mechanical properties and DC electrical conductivity was reported on that study.

The main objective of the present chapter is to review and expand the work done on the transport properties of thermoplastics filled with VGCF. Injection molded composites and composites processed in a fibrillar form are examined. To assess their practical interest, these composites are compared with systems produced in the same conditions, but using PAN-based carbon fibers as conductive fillers.

2. Experimental

Montell's Moplen F30G polypropylene, VGCF and commercial polyacrylonitrile (PAN) based chopped carbon fibers were used to fabricate the composites. VGCF Pyrograf III® [21], supplied by Applied Sciences, Inc, without any surface treatment, were used in the form of pellets of loosely aggregated curved fibers, with average diameters of about 200 nm and lengths varying from a few to 60 microns.

The PAN fibers, TENAX HTA 513, had nominal diameters of 7 μm and average lengths after composite fabrication close to 200 μm.

Composite masterbatches containing 15 vol. % of VGCF were produced in a Leistritz 30.34 twin-screw extruder and then granulated. Batches with fiber volume fractions of 10 and 5% were subsequently obtained by diluting with appropriate amounts of unfilled PP. A portion of this material was injection molded into rectangular bars of 10 x 12 x 1.5 mm. In these bars, the carbon fibers retained a preferential orientation along the 10 mm (flow) direction.

The remaining portion of material was extruded into circular monofilaments on a homemade spinneret at the IRC in Polymer Science & Technology, University of Leeds, UK. Three parameters: melt temperature, spinning rate and winding rate, were varied. By combining the latter two rates, samples drawn with different draw ratios (DR) could be obtained. The actual DR of the as-spun monofilaments was determined from the ratio of the linear density of each sample to the linear density of the sample spun without winding. The spinning conditions used allowed the stable processing of geometrically homogeneous samples. Subsequently, these samples were drawn at 130 °C in an Instron 4505 universal testing machine, at a gauge length of 25 mm and a crosshead speed of 100 mm/min. This speed was found to maximize the draw ratios. The actual DR was determined from the ratio of the linear density of the drawn samples to that of the as-spun ones.

Tensile tests both of the as-spun and drawn samples were carried out at room temperature, in the same tensile testing machine, at a crosshead speed of 50 mm/min and a gauge length of 50 mm.

The thermal conductivity was calculated from thermal diffusivity measurements at room temperature with a photothermal beam deflection set-up [22].

The electrical resistance of the composite monofilaments was measured along their axis by the two-point method using a Keithely 617 electrometer with a built-in voltage source. The bulk resistivity was calculated by

$$\rho = R\frac{S}{L} \qquad (1)$$

where ρ is the bulk resistivity, R is the measured resistance, S the cross-section area, and L the distance between electrodes. Values of the resistivity were measured at an applied voltage of 0.1 V and averaged over at least five specimens.

The resistance was determined as a function of the applied field from 0.1 to 100 V and presented as a dimensionless variable, $r(V)$, by dividing by R_0, the value corresponding to the lowest applied voltage:

$$r(V) = \frac{R(V)}{R_0} \qquad (2)$$

3. Results and discussion

3.1. INJECTION MOULDED COMPOSITES

The effect of the fiber loading on the bulk electrical resistivity is shown in Figure 1.

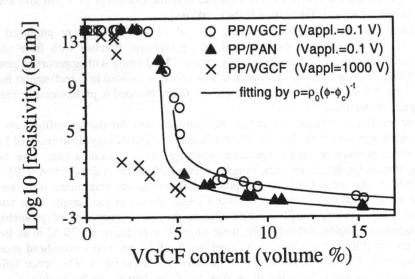

Figure 1. Dependence of the electrical resistivity of the PP/VGCF and PP/PAN composites on fiber content

PP/VGCF and PP/PAN composites are essentially similar, exhibiting typical percolation behavior. The resistivity decreases with increasing fiber loading, especially between 4 and 5% in volume. Other measurements, however, have shown differences in the electrical behavior of the two composites. In particular, the current-voltage characteristics of PP/VGCF are found to be non-linear even at high fiber loading, while those of PP/PAN obey Ohm's law under those conditions. As a result of this non-linearity, the percolation threshold for the PP/VGCF composite depends on the applied field. As can be seen in Figure 1, this effect was especially pronounced at the highest applied voltage, where the conductivity threshold occurs between 2 and 3 vol.%.

Another interesting peculiarity of the PP/VGCF samples was their time dependent resistivity behavior that could be observed in a cyclic mode. Although the current varied reversibly with the applied field, when the voltage changed the system needed some time to settle to a new resistivity.

The percolation threshold of the PP/VGCF composites (4 to 5 vol. %) is higher than that reported for polymers filled with anisotropic single crystal graphite flakes [17], where the filler particles had a high degree of graphite perfection. The percolation threshold of those systems was below 2 vol.%, even at low applied fields. Heremans *et al.* showed that the transport properties of carbon fibers correlate well with their structural characteristics [23]. As the as-grown VGCF have a low degree of crystalline perfection [9], they may have a comparatively high resistivity. This, in turn, can affect the percolation threshold of their composites.

The data in Figure 1 can be fitted to a power law that describes the relation between the resistivity of the composite and the filler concentration in the vicinity of the percolation threshold [10]:

$$\rho = \rho_0 (\phi - \phi_c)^{-t} \tag{3}$$

where: ρ is the composite resistivity, ρ_0 the resistivity of conductive filler, ϕ the filler volume fraction, ϕ_c the percolation threshold, and t the conductivity critical exponent. The results of the least-square fit are: $\rho_0 = 4.5 \cdot 10^{-5}$ Ωm, $\phi_c = 0.048$, t = 3.2 for the PP/VGCF composite and $\rho_0 = 7.5 \cdot 10^{-6}$ Ωm, $\phi_c = 0.041$, $t = 2.9$ for the PP/PAN composite. The resistivity of the PAN based fibers is in close to the value presented by the manufacturer ($5 \cdot 10^{-6}$ Ωm). Additionally, the values of the critical exponent agree with the mean-field value $t = 3$, typical of polymer-carbon-black composites [24].

The difference in the behavior suggests that the conductivity mechanisms of the two composites may be distinct. Indeed, as mentioned above, PP/PAN samples with fiber concentrations above percolation always obey Ohm's law at room temperature and do not show time dependent behavior. This means that the conductive paths through the sample are supplied by contacting fibers, with the electrons being the only charge carriers. On the other hand, the non-linearity of the I-V relationship for PP/VGCF should be mainly associated with the tunneling mechanism [17]. The tunneling barrier and the ionic conductivity mechanisms imply that, even at high fiber loading, the fibers are separated by polymer and do not touch each other directly. This morphology should contribute to the rise of the percolation threshold. However, to reach more definite conclusions, a study of the resistivity as a function of temperature is required.

The thermal conductivity of the composites, normalized with that of the unfilled polymer, is shown in Figure 2.

The conductivities of PP/VGCF and PP/PAN are similar, except at the highest fiber concentration. Although the conductivity increases substantially with fiber content in both systems, a threshold concentration is not observed. This is because heat conduction depends on the phonon mean free path and is limited by phonon scattering effects. Therefore is not associated with tunneling phenomena or physical contacts between fibers.

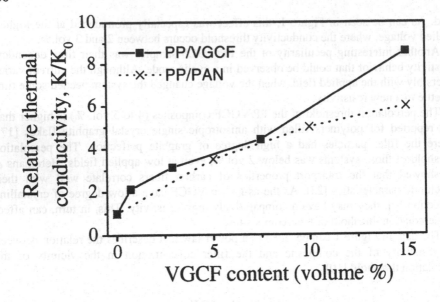

Figure 2. Dependence of the thermal conductivity of the PP/VGCF and PP/PAN composites on fiber content

3.2. AS-SPUN COMPOSITES

Table 1 summarizes the mechanical and electrical properties of the PP-VGCF composites extruded and spun in different conditions. SEM analysis showed that the fibers were distributed homogeneously in the matrix, although there was some preferential orientation along the extrusion direction.

The 5 vol. % composites were as flexible as the unfilled polymer. At higher concentrations the composites tended to be more brittle, but could still be wound easily. From 200 to 250°C, the effect of the spinning temperature on the processability and mechanical properties was negligible at all concentrations except at 15 %, where spinning instability was observed above 240°C.

The reinforcement effect of VGCF was assessed from the tensile properties of the spun samples, again normalized to those of the unfilled polymer spun under the same conditions. In Figure 3 it can be seen that the increase of VGCF concentration increases both the tensile strength and stiffness, but the dependence is complex due to the effect of draw-ratio (DR).

At low DR (1-1.5) the relative strength and stiffness increase with both the orientation and fiber content. Consequently, the highest values of the relative properties ($\sigma/\sigma_0 = 3.5$; $E/E_0 = 4.7$) were obtained for the sample filled with 15 vol. % VGCF spun to a DR= 1.5.

The properties of the most oriented samples exhibit this behavior only at 5 vol. % VGCF. At 10 and 15% the efficiency of the reinforcement decreases, but remains always higher than unity.

The electrical properties of the spun composites also depend on both fiber volume fraction and draw ratio (Table 1).

TABLE 1. Processing conditions and properties of the as-spun composites

VGCF fraction (%)	Spinning temperature (°C)	Draw ratio	Tensile strength (MPa)	Young's Modulus (GPa)	Strain at break (%)	DC resistivity (Ωm)
0	220	1.0	17.2[1]	1.15		
0	220	1.5	17.7[1]	1.2	>500	>10^{13}
0	220	3.5	24.5[1]	1.8		
5	200	1.5	56.9[1]	3.5	9-530	$2.7 \cdot 10^{10}$
5	220	1.0	49.3[1]	3.2	>500	$6.2 \cdot 10^{9}$
5	220	1.5	53.7	3.5	5.0	$3.5 \cdot 10^{10}$
5	220	3.5	77.1	6.3	33.2	$1.1 \cdot 10^{11}$
5	250	1.5	52.6	3.3	4.9	$7.1 \cdot 10^{9}$
10	200	1.5	57.5	4.1	2.4	$2.8 \cdot 10^{0}$
10	220	1.0	54.0	4.2	2.6	$6.8 \cdot 10^{-1}$
10	220	1.5	57.6	4.7	2.4	$1.2 \cdot 10^{0}$
10	220	3.5	45.1	5.2	16.9	$1.0 \cdot 10^{1}$
10	250	1.5	54.8	3.5	3.3	$1.25 \cdot 10^{0}$
15	200	1.5	60.5	5.4	2.0	$6.1 \cdot 10^{-2}$
15	220	1.0	56.2	4.9	2.2	$2.2 \cdot 10^{-2}$
15	220	1.5	62.4	5.4	2.3	$6.0 \cdot 10^{-2}$
15	220	3.5	42.4	4.6	1.9	$9.3 \cdot 10^{-2}$
15	245	1.5	41.1	4.4	1.6	$5.4 \cdot 10^{-2}$

1) Stress at yield point

The DC resistivity of the composite with 5 vol. % VGCF is several orders of magnitude lower than that of unfilled PP, but still higher than that of the corresponding injection molded composite (Figure 1). In other words, 5 vol. % VGCF is not enough to allow the percolation of charge carriers, whereas the percolation threshold of the injection molded composites was about 4 %. This results from the higher orientation of VGCF in the spun monofilaments, due to the unavoidable effect of matrix orientation. In fact, percolation theory predicts a strong dependence of the percolation threshold on the orientation of anisotropic filler particles. Therefore, it should be expected that even a little increase in the orientation might shift the onset of percolation to higher concentrations.

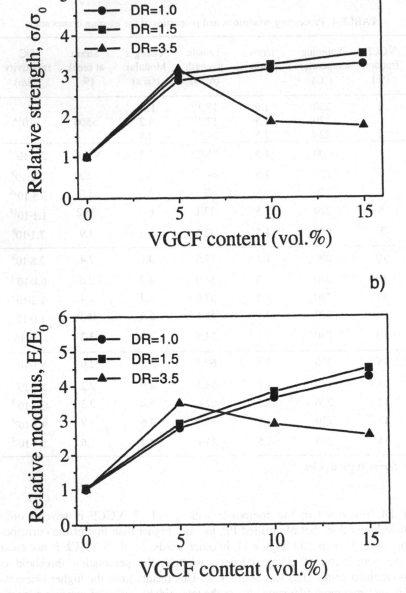

Figure 3. Dependence of the mechanical properties of the as-spun composites on fiber content: a) strength; b) modulus

The samples with 10% fibers were already above the critical concentration. However, their resistivity was still relatively high and went up with draw ratio, indicating a relatively weak conductive network that can be destroyed by further stretching. Only at 15 vol.% VGCF was the resistivity of the composites virtually independent on the orientation of the fibers. These composites were conductive at all spinning conditions, and, as shown below, subsequent hot drawing to high draw ratios did not destroy their conductive network.

The influence of the applied field in the shape of the *relative resistance versus applied voltage* curves is shown in Fig. 4.

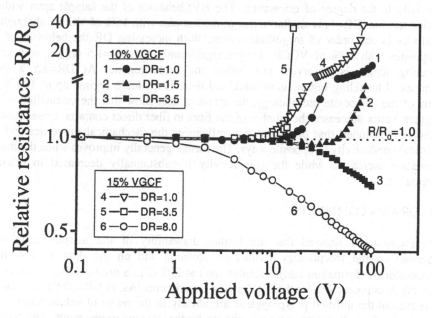

Figure 4. Relative resistance versus voltage curves for the VGCF/PP composites

It can be observed that none of the samples display a linear behavior over all applied voltages and that the actual behavior depends both on fiber concentration and DR. Some additional qualitative insight into the morphology of the composites can be derived from the analysis of these curves. It is well known that, in a real composite, the variation of the resistivity with the applied field is much more complicated than traditional percolation theory would suggest. This is mainly due to a variety of conduction mechanisms that determine its electrical response to an applied field.

It is well documented that four regions with distinct behaviors of $r(V)$ can be detected in the *applied field - filler concentration* plane [25]. These regions, or regimes, are: 1) the *linear or ohmic conductivity region* where $r \approx 1$, i.e., the resistance is independent of the applied field; 2) the *tunneling region*, where $dr/dV < 0$; 3) the *Joule regime*, reached when the field is sufficiently high and resistance begins to increase ($dr/dV > 0$) and 4) the *breakdown regime*, where the resistance increases steeply ($dr/dV \gg 0$). In terms of morphology, *linear conductivity* implies that there are enough fibers touching each other to allow the passage of the current throughout the composite. Conversely, for the tunneling regime to occur, the resistance of fibers separated by a thin polymer layer must predominate over the resistance of those in direct contact. Hence, the main contribution to the conductivity is the tunneling of electrons through polymer gaps. The Joule and breakdown regimes arise when changes in morphology caused by local overheating occur inside the bulk composite.

In Fig. 4 it can be seen that in the spun composites with 15 vol. % VGCF the number of fiber to fiber contacts is sufficient to avoid the appearance of the tunneling regime. As a result, the samples have a low resistivity and the Joule and breakdown regimes follow each other right after the linear regime, at relatively low applied voltages. The composites with 10% VGCF have a high sensitivity to the processing conditions, especially to the degree of orientation. The $r(V)$ behavior of the sample spun without applied tension (DR=1) is similar to that of the samples with 15% of VGCF, although its resistivity is one order of magnitude higher. With increasing DR the behavior of the composites with 10% of VGCF changes significantly. At DR=1.5 only a very small tunneling region is observed, just before the Joule regime. At DR=3.5 a more pronounced tunneling regime is noticed, and there is no Joule regime up to 100 V. In terms of the composite morphology, the above results show that the orientation of the polymer matrix decreases the number of the fiber to fiber direct contacts. Consequently, it can be concluded that the orientation influences the mechanical properties and the electrical conductivity in opposite ways. The former generally improves when the matrix orientation increases, while the conductivity is substantially decreased in oriented composites.

3.3 DRAWN COMPOSITES

Preliminary trials showed that the further drawability of the as-spun composites depended on the morphology formed on spinning and on the drawing conditions (temperature, deformation rate). Samples spun at 220°C to a nominal draw ratio of 1.5 were taken as precursors for the hot drawing experiments. As, at DR=10 the mechanical properties of the unfilled polypropylene are already in the range of values required for most applications, this draw ratio was chosen for the drawing experiments. The samples with 5 and 10 vol. % VGCF deformed via neck formation, whereas the sample with 15 vol. % deformed uniformly and had a lower drawability. As a result, in this case, it was only possible to reach a draw ratio of about 8. Table 2 presents the tensile strength and Young's modulus of the drawn composites.

TABLE 2. Properties of the drawn composites

VGCF fraction (vol.%)	Draw ratio	Tensile strength (MPa)	Young's modulus (GPa)	Strain at break (%)	DC resistivity (Ωm)
0	10	530	8.3	23.7	$>10^{13}$
5	10	717	14.9	10.3	10^{13}
10	10	537	14.1	9.3	$5.3 \cdot 10^9$
15	8	306	11.2	6.9	$5.2 \cdot 10^1$

The mechanical properties of the drawn composites, again normalized by the correspondent values of the properties of the unfilled polymer, spun and drawn with the same conditions, are plotted in Fig. 5. As mentioned above, the draw ratio of the composite with the highest VGCF loading is only 8.

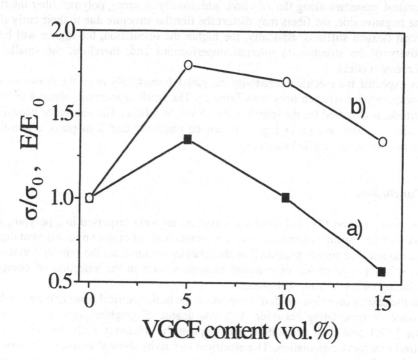

Figure 5. Dependence of the mechanical properties of the drawn composites on fiber content: a) strength; b) modulus

As can be seen in the Figure and also in Table 2, the increase of the relative strength and modulus is lower than that of the as-spun composites, but their absolute values are significantly higher. The reinforcement effect reaches a maximum at 5 vol. % VGCF. At higher loading, the relative strength decreases gradually, while the relative modulus goes down more markedly, but remaining always higher than unity.

There is insufficient data on the structure of the drawn composites to give a quantitative description of the observed reinforcement effect. However, the tendency of the relative mechanical properties to decline with VGCF content is consistent with existing concepts about oriented polymer structures. As commented before, most semi-crystalline polymers form the so-called fibrillar structure when stretched to a high extension. A principal feature of this structure is a fairly regular alternation of partially chain-folded (or extended chain) crystals and intervening amorphous layers. In addition, the stacked crystals are incorporated into microfibrils, which are units 50 nm in diameter and a high aspect ratio (>100). In turn, the microfibrils are united in fibrils, i.e., bundles of microfibrils, well oriented along the draw direction, held together by tie molecules and Van der Walls forces. A typical single fibril has a length of the order of the hundreds of microns, and a diameter of about 0.5 μm, which decreases upon orientation. This structure itself is a composite-like structure, whose formation gives rise to the so-called *self-reinforcement* of a polymer. It can be imagined that the VGCF, due to their small dimensions and shape, can be easily incorporated in the fibrillar structure, improving the mechanical performance. The reinforcing effect assumes that the fibers have high mechanical properties along the axis and, additionally, a strong polymer-fiber interface. On the negative side, the fibers may distort the fibrillar structure due to their curly shape and high flexural stiffness. Basically, the higher the orientation, the higher will be the sensitivity of the structure to internal imperfections and, therefore, the smaller the reinforcement effect.

As expected, the electrical resistivity increased dramatically in all the drawn samples with respect to the as-spun ones (see Table 2). The smallest increase, about 3 orders of magnitude, is observed for the sample with 15 vol. % VGCF. The resistance behavior of this sample is also shown in Fig. 4. It can be observed that it displays a non-linear behavior even at low applied voltages.

4. Conclusions

Vapor grown carbon fibers of submicron dimensions were dispersed in a polypropylene matrix and the resulting compounds were injection molded or drawn to different degrees of orientation. The results presented in this chapter demonstrate the potential that exists for developing a spectrum of material properties through the selection of composite constituents and processing techniques.

In the case of injection molded composites, the bulk electrical conductivity exhibits a characteristic percolating behavior. The low degree of graphite perfection of the as-grown VGCF and their orientation may explain the relatively high fraction of fibers required to achieve percolation. The electrical resistivity showed non-linearity and time dependent effects.

The thermal conductivity displayed by the composites is in agreement with the predictions of the effective medium theory.

The efficiency of VGCF in enhancing the mechanical properties of fibrillar composites is greatly dependent on their volume fraction. It seems that only up to 5% can the fibers be embedded into the oriented matrix without disturbing its structure. Apart from increasing the number of structural defects that affect the drawability of the composites, a larger amount of filler may create topological limitations to its own mobility during processing. On the other hand, these geometrical constraints help to maintain a continuous network of fibers, which provides the electrical conductivity. As a result, in the case of the highly drawn composites, only those with 15 % VGCF displayed low resistivity.

The spun composites show a higher degree of mechanical reinforcement than injection molded samples. Their electrical properties vary widely, depending on filler concentration and matrix orientation. The mechanical and electrical properties vary in opposite ways with the degree of orientation of the composite. However, by manipulating structural parameters, materials with unique combinations of properties can be obtained. In particular, a composite spun and then stretched to a draw ratio of 8, showed a tensile strength of 306 MPa, a Young's modulus of 11.2 GPa and a DC resistivity of 52 Ωm.

5. Acknowledgments

The authors are indebted to Applied Sciences, Inc. (Cedarville, Ohio, USA) for providing the Vapor Grown Carbon Fibers. The contribution of Dr. F. J. Macedo (Department of Physics, Universidade do Minho, Portugal) for the measurement and interpretation of the thermal conductivity data is also gratefully acknowledged.

6. References

1. Tibbetts, G.G., Doll, G.L, Gorkiewicz, D.W., Moleski, J.J., Perry, T.A., Dasch, C.J. and Balogh, M.J. (1993) Physical properties of vapor-grown carbon fibres, *Carbon* **31**, 1039-1047.
2. Tibbetts, G. G. and Devour, M. G. (1986), *U.S. Patent* No. 4,565,684
3. Heremans, J. (1985) Electrical conductivity of vapor-grown carbon fibres, *Carbon*, **23**, 431-436.
4. Tibbetts, G.G., Gorkiewicz, D.W. and Alig, R.L. (1993) A new reactor for growing carbon fibers from liquid- and vapor-phase hydrocarbons, *Carbon* **31**, 809-814.
5. Carneiro, O.S., Covas, J.A., Bernardo, C.A., Caldeira, G., van Hattum, F.W.J., Ting, J.-M., Alig, R.L. and Lake, M.L. (1998) Production and assessment of vapour grown carbon fibres-polycarbonate composites, *Compos. Sci. Techn.* **58**, 401-407.
6. Van Hattum, F.W.J., Bernardo, C.A., Finnegan, J. G., Tibbetts, Alig, R.L. and Lake, M.L. (1999) A study of the thermomechanical properties of carbon fiber-polypropylene composites, *Polymer Composites* **20**, 683.
7. Zhang, C., Yi, X., Yui, H., Asai, Sh. and Sumita, M. (1998) Morphology and electrical properties of short carbon fibre-filled polymer blends: high density polyethylene/ poly(methyl methacrylate), *J. Appl. Polym. Sci.* **69**, 1813-1819.
8. Patton, R.D., Pittman, C. U. Jr., Wang, L. and Hill, J. R. (1999) Vapor grown carbon fiber composites with epoxy and poly(phenylene sulfide) matrices, *Composites, Part A*, **30**, 1081-1091.
9. Gordeyev, S.A., Macedo, F.J., Ferreira, J A., Van Hattum, F.W.J. and Bernardo, C.A. (2000) Transport properties of polymer-vapour grown carbon fibre composites, *Physica B* **279**, 33-36.
10. Stauffer, D. and Aharony, A. (1994) *Introduction to Percolation Theory*, Taylor & Francis, London.
11. Balberg, I., (1987) Recent developments in continuum percolation, *Phil. Magazine B*, **56**, 991-1003.
12. Sichel, E.K. (1982) *Carbon Black in Polymer Composites*, Marcel Dekker, New York.
13. Ponomarenko, A.T., Shevchenko, V.G. and Enikolopyan, N.S. (1989) Formation processes and properties of conducting polymer composites, in N.S. Enikolopyan (ed.), *Filled Polymers*, Springer-Verlag, Berlin, 125-147.
14. Ketjenblack® EC.(1998) *Thecnical Bulletin on Polymer Additives*, Akzo Nobel.
15. Chung D. D. L., (1994) *Carbon fibre composites*, Butterworth-Heinemann, Boston.
16. Balberg, I. et al (1984) Excluded volume and its relation to the onset of percolation, *Phys.Rev.B* **30**, 3933-3943.
17. Celzard, A., McRae , E., Mareche , J.F., Furdin, G., Dufort, M.C. and Deleuze, C. (1996) Composites based on micron-sized exfoliated graphite particles: electrical conduction, critical exponents and anysotropy, *J. Phys. Chem. Solids* **57**, 715-718.
18. Geil, P. H., (1963) *Polymer Single Crystals*, Interscience Publ., New York.
19. Marichin, V.A. and Myasnikova L.P. (1977) *Supermolecular Structure of Polymers*, Khimiya, Leningrad.

314

20. Gordeyev, S.A., Ferreira, J.A., Bernardo, C.A. and Ward, I.M., Tensile and electrical properties of oriented polypropylene filled with VGCF, submitted to Appl. Phys. Letters.
21. Pyrograph III® (1998) *Technical Bulletin*, Applied Sciences, Inc., Cedarville, Ohio.
22. Macedo, F.J., Ferreira, J.A., Van Hattum, F.W.J. and Bernardo, C.A. (1999) Thermal diffusivity measurements of vapour grown carbon fibre composites using the optical beam deflection technique, *Journal of Materials Processing Technology* **92-93**, 151.
23. Heremans, J., Rahim, I. and Dresselhaus, M.S. (1985) Thermal conductivity and Raman spectra of carbon fibers, *Phys.Rev. B* **32**, 6742-6747.
24. Heaney, M.B. (1995) Measurement and interpretation of non-universal critical exponents in disordered conductor-insulator composites, *Phys. Rev.* **52**, 12477-12480.
25. Mukherjee, C.D., Bardhan, K.K. and Heaney, M.B. (1999) Predictable electrical breakdown in composites, *Phys. Rev. Lett.* **83**, 1215-1218.

PROPERTIES AND APPLICATIONS OF CARBON NANOTUBES

Materials Science Aspects

PULICKEL M. AJAYAN[1] AND RÓBERT VAJTAI[2]
*[1]Department of Materials Science and Engineering,
Rensselaer Polytechnic Institute, Troy, NY 12180-3590, USA
[2]Department of Experimental Physics,
University of Szeged, H-6720 Szeged, Dóm tér 9, Hungary*

Abstract. The discovery of fullerenes showed how structures of sp^2 carbon built based on simple geometrical principles can result in new symmetries that can have fascinating and useful properties. Carbon nanotube is the most striking example. The combination of structure, topology, and dimension create a host of physical properties in nanotubes that are paralleled by few known materials. The combination of high strength, low density, high conductivity, chemical inertness, and low-dimensionality makes nanotubes excellent candidate material for many practical applications. Here we will describe some of these applications of nanotubes, for example, in energy storage, as field emitters, as fillers in novel polymer based composites, and as templates for the creation of unique nanowires. We will discuss recent strategies and challenges in the synthesis of nanotubes tailored for specific applications and their future prospects.

1. Introduction

Carbon bonds in different ways to create structures with entirely different properties. Graphite and diamond, the two well-known solid phases of carbon, are evidences for this. The difference is in the hybridization of the four valence electrons in carbon. The sp^3 bonding creates four symmetric bonds resulting in diamond. When three of these electrons are covalently shared between neighbors in a plane and the fourth electron is allowed to be delocalized among all atoms, the resulting material is graphite. This second (sp^2) type of bonding results in a layered structure with strong in-plane bonds and weak bonds between planes; hence graphite is strongly anisotropic in its physical properties. These two different structures of carbon, both pure as well as combinations of the two bonding types, have been used widely in great many applications; from the transformation iron into steel, to making cutting tools, to creating high performance polymer composites for use in the aircraft industry.

The discovery of fullerenes [1] provided exciting insights into carbon nanostructures and how architectures built from sp^2 carbon units based on simple geometrical principles can result in new structures that have fascinating and useful

315

L.P. Biró et al. (eds.), Carbon Filaments and Nanotubes: Common Origins, Differing Applications, 315–330.
© 2001 *Kluwer Academic Publishers. Printed in the Netherlands.*

properties. Carbon nanotube [2] is an excellent example of such architecture. About a decade after the discovery of carbon nanotubes, the new knowledge available in this field indicates that nanotubes may be used in a number of practical applications. There have been great improvements in synthesis techniques and gram quantities of nanotubes can now be produced by electric arc discharge, laser ablation and catalytic chemical vapor deposition. Theoretical modeling provided strong basic knowledge for experimentalists and this synergy has helped the rapid expansion of this field [3-10].

Carbon nanotubes were probably being generated, without being identified, in different high temperature carbonaceous environments over the last several decades, but they were discovered and systematically analyzed in detail by Sumio Iijima in 1991 [2], while studying the surfaces of graphite electrodes used in fullerene synthesis. His observation and analysis of the nanotube structure started a new direction in carbon research, which complemented the excitement and activities in fullerene research and industrial applications of carbon fibers. If one compares the conventional planar graphite-related structures, such as micron-size carbon fibers, and the nanotubes, the uniqueness of the nanotubes arises from its near flawlessness in their structure and the high predictability of their properties. The novel electronic properties, based on helicity (local symmetry) and dimension, lead to fascinating electronic device applications and this will be specially reviewed in another ASI lecture here [11].

Another sign of uniqueness of nanotubes is topology, or the closed nature of individual nanotube shells; when individual layers wrap around into seamless structures, certain aspects of the anisotropic properties of graphite disappear, making the structure remarkably different from graphite. The combination of size, structure and topology endows nanotubes with important mechanical, electronic and surface properties, and the applications based on these properties form the central topic of this chapter.

The structure of nanotubes remains distinctly different from traditional carbon fibers that have been industrially used (e.g., as reinforcements in rackets, airplanes, space shuttles or special bikes) [12]. Most importantly, nanotubes, for the first time represent the idealized, perfect and ordered, carbon fiber. This places them along with molecular fullerene species in a special category of prototype macromolecular material or along with a well defined crystal. We have to describe the difference between the two varieties of nanotubes, which differ in the arrangement of their graphene cylinders, namely multi-walled nanotubes (MWNTs) and single walled nanotubes (SWNTs). MWNTs are collections of several concentric graphene cylinders and are larger structures compared to single-walled nanotubes (SWNTs), which are individual cylinders of 1-2 nm diameter (see Fig. 1). The former can be considered as mesoscale graphite system, in between conventional carbon fibers and SWNTs, whereas the latter is truly a single large molecule. However, SWNTs also show a strong tendency to bundle up into ropes consisting of aggregates of several tens of individual tubes organized into one-dimensional triangular lattices.

Materials Science applications of nanotubes include the use of nanotube arrays in field emitting devices, individual MWNTs and SWNTs attached to the end of an atomic force microscope (AFM) tip for use as nanoprobes, MWNTs as a supports in heterogeneous catalysis, SWNTs as good media for hydrogen storage, and nanotubes as reinforcements in composites. Some of these may become real marketable applications in the near future, but others need further modification and optimization. The potential

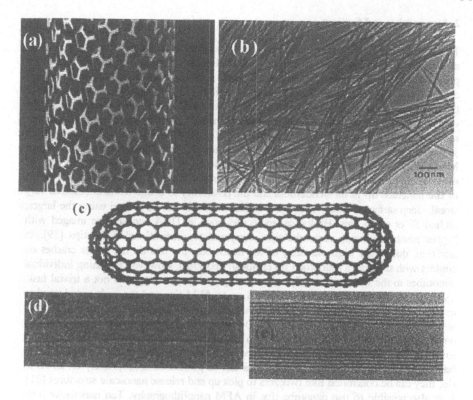

Figure 1. (a) Schematics of hexagons in the wall of a nanotube layer; (b) TEM image of purified MWNT sample. Most of the polyhedral nanoparticles have been removed by oxidation in air at 750 °C. The nanotubes in such oxidized samples are all open [6]; (c) SWNT schematic, showing the capping of the ends; HRTEM pictures of a 1.5 nm diameter singlewalled nanotube (d) and a multiwalled nanotube (e).

for nanotube based devices go far in to future nanotechnology and electronics based applications, but these latter topics will be dealt by other authors in this book.

The best methods yet to produce defect-free nanotubes with acceptable yields are based on the electric arc [13,14] and laser ablation processes [15]. The product prepared by these techniques has to be purified using chemical and physical separation methods (such as filtration). We are fully aware that the presently available synthesis and processing techniques are not scalable to make the industrial quantities needed for most applications, and this is a real bottleneck in nanotube R&D. In recent years, experimental and innovative work has focused on developing chemical vapor deposition techniques using catalyst particles and hydrocarbon precursors to grow nanotubes [16]; such techniques have been used earlier to produce hollow nanofibers of carbon in large quantities [12]. In the catalytic CVD nanotube production the great drawback is the inferior quality of the structures that contain gross defects (twists, tilt boundaries etc.), particularly because the structures are created at much lower temperatures (600-1000 °C) compared to the arc or laser processes (2000 °C). But CVD can be used to grow highly oriented nanotube arrays by pre-patterning the catalysts on selected substrates.

2. Nanoprobes and Sensors

The most coveted applications of single nanotubes may eventually appear in molecular electronics and nanoprobes. The small and well-defined dimensions of the nanotubes produce interesting results. The extremely small sizes, high electrical and thermal [17] conductivity, high mechanical strength and flexibility of nanotubes make them one of the best candidates to be used as nanoprobes. Nanotube probes have been found to be useful in a variety of applications, such as high-resolution imaging, nanolithography, nanoelectrodes, sensors and field emitters. Use of single MWNT attached to the end of a scanning probe microscope tip for imaging has already been demonstrated [18]. Since MWNT tips are conducting, they can be used in STM, AFM instruments as well as other scanning probe techniques, such as electrostatic force microscopy. The advantage of the nanotube tip is its slenderness and the possibility to image features (such as very small, deep surface cracks), which are almost impossible to be probed using the larger, etched Si or metal tips. Biological molecules such as DNA can also be imaged with higher resolution using nanotube tips, compared to conventional STM tips [19]. In addition, due to the high elasticity of the nanotubes, the tips do not suffer crashes on contact with substrates (nanotube tip elastically buckles). However, attaching individual nanotubes to the conventional tips of scanning probe microscopes is not a trivial task. Typically bundles of nanotubes are pasted on to AFM tips and the ends are cleaved to expose individual nanotubes. Successful attempts have been made to grow individual nanotubes directly onto Si tips using CVD [20], in which case the nanotubes are firmly anchored to the probe tips. In addition to the use of nanotube tips for high-resolution imaging, it is also possible to use nanotubes as active tools for surface manipulation. It has been shown that if a pair of nanotubes can be positioned appropriately on an AFM tip, they can be controlled like tweezers to pick up and release nanoscale structures [21]. It is also possible to use nanotube tips in AFM nanolithography. Ten nanometer lines have been written on oxidized silicon substrates using nanotube tips at high speeds.

Chemical functional groups can be attached to the tips of nanotubes and these modified nanotubes can then be used as molecular probes, with potential applications in chemistry and biology. These tips can be used for chemical and biological discrimination on surfaces [22]. Functionalized nanotubes were used as AFM tips to perform local chemistry, to measure binding forces between protein-ligand pairs and for imaging chemically patterned substrates. The chemical functionalization of nanotubes can set the stage for nanotube-based molecular engineering and many new chemical and biological nanotechnology applications.

Actuators and miniaturized chemical sensors have been constructed using sheets of SWNT. A few volts of electricity applied to strips of polymer-nanotube films in an electrolyte can move the films electromechanically [23]. This interesting behavior of nanotube sheets in response to applied voltage suggests several other applications, including nanotube-based micro-cantilevers for medical catheter applications and as novel substitutes for higher temperatures ferroelectrics. Recent research has also shown that nanotubes can be used as chemical sensor [24]. The electrical resistivities of SWNTs were found to change sensitively on exposure to environments containing gas molecules such as NH_3 and O_2. By monitoring the change in electrical resistance of nanotubes, the presence of gases could be precisely monitored; the response times of nanotube sensors are an order of magnitude faster than present solid-state sensors.

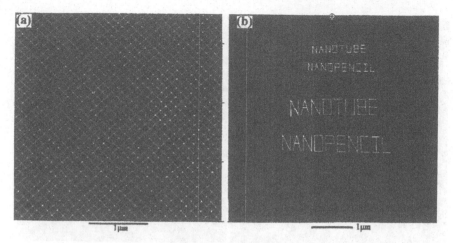

Figure 2. (a) AFM image of 2 nm tall, 10 nm wide, and 100 nm spaced silicon-oxide (light) lines fabricated by a nanotube tip. Bias voltage= -9 V; speed 510 mm/s; cantilever amplitude during lithography 50-55 nm; cantilever amplitude59.6 nm. (b) Silicon oxide "words" written by the nanotube tip.

Figures and the original figure caption adapted from [25]

3. Electron Emitting Properties and Applications

Electron field emission materials have been investigated extensively for technological applications, such as flat panel displays, electron guns in electron microscopes, power amplifiers for microwaves [26,27] etc. For technological applications, electron emissive materials should have low threshold emission fields and should be stable at high current density. In order to minimize the electron emission threshold field, it is desirable to have emitters with a low work function and a large field enhancement factor. The work function is an intrinsic materials property, while the field enhancement factor depends mostly on the geometry of the emitter. Carbon nanotubes have the right combination of properties - nanometer-size diameter, structural integrity, high electrical conductivity, and chemical stability - that make good electron emitters. Electron field emission from carbon nanotubes was first demonstrated in 1995 [28], and has since been studied intensely on various carbon nanotube materials. It is seen from experimental observations that electron emission is enhanced by a few localized electronic states [29, 30] partly due to the presence of topological defects (five member rings) present at the nanotube tips. Compared to conventional emitters, carbon nanotubes exhibit a lower turn-on electric field (threshold field for 10 mA/cm^2 current density), as 2-3 V/μm for random SWNT films and 3-5 V/μm for random and aligned MWNTs [31]. These values for the threshold field are all significantly better than those from conventional emitters such as the Mo and Si tips, which have a threshold electric field of 50-100 V/μm. The low threshold field for electron emission observed in carbon nanotubes is seen as a direct result of the large field enhancement factor rather than a reduced electron work

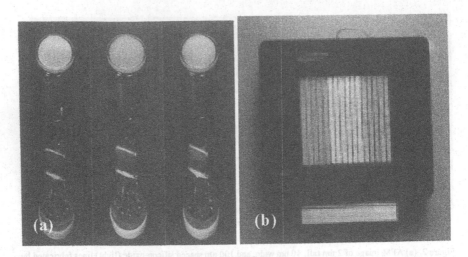

Figure 3. Prototype electron emission devices based on carbon nanotubes. (a) Demonstration field emission light source using carbon nanotubes as the cathodes (fabricated by Ise Electronic Co., Japan) {adapted from [32]} Fluorescent materials are ZnS:Cu,Al for green, Y₂O₃:Eu for red and ZnS:Ag for blue. The typical anode currents and voltages are 200 µA and 10 kV, respectively;
(b) A prototype 4.5" field emission display using carbon nanotubes fabricated by Samsung (image adapted from [33], originally provided by Dr. W. Choi of Samsung Advanced Institute of Technology)

function [34]. SWNTs have a higher degree of structural perfection than either MWNTs or CVD-grown materials and have a capability for achieving higher current densities and have a longer lifetime [31]. Stable emission above 20 mA/cm² has been demonstrated in SWNT films deposited on Si substrates [35].

Only a few years after the discovery of these promising properties some prototype devices have made, namely cathode ray lighting tubes, field emission displays, and gas-discharge tubes. Ise Electronic Co. has fabricated cathode ray lighting elements with carbon nanotube materials as the field emitters in Japan [32] (fig. 3a). In recent models, nanotubes are screen-printed onto the metal plates serving as a cathode, while a phosphor screen is printed onto the inner surfaces of the glass tube. Different colors can be obtained by using different fluorescent materials. The luminance of the phosphor screens measured from nanotube emitters is about twice that of conventional cathode ray tube (CRT) lighting elements operated under similar conditions [33]. A matrix-addressable diode flat panel display (demo) was constructed at Northwestern University, consisting of nanotube-epoxy stripes on the cathode glass plate and phosphor-coated indium-tin-oxide (ITO) stripes on the anode plate. The 4.5 inch diode-type field emission display shown in Fig. 3(b) has been fabricated by Samsung with SWNT stripes on the cathode and phosphor-coated ITO stripes on the anode running orthogonally to the cathode stripes [33].

As gas discharge tube (GDT) protectors, devices with nanotube covered cathodes are strong competitors; these could be used to protect against transient over-voltages in telecom circuits [36]. Prototype GDT devices using carbon nanotube coated electrodes have recently been fabricated and tested [37]. Molybdenum electrodes with various interlayer materials were coated with single-walled carbon nanotubes and analyzed for both electron field emission and discharge properties. A mean DC breakdown voltage of 448.5 V and a standard deviation of 4.8 V over 100 surges were observed in nanotube-based GDTs with 1 mm gap spacing between the electrodes. The breakdown reliability is a factor of 4-20 better and the breakdown voltage is ~30% lower than the two commercial products measured as a reference. The enhanced performance shows that nanotube-based GDTs can be attractive over-voltage protection units in advanced telecom networks.

4. Tailoring Nanotube Growth for Applications

Although the strategies of CVD growth and properties of CVD grown nanotubes and fibers will be discussed in a separate chapter [38] we will describe here some of the recent approaches used in the synthesis of nanotubes to accomplish specific materials science applications.

There are several, independent methods so far designed for growing nanotubes. Carbon (and non-carbon) nanotubes are synthesized in dc electric arcs [2,13,39-41], by laser evaporation/ablation of different targets [15,42], in soot of ethylene, benzene and acetylene flames [43,44], and by catalytic decomposition of organic compounds [45-49]. Certain catalysts (transition metals and their combinations) are used in all these techniques, particularly for the growth of singlewalled nanotubes. The electric arc and laser methods are inherently not scalable; however these techniques are used routinely to make gram quantities of nanotubes. The as-produced nanotubes (singlewalled nanotubes in particular) needs to be purified extensively, via chemical and filtration techniques; nanotube samples with greater than 99% purity is now commercially available in small quantities. CVD is scalable and there already exists industrial scale production of nanotubes, which are defective, nevertheless dimensionally similar to the nanotubes discussed above. CVD can also be used to create oriented nanotube arrays on substrates, by first patterning the catalyst particles on the substrates. Recently, for large area applications (flat-panel displays, catalyzers, etc.) this latter method is being increasingly utilized. The catalyst covering of the substrates can be produced by lithography, but several other methods can also be used, e.g. one can build the structure from nanometer size catalyst nanoparticles dot by dot [50], CVD on patterned catalyst islands [51] or by microcontact printing for patterning of larger areas with medium resolution [52]. Micropatterns of carbon nanotubes on a wide range of substrates have been generated by patterned growth of aligned nanotubes on surfaces pre-patterned with organic material (e.g. n-hexane plasma polymer) [53]. There have also been reports of autocatalytic growth of high aspect carbon nanostructures on graphite substrates by controlling the CVD process [54]. Many of these recent results point to the flexibility and power of the CVD technique, combined with micro- to nano-lithography. Ultimately, by tailoring the positioning of catalyst particles on substrates and controlling the CVD conditions, one hopes to achieve precise control of the nanotube architectures (orientation, position, density, and joining) that can be grown on various substrates. Such control will be needed if multiscale device structures are to be created from nanotubes for electronic applications.

322

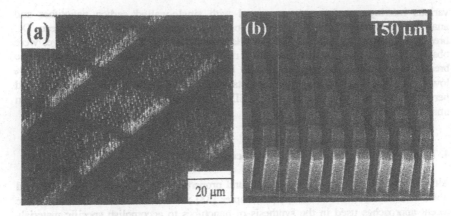

Figure 4. (a) micrograph showing growth of carbon nanotube obelisks on an array of submicron nickel dots: {adapted from [50]}; (b) the towers, used in electron field emission studies, are composed of many multi-walled nanotubes {taken from [51]}. Such ordered nanotube architectures can now be synthesized by several CVD methods on various substrates. The production of these structures could be scaled up in straightforward ways. Such nanotube materials can then find a variety of applications in electron field emission, flat-panel displays, large-scale scanning force probes, electrochemcial sensors and electrodes[51];

5. Energy Storage

Graphite, glassy carbon and carbon fiber electrodes have been used for decades in fuel cells, battery and several other electrochemical applications [55]. Nanotubes have smooth surface topology and perfect surface specificity. The rate of electron transfer, which ultimately determines the efficiency of fuel cells, depends on the structure and morphology of the carbon material used in the electrodes. Compared to conventional carbon electrodes, the electron transfer kinetics takes place fastest on nanotubes, following ideal Nernstian behavior [56]. Their performance has been found to be superior to other carbon electrodes in terms of reaction rates and reversibility in the case of several electrochemical reactions including some bio-electrochemical reactions [57]. Pure MWNTs and MWNTs deposited with metal catalysts (Pd, Pt, Ag) have been used to electro-catalyze oxygen reduction, which is important for fuel cells [58-59]. The properties of catalytically grown carbon nanofibers (which are basically defective nanotubes) have also been seen desirable for high power electrochemical capacitors [60].

The basic working mechanism of rechargeable lithium batteries is electro-chemical intercalation and de-intercalation of lithium between their two electrodes. Current lithium batteries use carbon materials (graphite or disordered carbon) as the anodes. It is desirable to have batteries with a high energy capacity, fast charging time and long cycle time. In carbon nanotubes a higher Li capacity may be obtained if all the interstitial sites (inter-shell van der Waals spaces, inter-tube channels, and inner cores) are accessible for Li intercalation. Electrochemical intercalation of MWNTs [61] and SWNTs [62] has been experimentally investigated. Electrochemical intercalation was studied using arc-discharge-grown MWNT sample using an electrochemical cell with a carbon nanotube film and a lithium foil as the two working electrodes [62]. A reversible capacity (C_{rev}) of 100-640mAh/g has been reported, depending on the sample processing conditions [61,62]. Studies [63-65] have also demonstrated that alkali metals can be intercalated into the inter-shell spaces within the individual MWNTs through defect sites. Single-walled nanotubes are shown to have both high reversible and irreversible capacities. Two groups reported 400-650mAh/g reversible and ~1000 mAh/g irreversible capacities in SWNTs. The exact locations of the Li ions in the intercalated SWNTs are unknown. The large irreversible capacity could be related to the large surface area of the SWNT films (~300 m^2/g by BET characterization) and the formation of a solid-electrolyte-interface.

6. Gas Storage, Usage as Template Material

The well-defined room inside the nanotubes offers the possibility to fill it with different materials. The goal may be the storage of the filling material, for example hydrogen, drug molecules and similar applications, or to shape the material (template based fabrication of one-dimensional nanowires). Early calculations suggested that strong capillary forces exist in nanotubes, strong enough to hold gases and fluids inside them [66]. Extraordinarily high and reversible hydrogen adsorption in SWNT-containing materials [67-70] and graphite nanofibers [71,72] has been reported. The basic mechanism(s) of hydrogen storage in these materials is not understood, and experimental data measured in different groups need to be verified on well-characterized materials under controlled conditions. Measurements performed on large samples of unpurified materials showed a hydrogen storage capacity of 4.2 wt% when the samples were exposed to 10 MPa hydrogen at room temperature. About 80% of the absorbed H_2 could be released at room temperature [69], thus achieving/exceeding the benchmark of 6.5 wt% H_2 to system weight ratio set by the Department of Energy. At this point it is still not clear whether carbon nanotubes will have real technological relevance in the hydrogen storage applications area. Some scientific studies also show that the diameter of the nanotubes is critical for storage of gases such as hydrogen. In recent years there has been partially successful attempts to tailor the diameter of nanotubes during synthesis and the hydrogen storage results on larger diameter tubes (>5 nm) have been very promising.

The first experimental proof on usage of nanotubes as templates was demonstrated in 1993, by the filling and solidification of molten metal inside the channels of MWNTs [74]. Wires as small as 1.2 nm in diameter were fabricated by this method, inside nanotubes. Since then, nanotubes have been extensively studied as templates to create nanowires of various compositions and structures (see Fig. 5).

The critical issue in the filling of nanotubes is the wetting characteristics of nanotubes, which seems to be quite different from that of planar graphite, because of the curvature of the tubes. Wetting of low melting alloys and solvents occurs quite readily in the internal high curvature pores of MWNTs and SWNTs. In the latter, since the pore sizes are very small, filling is more difficult and can be done only for a selected few compounds. Liquids such as organic solvents wet nanotubes easily and it has been

324

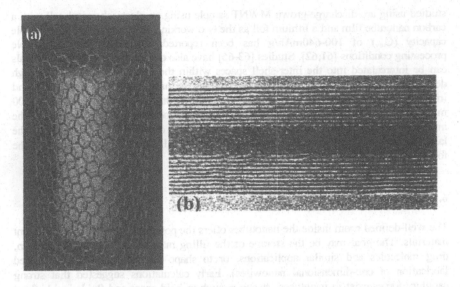

Figure 5. Illustrations to show the use of nanotubes as templates. (a) is a schematic that shows the filling of the empty one-dimensional hollow core of nanotubes with foreign substances. (b) TEM image of a filled MWNT, its cavity filled uniformly with lead oxide. The filling was achieved by capillarity [73].

proposed that interesting chemical reactions could be performed inside nanotube cavities [75]. Filled nanotubes can also be synthesized in situ, by encapsulation during the growth of nanotubes in an electric arc or laser ablation. These techniques produce encapsulated nanotubes with carbide nanowires (e.g., transition metal carbides) inside [76,77]. Multi-element nanotube structures consisting of multiple phases (e.g., coaxial nanotube structures containing SiC, SiO, BN and C) have been successfully synthesized by reactive laser ablation [78]. Similarly, post-fabrication treatments can be used to create heterojunctions between nanotubes and semiconducting carbides [79,80]. There are other ways in which pristine nanotubes can be modified into composite structures. Chemical functionalization can be used to build macro-molecular structures from fullerenes and nanotubes. Decoration of nanotubes with metal particles has been achieved for different purposes, most importantly for use in heterogeneous catalysis [81].

7. Nanotubes as Reinforcements

The mechanical behavior of carbon nanotubes is exciting since nanotubes are seen as the "ultimate" carbon fiber ever made. Nanotubes should then be ideal candidates for structural applications. Carbon fibers conventionally have been used as reinforcements in high strength, light weight high-performance composites; one can typically find these in a range of products ranging from expensive tennis rackets to aircraft body parts.

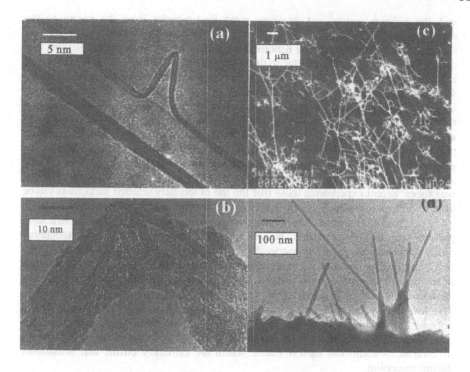

Figure 6. (a) TEM image of a buckled MWNT in a polymer film due to compressive stresses. [6] (b) HRTEM showing the deformation of individual layers during the buckling of an individual MWNT. (c) SEM image of a typical fracture surface in a SWNT-epoxy resin composite [6]. Notice the curved network of the nanotube ropes in the composite. (d) Pull-out of MWNTs at a broken surface of a nanotube-PPV composite. Notice a thin layer of polymer web that joins the nanotubes together {figure taken from [89]}.

Early theoretical work and recent experiments on individual nanotubes (mostly MWNTs) have confirmed that nanotubes are one of the stiffest structures ever made [82-85]. Since carbon-carbon covalent bonds are one of the strongest in nature, a structure based on the perfect arrangement of these bonds oriented along the axis of nanotubes would produce an exceedingly strong material. Theoretical studies have suggested that SWNTs could have a Young's modulus as high as 1 TPa [83], which is the in-plane value of defect free graphite. Although the theoretical estimate for the tensile strength of individual SWNTs is about 300 Gpa, the observed tensile strength value corresponded to < 60 GPa. High temperature plastic behavior in nanotubes is also interesting, as seen from simulations [86]. It is suggested that pairs of 5-7 (pentagon-heptagon) pair defects, called a Stone-Wales defect [87] in sp^2 carbon systems, are created under high strain in the nanotube lattice. The motion of these defects in nanotubes leads to a step-wise diameter reduction (localized necking).

The obvious application of nanotubes based on their mechanical properties will be as structural reinforcements in composite materials. The efficiency of nanotube reinforcements will however depend on creating a good interface between nanotubes and the polymer matrix and attaining good load transfer from the matrix to the nanotubes, during loading. This is clearly not a trivial task. There have been conflicting reports about the strength of the nanotube-polymer interface. Most studies suggest that the load transfer is not efficient due to the sliding of individual layers in MWNTs and of individual nanotubes in SWNT ropes. It has been suggested that there two steps to surmount before nanotubes can become efficient load carriers; one, to create a strong interface between nanotubes and the polymer by chemically modifying and designing the nanotube surface and two, the nanotube aggregates (ropes) have to be broken down and dispersed or cross-linked internally so that sliding of individual tubes or layers do not happen during load transfer. There are also other challenges that relate to the processing of nanotube polymer composites. Achieving good dispersion of the nanotubes in the matrix is a formidable task, especially at higher volume fractions of the nanotubes. Recent experiments show that the use of specially selected surfactants can be used to disperse nanotube in epoxy matrices without affecting the matrix properties.

There are incontestable advantages of using carbon nanotubes in structural polymer (e.g., epoxy) composites. Nanotube reinforcements will increase the toughness of the composites by absorbing energy during their highly flexible elastic behavior. This will be especially important for nanotube-based ceramic matrix composites. An increase in the fracture toughness on the order of 25% has been seen in nano-crystalline alumina nanotube (5% weight fraction) composites, without compromising on hardness [88]. The low density of the nanotubes will clearly be an advantage for nanotube-based polymer composites, in comparison to short carbon fiber reinforced (random) composites. Nanotubes also has the advantage of being highly elastic during processing, in contrast to traditional carbon fibers, which are extremely brittle, and breakdown during processing.

There are also novel applications that are being developed for nanotube polymer systems. One such system is photoactive polymers which are filled with nanotubes. Recently, it has been demonstrated in a conjugated luminescent polymer, poly(m-phenylenevinylene-co-2,5-dioctoxy-p-phenylenevinylene) (PPV), filled with MWNT and SWNT [89]. Nanotube/PPV composites have shown large increases in electrical conductivity (by nearly eight orders of magnitude) compared to the pristine polymer, with little loss in photoluminescence/electro-luminescence yield. Such structures have been shown to be useful in organic light emitting device applications. The doping of nanotubes in the polymer can be used to change the emission properties of the polymer, for example, color tenability [90]. The increase in electrical conductivity, at low percolations using nanotube fillers, has also been used to prepare bioactive substrates for cell growth (osteo-integration). A biodegradable polymer (polylacticacid, PLA) with MWNT is used to provide electrical stimulus for enhanced adhesion and growth of cells and continued mineralization (bone formation) [91].

8. Conclusions and Challenges

Since their discovery in 1991, nanotubes have caught the fancy of researchers, industry and policy makers alike. It is now considered to play a major role in the ensuing nanotechnology revolution. Nanotube is an extremely versatile material with applications ranging from biomedical to electronic to structural areas. There is still a need to develop and optimize processing methods to make well defined (in terms of structure and dimensions) and well organized nanotubes, since this will have real impact on applications. For bulk applications, the quantities that can be made still fall far short of what industry would need and the procedures for purification are tedious and expensive. The market price of nanotubes has fallen over the last few years but is still too high (~$ 200 per gram for as produced material).

The discovery, research and development of nanotubes in the last decade have changed our view of carbon materials. The greatest virtue of the nanotube structure is that it can be entirely documented and presented from a unit cell picture, giving the material a molecular characteristic. This means that the properties of nanotubes can be predicted precisely, theoretically, and this can excite experimentalists to find new applications. The excitement in this field has prompted physicists, chemists and engineers to come together and work on some fascinating interdisciplinary problems and challenges associated with nanotubes. From the progress made in recent years, it seems inevitable that nanotubes will find its way to the marketplace in more than one route: nanoprobes, field emitting devices and composites could be such strong possible ways.

9. References

1. Kroto, H.W., Heath, J.R., O'Brien, S.C., Curl, S.C. and Smalley, R.E. (1985) C$_{60}$: Buckminsterfullerene, *Nature* 318, 162-163.
2. Iijima, S. (1991) Helical mirotubules of graphitic carbon, *Nature* 354, 56-58.
3. Dresselhaus, M.S., Dresselhaus, G. and Eklund, P.C. (1996) *Science of Fullerenes and Carbon Nanotubes*, Academic Press, New York.
4. Ebbesen, T.W. (1997) *Carbon Nanotubes: Preparation and Properties*, CRC Press, Boca Raton.
5. Saito, R., Dresselhaus, G. and Dresselhaus, M.S. (1998) *Physical Properties of Carbon Nanotubes*, Imperial College Press, London.
6. Ajayan, P.M. (1999) Nanotubes from carbon, *Chem. Rev.* 99, 1787-1799.
7. Dekker, C. (1999) Carbon nanotubes as molecular quantum wires, *Physics Today* 52, 22-28.
8. Yakobson, B.I. and Smalley, R.E. (1997) Fullerene nanotubes: C$_{1,000,000}$ and beyond, *American Scientist* 85, 324-337.
9. Ajayan, P.M. and Ebbesen, T.W. (1997) Nanometer size tubes of carbon, *Reports on Progress in Physics* 60, 1025-1062.
10. Dresselhaus, M.S., Dresselhaus, G., Eklund, P.C. and Saito, R. (1998) Nanotube connections, *Physics World*, January Issue, 33-34.
11. Dresselhaus, M.S. (2000) chapter in this ASI issue.
12. Dresselhaus, M.S., Dresselhaus, G., Sugihara, K., Spain, I. L., Goldberg, H. A. (1988) *Graphite Fibers and Filaments*, Springer-Verlag, New York.
13. Ebbesen, T.W. and Ajayan, P.M. (1992) Large scale synthesis of carbon nanotubes, *Nature* 358, 220-222.
14. Journet, C., Maser, W.K., Bernier, P., Loiseau, A., Lamy de la Chapelle, M., Lefrant, S., Deniard, P., Lee, R. and Fischer, J.E. (1997) Large-scale production of single-walled carbon nanotubes by the electric-arc technique, *Nature* 388, 756-758.
15. Thess, A., Lee, R., Nikdaev, P., Dai, H., Petit, P., Robert, J., Xu, C., Lee, Y.H., Kim, S.G., Rinzler, A.G., Colbert, D.T., Scuseria, G.E., Tomanek, D., Fischer, J.E. and Smalley, R.E. (1996) Crystalline ropes of metallic carbon nanotubes, *Science* 273, 483-487.
16. Li, W.Z., Xie, S.S., Qian, L.X., Chang, B.H., Zou, B.S., Zhou, W.Y., Zhao, R.A. and Wang, G. (1996) Large-scale synthesis of aligned carbon nanotubes, *Science* 274, 1701-1703.
17. Berber, S., Kwon, Y-K., Tomanek, D. (2000) Unusually high thermal conductivity of carbon nanotubes, *Phys. Rev. Lett.* 84, 4613-4616.

328

18. Dai, H.J., Hafner, J.H., Rinzler, A.G., Colbert, D.T. and Smalley, R.E. (1996) Nanotubes as nanoprobes in scanning probe microscopy, *Nature* **384**, 147150.
19. Wong, S.S., Harper, J.D., Lansbury, P.T. and Lieber, C.M. (1998) Carbon nanotube tips: High-resolution probes for imaging biological systems, *J. Am. Chem. Soc.* **120**, 603-604.
20. Hafner, J.H., Cheung, C.L. and Lieber, C.M. (1999) Growth of nanotubes for probe microscopy tips, *Nature* **398**, 761-762.
21. Kim, P. and Lieber, C.M. (1999) Nanotube nanotweezers, *Science* **286**, 2148-2150.
22. Wong, S.S., Joselevich, E., Woolley, A.T., Cheung, C. L. and Lieber, C.M. (1998) Covalently functionalized nanotubes as nanometre-sized probes in chemistry and biology, *Nature* **394**, 52-55.
23. Baughman, R.H., Cui, C., Zhakhidov, A.A., Iqbal, Z., Barisci, J.N., Spinks, G.M., Wallace, G.G., Mazzoldi, A., Rossi, D.D., Rinzler, A.G., Jaschinski, O., Roth, S. and Kertesz, M. (1999) Carbon nanotube actuators, *Science* **284**, 1340-1344.
24. Kong, J., Franklin, N.R., Zhou, C., Chapline, M.G., Peng, S., Cho, K. and Dai, H. (2000) Nanotube molecular wires as chemical sensors, *Science* **287**, 622-625.
25. Dai, H., Franklin, N., Jie, H. (1998) Exploiting the properties of carbon nanotubes for nanolithography, *Appl. Phys. Lett.* **73**, 1508-1510.
26. Fowler, R.H. and Nordheim, L.W. (1928) Electron emission on intense electric fields, *Proc. R. Soc. London* **A119**, 173-181.
27. Brodie, I. and Spindt, C.A. (1992) Vacuum microelectronics, *Adv. Electronics & Electron Phys.*, **83**, 1-95.
28. Rinzler, A.G., Hafner, J.H., Nikolaev, P., Lou, L., Kim, S.G., Tomanek, D., Colbert, D. and Smalley, R.E. (1995) Unraveling nanotubes: Field emission from an atomic wire, *Science* **269**, 1550-1553.
29. Bonard, J.-M., Stöckli, T., Maier, F., de Heer, W.A., Châtelain, A., Salvetat, J.-P. and Forró, L. (1998) Field-emission-induced luminescence from carbon nanotubes, *Phys. Rev. Lett.* **81**, 1441-1444.
30. Bonard, J.-M., Salvetat, J.-P., Stöckli, T., Forró, L., Châtelain A. (1999) Field emission from carbon nanotubes: perspectives for applications and clues to the emission mechanism, *Appl Phys A* **69**, 245-254.
31. Bower, C., Zhou, O., Zhu, W., Ramirez, A.G., Kochanski, G.P., Jin, S. in J.P. Sullivan, J.R. Robertson, B.F. Coll, T.B. Allen and O. Zhou (eds.) *Amorphous and Nanostructured Carbon,* Mater. Res. Soc. (in press).
32. Saito, Y., Uemura, S. and Hamaguchi, K. (1998) Cathode ray tube lighting elements with carbon nanotube field emitters, *Jpn. J. Appl. Phys.* **37**, L346-348.
33. Choi, W. B., Chung, D. S., Kang, J. H., Kim, H. Y., Jin, Y. W., Han, I. T., Lee, Y. H., Jung, J. E., Lee, N. S., Park, G. S. and Kim, J. M. (1999) Fully sealed, high-brightness carbon-nanotube field-emission display, *Appl. Phys. Lett.* **75**, 3129-3131.
34. Suzuki, S., Bower, C., Watanabe, Y. and Zhou, O., Unpublished results.
35. Zhu, W., Bower, C., Zhou, O., Kochanski, G., Jin, S. (1999) Large current density from carbon nanotube field emitters, *Appl. Phys. Lett.* **75**, 873-875.
36. Standler, R. (1989) *Protection of Electronic Circuits from Over-voltages,* John Wiley & Sons, New York.
37. Rosen, R., Simendinger, W., Debbault, C., Shimoda, H., Fleming, L., Stoner, B., Zhou, O. (2000) Application of carbon nanotubes as electrodes in gas discharge tubes, *Appl. Phys. Lett.* **76**, 1668-1670.
38. Kónya, Z. (2000) chapter in this ASI issue.
39. Iijima, S., Ichihashi, T. (1993) Single-shell carbon nanotubes of 1-nm diameter, *Nature* **363**, 603- 605.
40. Bethune, D.S., Kiang, C.H., De Vries, M.S., Gorman, G., Savoy, R., Vazquez, J, Beyers, R., (1993) Cobalt-catalysed growth of carbon nanotubes with single atomic layer walls, *Nature* **363**, 605- 607.
41. Gamaly, E.G., Ebbesen, T.W. (1995) Mechanism of carbon nanotube formation in the arc discharge, *Phys. Rev. B* **52**, 2083-2089.
42. Puretzky, A.A., Geohegan, D.B., Fan, X., Pennycook, S.J., (2000) Dynamics of single-wall carbon nanotube synthesis by laser vaporization, *Appl Phys A* **70**, 153-160.
43. Chowdhury, K.D., Howard, J.B., Vandersande, J.B. (1996) Fullerenic nanostructures in flames, *J. Mater. Res.* **11**, 341-347.
44. Richter, H., Hernádi, K., Caudano, R., Fonseca, A., Migeon,. H.-N., Nagy, J.B., Schneider, S., Vandooren, J., Van Tiggelen. P.J. (1996) Formation of nanotubes in low pressure hydrocarbon flames, *Carbon* **34**, 427-429.
45. Amelinckx S., Zhang X.B., Bernaerts D., Zhang X.F., Ivanov V., Nagy J.B. (1994) A formation mechanism for catalytically grown helix shaped graphite nanotubes, *Science* **265**, 635-639.
46. Terrones, M., Grobert, N., Olivares, J., Zhang, J.P., Terrones, H., Kordatos, K., Hsu, W.K., Hare, J.P., Townsend, P.D., Prassides, K., Cheetham, A.K., Kroto, H.W., Walton, D.R.M., (1997) Controlled production of aligned-nanotube bundles, *Nature* **388**, 52-55.

47. Kong, J., Cassell, A.M. and Dai, H.J. (1998) Chemical vapor deposition of methane for single-walled carbon nanotubes, *Chem. Phys. Lett.* **292**, 567-574.
48. Colomer, J.-F., Bister, G., Willems, I., Konya, Z., Fonseca, A., Van Tendeloo, G . and Nagy, J.B. (1999) Synthesis of single-wall carbon nanotubes by catalytic decomposition of hydrocarbons, *Chem. Commun.* **14**, 1343-1344.
49. de Heer, W.A., Chatelain, A. and Ugarte, D. (1995) A carbon nanotube field-emission electron source, *Science* **270**, 1179-1180.
50. Ren, Z.F., Huang, Z.P., Wang, D.Z., Wen, J.G., Xu, J.W., Wang, J.H., Calvet, L.E., Chen, J., Klemic, J.F. and Reed, M.A. (1999) Growth of a single freestanding multiwall carbon nanotube on each nanonickel dot, *Appl. Phys. Lett.* **75**, 1086-1088.
51. http://www-chem.stanford.edu/group/dai/Projects/fieldemission.html
52. Kind, H., Bonard, J-M., Emmenegger, C., Nilsson, L-O., Hernadi, K., Maillard-Schaller, E., Schlapbach, L., Forro, L., Kern, K. (1999) Patterned films of nanotubes using microcontact printing of catalysts, *Adv. Mater.* **11**, 1285-1289.
53. Chen, Q. and Dai, L. (2000) Plasma patterning of carbon nanotubes, *Appl. Phys. Lett.* **76**, 2719-2721.
54. Ajayan, P.M., Nugent, J.M., Siegel, R.W., Wei, B., Kohler-Redlich, P. (2000) Growth of carbon micro-trees - Carbon deposition under extreme conditions causes tree-like structures to spring up, *Nature* **404**, 243-243.
55. McCreery, R.L. (1991) in A.J. Bard (ed.), *Electroanalytical Chemistry* Vol. 17, Marcel Dekker, New York.
56. Nugent, J., Santhanam, K.S.V., Rubio, A. and Ajayan, P.M., Unpublished results.
57. Britto, P.J., Santhanam K.S.V., and Ajayan, P.M. (1996) Oxidation of dopamine at carbon nanotube electrodes, *Bioelectrochemistry and Bioenergetics* **41**, 121- 126.
58. Britto, P.J., Santhanam, K.S.V., Rubio, A., Alonso, J.A., Ajayan, P.M. (1999) Improved charge transfer at carbon nanotube electrodes, *Adv.Mater.* **11**, 154-157.
59. Che, G., Lakshmi, B.B., Fisher, E.R. and Martin, C.R. (1998) Carbon nanotubule membranes for electrochemical energy storage and production, *Nature* **393**, 346-349.
60. Niu, C., Sichel, E.K., Hoch, R., Moy, D. and Tennent, H. (1997) High power electrochemical capacitors based on carbon nanotube electrodes, *Appl. Phys. Lett.* **70**, 1480-1482.
61. Frackowiak, E., Gautier, S., Gaucher, H., Bonnany, S., Beguin, F. (1999) Electrochemical storage of lithium multiwalled carbon nanotubes, *Carbon* **37**, 61-69.
62. Gao, B., Kleinhammes, A., Tang, X.P., Bower, C., Fleming, L., Wu, Y., Zhou, O. (1999) Electrochemical intercalation of single-walled carbon nanotubes with lithium, *Chem. Phys. Lett.* **307**, 153-157.
63. Zhou, O., Fleming, R.M., Murphy, D.W., Chen, C.T., Haddon, R.C., Ramirez, A.P. and. Glarum, S.H. (1994) Defects in carbon nanostructures, *Science* **263**, 1744-1747.
64. Suzuki, S. and Tomita, M. (1996) Observation of potassium intercalated carbon nanotubes and their valence band excitation spectra, *J. Appl. Phys.* **79**, 3739 -3743.
65. Suzuki, S., Bower, C. and Zhou, O. (1998) In-situ TEM and EELS studies of alkali-metal intercalation with single-walled carbon nanotubes, *Chem. Phys. Lett.* **285**, 230-234.
66. Pederson, M.R. and Broughton, J.Q. (1992) Nanocapillarity in fullerene tubules, *Phys. Rev. Lett.* **69**, 2689-2692.
67. Dillon, A.C., Jones, K.M.,. Bekkedahl, T.A., Kiang, C.H., Bethune, D.S. and Heben, M.J. (1997) Storage of hydrogen in single-walled carbon nanotubes, *Nature* **386**, 377-379.
68. Chen, P., Wu, X., Lin, J., Tan, K.L. (1999) High H_2 uptake by alkali-doped carbon nanotubes under ambient pressure and moderate temperatures, *Science* **285**, 91-93.
69. Liu, C., Fan, Y.Y., Liu, M., Cong, H.T., Cheng, H.M., Dresselhaus, M.S. (1999) Hydrogen storage in single-walled carbon nanotubes at room temperature, *Science* **286**, 1127-1129.
70. Nutzenadel, C., Zuttel, A., Chartouni, D., Schlapbach, L. (1999) Electrochemical storage of hydrogen in nanotube materials, *Electrochemical and Solid-state Letters* **2**, 30-32.
71. Chambers, A., Park, C., Baker, R. T. K. and Rodriguez, N. M. (1998), Hydrogen storage in graphite nanofibers, *J. Phys. Chem. B*, **102**, 4253-4256.
72. Dresselhaus, M.S., Williams, K.A., Eklund, P.C. (1999) Hydrogen adsorption in carbon materials, *MRS Bulletin* **24**, 45-50.
73. Tsang, S.C., Chen, Y.K., Harris P.J.F. and Green, M.L.H. (1994) A simple chemical method of opening and filling carbon nanotubes, *Nature* **372**, 159-162.
74. Ajayan, P.M. and Iijima, S. (1993) Capillarity induced filling in carbon nanotubes, *Nature* **361**, 333-334.
75. Dujardin, E., Ebbesen, T.W., Hiura, T. and Tanigaki, K. (1994) Capillarity and wetting of carbon nanotubes, *Science* **265**, 1850-1852.

330

76. Ruoff, R.S., Lorents, D.C., Chan, B., Malhotra, R. and Subramoney, S. (1992) Single crystal metals encapsulated in carbon nanoparticles, *Science* **259**, 346-348.
77. Guerret-Plecourt, C., Le Bouar, Y., Loiseau, A. and Pascard, H. (1994) Relation between metal electronic structure and morphology of metal compounds inside carbon nanotube, *Nature* **372**, 761-763 .
78. Zhang, Y., Suenaga, K., Colliex, C. and Iijima, S. (1998) Coaxial nanocable: Silicon carbide and silicon oxide sheathed with boron nitride and carbon, *Science* **281**, 973-975.
79. Hu, J.T., Min, O.Y., Yang, P.D., Lieber, C.M. (1999) Controlled growth and electrical properties of heterojunctions of carbon nanotubes and silicon nanowires, *Nature* **399**, 48-51.
80. Zhang, Y., Ichihashi, T., Landree, E., Nihey, F., Iijima, S. (1999) Heterostructures of single-walled carbon nanotubes and carbide nanorods, *Science* **285**, 1719-1722.
81. Planeix, J.M., Coustel, N., Cog, B., Brotons, V., Kumbhar, P.S., Dutartre, R., Geneste, P., Bernier, P. and Ajayan, P.M. (1994) Application of carbon nanotubes as supports in heterogeneous catalysis, *J. Am. Chem. Soc.* **116**, 7935- 7936 .
82. Overney, G., Zhong, W. and Tomanek, D. (1993) Structural rigidity and low frequency vibrational modes of long carbon tubules, *Z. Phys. D* **27**, 93-96.
83. Yakobson, B.I., Brabec, C.J. and Bernholc, J. (1996) Nanomechanics of carbon tubes: Instabilities beyond linear response, *Phys. Rev. Lett.* **76**, 2511-2514.
84. Treacy, M.M.J., Ebbesen, T.W. and Gibson, J.M. (1996) Exceptionnaly high Young's modulus observed for individual carbon nanotubes, *Nature* **381**, 678-680.
85. Wong, E.W., Sheehan, P.E. and Lieber, C.M. (1997) Nanobeam mechanics: Elasticity, strength, and toughness of nanorods and nanotubes, *Science* **277**, 1971-1975.
86. Yakobson, B.I (1998) Mechanical relaxation and "intramolecular plasticity" in carbon nanotubes, *Appl. Phys. Lett.* **72**, 918-920.
87. Stone, A.J. and Wales, D.J. (1986) Theoretical studies of icosahedron C_{60} and some related species, *Chem. Phys. Lett.* **128**, 501-503.
88. Chang, S., Doremus, R.H., Ajayan, P.M. and Siegel, R. W. (2000) unpublished results.
89. Curran, S.A., Ajayan, P.M., Blau, W.J., Carroll, D. L., Coleman, J.N., Dalton, A.B. Davey, A.P., Drury, A., McCarthy, B., Maier, S. and Strevens, A (1998) Composite from poly(m-phenylenevinylene-co-2,5-dioctoxy-p-phenylenevinylene) and carbon nanotubes: A novel material for molecular optoelectronics, *Adv. Mater.* **10**, 1091-1093.
90. Carroll, D. L.., Unpublished results.
91. Bizios, R. and Ajayan, P. M., Unpublished results.

NOVEL APPLICATIONS OF VGCG INCLUDING HYDROGEN STORAGE

MAX L. LAKE
Applied Sciences, Inc.
P.O. Box 579, Cedarville, OH 45314, USA
www.apsci.com

Abstract. From a variety of synthesis studies of vapor grown carbon fibers (VGCF), two forms of discontinuous fibers have emerged as leading candidates for commercial application. The first is made to have a diameter comparable to conventional carbon fibers by coating the original graphite filament with pyrolytic carbon. Composites of this fiber have been fabricated with exceptionally high values of thermal and electrical conductivity, but modest values of strength and modulus. This combination of properties is assumed to derive from the fact that, while the pyrolytic carbon coating is of high purity and graphitizes well, defects incorporated into the graphite lattice are proportional to the volume of the fiber. These defects lead to a decrease in strength and modulus at larger fiber diameters. The second form of VGCF, that has higher potential for low cost and large volume production, is a nanofiber that has a diameter on the order of 100 nanometers and a length of about 100 microns. Methods for fabricating composites from these materials are under development, and show promise in a number of large volume applications.

Models describing the strength and modulus of other short fiber reinforced composites suggest that this form of fiber could be used to produce excellent engineering composites with modulus several times in excess of aluminum. This level of reinforcement implies that critical degrees of two-dimensional orientation are achievable, as well as a proper fiber/matrix interface. This premise, coupled with the prospect of a production cost comparable to that of glass fibers provides the basis of a potential revolution in the field of carbon fiber reinforced composites, wherein the worldwide demand would increase by over an order of magnitude. Some new and quite promising applications of vapor grown carbon nanofibers are reported in this chapter.

1. Introduction

Carbon nanofibers are touted for an astonishing variety of applications in fields ranging from engineering materials to chemical and biomedical processing to semiconductor devices, and even to hydrogen storage. This optimism is derived from the unique and extreme values of their physical properties, including their size, stiffness, and electrical and thermal conductivity. To this point, the latter application of hydrogen storage has yet to be demonstrated as feasible. While the theoretical predictions are motivating, practical experience has been discouraging (details are given near the end of the paper). However, substantial progress has been made toward practical reinforcement of engineering

L.P. Biró et al. (eds.), Carbon Filaments and Nanotubes: Common Origins, Differing Applications, 331–341.
© 2001 *Kluwer Academic Publishers. Printed in the Netherlands.*

materials to provide both mechanical reinforcement and an added measure of electrical enhancement not available from alternative reinforcement agents.

2. The Demand for Nanofiber Reinforced Composites

As a practical matter, the more significant applications impacting society and commerce will be those for which high volume of nanofibers is required. For perspective, an assessment of current and forecast composite materials utilization assists in establishing the potential scale of economic impact.

The annual market for composite materials is estimated to be in excess of $10 billion at the beginning of the 21st century, with annual growth exceeding 5% per year, as composites penetrate the automotive, construction and civil engineering, marine, and electronics industries. The Society of the Plastics Industry's Composites Institutes reports that composite shipments reached 1.63 million tons for 1998, representing a 5.5% increase over 1997, and a 53% increase since 1991. Drivers of this growth included sales of automobiles and new and existing homes, as well as other critical factors which affected the economy, including consumer spending, interest rates and infrastructure spending. Material substitution is the key mechanism in the growth of the composites industry. Benefits from composites, which underpin this growth, include high strength, design flexibility, corrosion resistance, lightweight, dimensional flexibility, high dielectric strength, parts consolidation, finishing, low tooling costs, and availability of standard shapes.

Figure 1 illustrates the utilization of composites by industry sector. The largest consumer of composites is the transportation sector, which utilized 516.6 million kilograms in 1998 or one third of the total composite usage for that year. The Automotive Composites Alliance predicts a further 27% increase in the use of reinforced thermosets in the next five years.

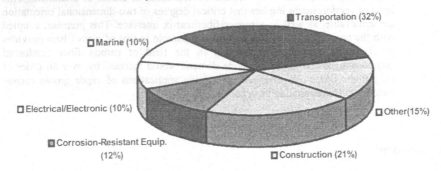

Figure 1. Utilization of composites by industry sector

Construction applications increased by 7.2% from 1997 to 1998, due to strong consumer demand and low interest rates. In fact, construction and civil engineering applications are considered to be the largest untapped opportunity for composites. Composite materials are continuing to gain market share for applications ranging from wood decks, where pressure treated wood currently dominates the market, to composite bathtubs, where they have already achieved 60% of the market.

Marine and electronics applications each represent approximately 10% of composite usage. Although these have experienced less robust growth than transportation and construction, continued growth at around 3.5% is anticipated for both industries. The remainder of composite usage is distributed among appliance and business equipment, corrosion resistant applications, consumer goods, aircraft, and other miscellaneous use.

An emerging element of the composites market is the so-called "engineered nanostructures", where improved composite properties are achieved with carbon nanofibers and other nanoscale reinforcements. Vigorous R&D activity to perfect applications is currently underway in academic and industry laboratories. These efforts will depend on the development of high efficiency methods of production of single-walled and multi-walled carbon nanotubes, and carbon nanofibers. Such materials will initially be used in large volumes to enhance the properties of conventional composite materials, such as adding electrical conductivity to glass fiber composites, thus expanding the number of applications that can be supported by a given composite system.

3. Engineered Polymers

An early commercial application of carbon nanofibers is for engineered polymers. Composites synthesized with discontinuous fibrous reinforcements can be produced with an inherently low processing cost. In equivalent production volumes, carbon nanofibers are projected to have a cost comparable to E-glass on a per-pound basis, yet to possess properties, which far exceed those of glass and are equal to or exceed those of much more costly PAN-based and pitch-based carbon fibers.

Because they are a discontinuous reinforcement, carbon nanofibers can be incorporated into commercially available thermoplastics, thermosets and elastomers and can be used directly in existing high volume molding processes without any significant new manufacturing development. Due to their extraordinary intrinsic properties, particularly elastic modulus, nanofibers are expected to enable a reduction in the material required to produce a given stiffness, thus providing net weight and cost savings.

3.1 MECHANICAL PROPERTIES

While mechanics of continuous reinforcements have been reasonably well studied over the past thirty years, there have been fewer stimuli to study the corresponding properties of short fiber reinforced composites. This knowledge is required to optimize interface and pre-form development, and for establishing design rules for composite synthesis. Thus the database required to predict the mechanical properties of discontinuous-fiber reinforced composites from the properties of the matrix, the fibers, the fiber/matrix interface, and the fiber architecture is not well developed. However, there do exist treatments of short fiber reinforcements from which insights can be drawn.

The earliest derivation of a formula to explain short-fiber reinforcement is due to Cox, who wanted to understand the elastic properties of paper [1]. His work was later extended by Baxter to include general fiber architecture [2]. This theory predicts that the modulus of a composite, E_c, can be determined from the fiber and matrix moduli, E_f and E_m respectively, and the fiber volume fraction, V_f, by a variation of the familiar rule of mixtures, that incorporates the fibers' degree of orientation, θ, and aspect ratio, l/d:

$$E_c = E_m V_m + E_f V_f \, g(l/d) \, f(\theta) \qquad (1)$$

where the functions, f and g, take on values between 0 and 1. The former is small for stubby particles, but increases rapidly with aspect ratio. The latter is dependent upon θ, and is greatest for uniaxial alignment. As shown in the previous chapter of this book, Baxter's theory predicts that if carbon nanofibers - which have very high aspect ratios - can be orientated in two dimensions, the resulting composite could be several times as stiff as a glass reinforced composite. For example, a panel with only 30 vol. % fibers would exhibit stiffness equivalent to aluminum.

3.2 AUTOMOTIVE APPLICATIONS

3.2.1 *SMCs*
The major attributes of sheet molding compounds (SMCs) are low tooling costs compared to steel, and the ability to form complex shapes, leading to increased design degrees-of-freedom and ultimate part reduction. E-glass reinforcements impart moderate modulus to the compound, but suffer from long curing cycles and from the fact that the compounds must be laminated to provide a suitable surface for painting. The potential advantages of carbon-nanofiber-reinforced sheet molding compounds lie in several areas. Nanofiber/polyester composites having equal stiffness with decreased fiber loading will reduce cost and weight. The small diameter of carbon nanofiber and low water content will allow as-molded parts to be painted without laminate layers, and without pinholes and pits. The high thermal and electrical conductivity of carbon nanofibers may also contribute to shorter, less-expensive curing cycles. Additionally, the low coefficient of thermal expansion of carbon nanofibers will ameliorate the thermal shrinkage problems experienced with current SMCs.

3.2.2 *Underhood Applications*
The range of glass fiber reinforced composites (GFRCs) applications gives a useful guide to the potential uses of VGCF composites. Nylon-based GFRCs constitute most of the underhood applications of plastics: fans, radiator parts, air conditioners, air filters, inlet manifolds, brake fluid reservoirs, air control valves, heater housings, windshield wiper reservoirs, gears, etc. All of these could potentially be replaced by carbon nanofiber composites. In addition, many applications that avoid glass fibers, either because they are electrically insulating, or because they are too coarse in size, could also be considered (e.g., electrical boxes and connectors, gas tanks, interior/console/dash surfaces, and body panel exterior surfaces).

3.2.3 *Plastics*
In polymer composites VGCF can improve stiffness, surface paintability, and impact resistance and ameliorate creep and shrinkage problems. In the automotive sector, the use of plastics for passenger vehicles and light trucks in North America in 1999 was approximately 1.8 billion kilograms [3]. Thermoplastic panels will probably be more desirable than thermosets, since they offer potential for recycling by melting. The fiber

modulus and surface area available for bonding the fiber to the matrices are many times that of glass and other commercial fibers. This creates the probability that less fiber will be required to attain similar modulus properties, which would influence the overall cost and would definitely create weight savings. In addition to modulus, creep properties at room and higher temperatures may be improved. These possibilities may move the thermoplastics industry into applications now occupied by more expensive thermoset processes. Another useful property for VGCF is that the axial coefficient of thermal expansion is slightly negative, while that of glass is higher than plastics. This property may reduce mold shrinkage, cracking, and stress problems associated with the higher coefficients of thermal expansion. In spite of all property improvements, the most important need for market penetration with carbon fiber is a ten-fold price reduction, to below $10/kg.

3.2.4 *Tires*
Partial replacement of carbon blacks with silica in tires has added greater wear resistance. However, this reduction has been at the expense of the electrical conductivity of the tire, leading to electrical isolation of the automobile from the ground. This undesirable outcome might be remedied by the addition of a small volume fraction of carbon nanofibers in the tread compound.

Rubber automobile tires reinforced with larger volume fractions of carbon nanofibers may ultimately improve other tire performance parameters, such as fuel efficiency, traction, and wear resistance. Carbon nanofibers are envisioned as a new dimension in rubber reinforcement, with promise to extend the engineering of the rubber compounds over what has been achieved with carbon black. The potential exists for partial replacement of carbon black with carbon nanofibers to reduce hysteresis losses, while maintaining approximately the same stiffness, traction, and strength. If the hysteresis improvement were to give the same decrease in rolling resistance, this would be equivalent to reducing the vehicle weight.

3.3 ELECTRICAL APPLICATIONS

3.3.1 *Static dissipation*
In the semiconductor industry, static dissipation is used to prevent damage to electronic devices during handling. Generally, plastic cases are used in these devices, which can build up static charges that may originate sparks. Sparks above 10-15 volts can render a semiconductor device unusable. Electrically conductive wafer and chip handling cassettes fabricated from high purity materials are used during the manufacturing process to transport the electronics components without risk of static discharge or contamination. Additional processing steps of wafer polishing, cutting, testing, packaging and assembly still provide ample opportunity for static damage to semiconductor materials. Even after assembly, boards are shipped in metal-coated plastics to prevent damage. Here, static dissipation composites may be used, and represent markets for carbon nanofibers.

3.3.2. *Batteries*
Graphitic systems are of interest in lithium ion batteries because they are easily intercalated. These power storage devices require an anode that is conducting, has high effective surface area, and the ability to be easily and reversibly intercalated with the Li

ions. Carbon nanofibers are prospective candidates for this use, due to the low cost and small diameter that results in a high relative surface area. Applications also exist in lithium batteries to enhance the conductivity of metal oxide cathodes.

3.3.3. *EMI Shielding*

There is a rapidly growing need for improved materials for shielding against electromagnetic interference (EMI), and containment of EM signals.

EMI can come from external sources, either natural, such as electrical storms, or artificial, such as transmitters. However, the main concern is EMI coming from other components of related or nearby electronic systems. This problem grows as more high-powered as well as more sensitive electronics are packaged into smaller spaces. It is particularly a problem when digital electronics are coupled with high power transmitters, as may occur in microwave systems or even cellular phone systems. Thus, there is a growing need for thin coatings that can help isolate some components of the systems from other components and from the outside world.

EMI shielding is used in almost all electronic devices. With the advent of small, portable devices such as laptop computers, carbon fibers have been increasingly utilized to provide the required properties of shielding, combined with lightweight and strength. Because other materials - some of which are less expensive - will provide adequate shielding, it is only when the manufacturer needs properties of reduced weight and increased strength that carbon fibers are selected. The trend in this market is for more portability and lighter weight, with the result that this market segment is growing faster than the total carbon fiber market in general - about 15% annually.

For EMI applications the fiber needs to act as an absorber/scatterer of radar and microwave radiation [4]. Carbon nanofibers have a combination of properties, which is particularly advantageous for this sort of application [5,6]. First, they have a moderate electrical conductivity. Some conductivity is necessary for a material to interact strongly with EM radiation, but too high a conductivity makes a material reflective. The moderate conductivity of nanofibers, similar to a poor metal, is ideal for absorption. Also advantageous is the high aspect ratio. Theory clearly shows that high aspect ratio of a particle leads to the best extinction efficiency (i.e., cross section per unit volume). The aspect ratio of nanofibers is typically in the range of 100 to 1000. Their specific dimensions are also useful. Radiation interacts most strongly with physical features that are comparable in size to its wavelength. Nanofibers are typically 50 to 100 microns long, an appropriate size for radar/microwave absorption. Finally, the efficiency is boosted by the small diameter. At the frequencies and conductivities of interest, the effects are all surface phenomena. Fibers of large diameter can have material too deep inside them to affect the process. At only about 0.1 microns in diameter, nanofibers allow all their material to participate in the absorption. Along this line, it is also important to note that the density of carbon nanofibers is only 2 g/cc, giving them a better specific conductivity than metal fillers, which are sometimes used in radar absorbing materials.

3.4 THERMAL APPLICATIONS

As electronic devices and actuators continue to shrink in size, the power density of these devices will continue to increase. As a consequence, thermal management is increasingly difficult, requiring new technologies to remove increasingly higher heat flux. Because the inherent thermal conductivity of well-ordered graphite is about 1950 W/mK, the ability to grow nanofibers at specific locations on semiconductor targets is essential. In addition, the nanofibers must be densely packed in order to maximize the amount of heat, which can be conducted

3.5 CEMENT

The market for improved cement or concrete materials is significant particularly for expensive structures, such as bridges and tunnels, where life cycle improvements are important. Research is being conducted on a variety of materials to reduce shrinkage, eliminate micro-cracking, enhance flexural strength and toughness and improve compressive strength for road and bridge repair.

Castable, high-strength cements are produced using Densification with Small Particles (DSP). Strong, dense materials are formed by casting mixtures of calcium silicate (Portland) cement and silica fume at very low water contents. The major hydration product of Portland cement, calcium silicate hydrate (commonly abbreviated C-S-H) gel, is a porous material, a composite of nanometer-sized particles and pores. Another approach for obtaining high strength materials are the Macro-Defect-Free (MDF) cements. MDF cements, a combination of inorganic cement and a water-soluble polymer, are processed under high-shear mixing on a two-roll mill. Chemical interactions between cement and polymer results in a strong, stiff composite.

Both types of cement can exhibit compression strength and modulus similar to steel, although tensile and flexural strengths are typically only about 10% that of steel and brittle behavior is usually observed. The microstructure of MDF and DSP is fine-grained with very low porosity, similar to conventional ceramic materials. The difference between MDF/DSP and conventional ceramics is that they solidify and gain structural integrity through hydration at room temperature rather than solidify through solid-state sintering at elevated temperatures. MDF and DSP are most often produced as hardened pastes, and do not typically include the addition of aggregates such as sand and gravel used in concrete. The aggregate in concrete contributes some degree of toughness by deflecting and arresting cracks. However, with little capacity in the microstructure to deflect or arrest cracks, MDF and DSP pastes exhibit nearly classic brittle behavior.

Fiber reinforcement is an established strategy for adding toughness to brittle materials. The scale of the fibers addresses different aspects of crack control. Typically long fibers (2-3 cm) are used in concrete to bridge cracks and provide closing stresses. The cracks initiate in their normal manner, but the long fibers serve to prevent propagation.

Carbon nanofibers have particular advantages for use in MDF and DSP cements because the extremely small diameter makes possible their inclusion in the mixing and processing stages. In addition, because the fibers are an entangled mass, it may be that their aggregate length is sufficient to be effective against crack propagation.

Preliminary samples show significantly improved machining characteristics compared to control samples without nanofibers, as shown in figures 2 and 3 below.

338

Figure 2. Nanofiber reinforced cement, molded, injection molded, and extruded into various complex shapes.

Empirically, it has been found that the addition of as little as one volume percent carbon nanofibers to cement can result in significantly modified properties. Tests of MDF and DSP with the carbon nanofibers have demonstrated that significant benefits are imparted to the material without compromising strength.

Figure 3. SEM micrograph of threads machined in nanofiber-reinforced cement. Very sharp structures, with less than 100 microns, can be cut into the material.

A newly identified emerging market is in wood composites. Composite wood is a large industry. The fast-growing replacement trees that the lumber industry has been planting yields poor quality wood. The industry is entertaining the possibility of combining wood with other materials to create a wood-like composite to meet construction volume and specifications. There is now opportunity to improve the structural properties of the manufactured plywood by the inclusion of carbon fibers in the outer glue layer. It has been determined that carbon fibers can be easily added to that layer to significantly improve the structural properties of the plywood. The primary barrier to such application is the cost of conventional carbon fiber. Because carbon nanofibers have potentially low cost, they would be natural candidates for this application.

4. Hydrogen Storage

Chambers, et al. [7], have made the claim that a single gram of carbon can absorb about 20 L (STP) of hydrogen at 120 atm. This corresponds to a mass absorption ratio of 1.76 grams of hydrogen, or an atomic ratio of about 10.6 atoms of hydrogen per atom of carbon. More recently, density-functional calculations by Lee and Lee [8] suggested that Single Wall Nanotubes have adsorption sites that would permit hydrogen storage of up to 14% by weight.

Attempts at ASI to reproduce the result of Chambers et al., have resulted in much lower absorption of hydrogen – only about 0.002 atoms of hydrogen per atom of carbon, or about four orders of magnitude lower than claimed. Moreover, we encountered potential error mechanisms that could yield erroneously high results. Hydrogen leakage can result in an apparent pressure drop in the system. Similarly, oxygen present in the system could in principle convert hydrogen to water, which would have a very small vapor pressure at ambient temperature, though the catalytic process by which this could happen has not been identified. In both cases, the result is an apparent hydrogen pressure drop in the sample vessel.

The procedure was to attempt to load the graphite lattice of carbon nanofibers at 13.79 MPa (136 atm), with retention of hydrogen maintained at a pressure of about 4.14 MPa (40.8 atm). The nanofibers used were produced at ASI but were of size similar to those used by Chambers, et al., and possessed a "herringbone" structure, in which the graphitic planes are actually oriented at an angle relatively to the axis. It was assumed that this feature might be essential to observing the above effect, although of course there can not be perfect assurance that these nanofibers are sufficiently similar to those used in reference [9].

To perform the test, approximately 1 gram of carbon nanofibers was soaked in 12 N hydrochloric acid for several hours at room temperature. The acid was filtered off using a fine mesh polyethylene screen and the remainder of the HCl adsorbed on the fibers was driven off at 75°C. The acid washed fiber sample was placed in the sample chamber, which was then heated to 150°C for several hours under vacuum.

Next, hydrogen at 12.16 MPa (120 atm) was admitted into the sample chamber. After sealing the chamber the pressure drop was recorded for a period of 24 hours. The original protocol did not include a leak check and in all cases using this protocol, a significant pressure drop was observed, which could be interpreted as adsorption of hydrogen. However, when a leak check step was instituted using pressurized helium, leaks were found in the system. After correcting the leaks, anomalous results were never obtained. The experience of this test suggests that helium leaks are common and can easily be misinterpreted as evidence of adsorption.

In a properly sealed system, it is possible to measure the sample size as well as the amount of absorbed hydrogen by using the ideal gas law. Specifically,

$$p_1 v_1 = p_2 v_2 \tag{2}$$

where p_1 and v_1 are the pressure and volume of the sample chamber, respectively; v_2 is the gas volume after it has expanded to the prechamber and auxiliary chamber, including the connecting gas lines, and p_2 is the corresponding pressure. To calibrate the system, a sample of known volume v_o is added to the sample chamber. Thus

$$p_1' v_1' = p_2' v_2' \tag{3}$$

where $v_1' = v_1 - v_o$ and $v_2' = v_2 - v_o$; p_1' and p_2' are the corresponding pressures.

Also, since the initial pressure is determined by the tank pressure,

$$p_1 (v_1 - v_o) = p_2' (v_2 - v_o). \tag{4}$$

and since

$$v_1 = \frac{p_2}{p_1} v_2, \tag{5}$$

then

$$v_2 = \frac{v_o (p_1 - p_2')}{(p_2 - p_2')}. \tag{6}$$

Thus v_1 and v_2 can be calculated from equations (5) and (6).

If hydrogen gas is used, and the pressure drops due to absorption, the ideal gas law can be used to estimate the amount of hydrogen missing from the gas phase:

$$\frac{p_1 v_1'}{T} = \frac{p_{1,f} v_1'}{T} + \frac{\Delta p_1 v_1'}{T} \tag{7}$$

where $p_{1,f}$ is the final pressure measurement, and Δp_1 is the difference between the initial and the final pressure measurements. In equation (7), the value v_o used to calculate v_1' is the weight of the fiber divided by its estimated density (not including the hollow core) of 1.95 g/cc. The number of moles of absorbed hydrogen can be calculated from

$$\Delta n_{H2} = \frac{\Delta p_1 v_1'}{RT} \tag{8}$$

where R is the ideal gas constant.

Table 1 summarizes typical results. The uncertainties quoted assume zero leakage, and so the possibility that micro-leaks could have resulted in overestimating the absorption cannot be excluded.

Table 1. Experimental values of H/C

Sample	Moles of H/C
CO_2 # 4 no acid wash	3.54×10^{-2} (+/-25.4%)
CO_2 # 4 acid washed	3.05×10^{-2} (+/-36.0%)
CO_2 # 7 acid washed	2.05×10^{-2} (+/-122%)
HT acid washed	1.18×10^{-2} (+/-132.7%)

Thus, under the conditions of these experiments, substantial absorption of hydrogen was not observed. It is then left to debate whether the attempted replication was adequate to duplicate the conditions of the original experiment.

Acknowledgements

This discussion was prepared with the benefit of the significant contributions of researchers at Applied Sciences, Inc., including R. Alig, D. Burton, G. Glasgow, J. Guth, J. Hager, R Jacobsen, E. Kennel, C. Tang and J-M. Ting, whose efforts and service are gratefully acknowledged. This work was sponsored in part by a NIST Advanced Technology Program under Agreement No. 70NANB5H1173, and the Ohio Coal Development Office under Grant No. CDO/D-96-3. The work on hydrogen storage was sponsored by NASA JSC Contract NAS9-98064.

References

1. Cox, H.L. (1952) Elasticity and Strength of Paper and other Fibrous Materials, *British Journal of Applied Physics* **3**, 72-79.
2. Baxter, W.J. (1992) An analysis of the Modulus and Strength of PYROGRAF/Epoxy Composites, Report No. PH-1717, General Motors Research Laboratories, Warren, MI.
3. Automobile Plastics Report (1999) published by Market Search Incorporated.
4. Bohren, C. and Huffman, D. (1983) *Absorption and Scattering of Light by Small Particles* Wiley, New York.
5. Jacobsen, R.L., Lake, M.L., Resetar-Racine, T. and Alexander Jr., R.W. (1995) Vapor Grown Carbon Fibers for Electromagnetic Scattering and Absorption, *Proceedings of the 22nd Biennial Conference on Carbon*, San Diego, 286-287.
6. Waterman, P.C. and Pedersen, J.C. (1992) Scattering by Finite Wires, *J. Appl. Phys.* **72**, 349-359.
7. Chambers, A., Park, C., Baker, R.T.K. and Rodriguez, N.M. (1998), Hydrogen Storage in Graphite Nanotubes, *J. Phys. Chem. B*, **102**, pp. 4253-4256.
8. Lee, S.M. and Lee, Y.H. (2000) Hydrogen Storage in Single-walled Carbon Nanotubes, *Appl. Phys. Lett.* **76**, 2877-2879.

ROUND TABLE 1

"Common and Disparate Elements in Filament and Nanotube Growth"

Notes taken by N. Grobert and D. Toma

Introduction by the chairman: G. G. Tibbetts

The starting point of the discussion was to distinguish nanotubes from carbon filaments. Briefly, the main characteristics of a nanotube are:

- An elongated structure with a small diameter and graphene layers parallel to the axis.
- The number of walls should be limited to one or only a few.
- Nanotubes should exhibit quantum size effects.

Filaments are characterized by:

- An elongated structure, which is multi-walled, quite thick, and usually possess a fishbone or a more complicated orientation of the graphene planes.
- Filaments do not exhibit quantum behavior.

Summary of the discussion

P. Bernier suggested that exact limits for the diameter should be given, e.g. up to 10 nm for nanotubes and ca. 100 nm for filaments. Nanotubes show an unusual conductivity (quantum behavior), which can be metallic or semiconducting and differs from the classical conductivity of macroscopic carbon systems. L. P. Biró added that the STS spectrum of thick carbon nanotubes (ca. 30 nm) is similar, if not identical, to that of graphite and does not reveal the characteristic features of nanotubes with diameters smaller than 30 nm. According to theory, the electronic band gap of a nanotube shrinks with increasing diameter and therefore this gap could be used to indirectly determine tube diameters.

M. Terrones noted that nanotubes are composed of highly crystalline concentric cylinders (up to 150 layers). Nanotubes can be classified as "quantum nanotubes" (less than 10 nm) and "nanotubes" (large diameter, highly crystalline cylinders), which are not filaments. Ph. Lambin stated that a semiconductor with a band gap E_g smaller than 25 meV (~300 K) behaves like a metallic conductor at room temperature. With $E_g = 2\gamma_0$ $d_{C-C}/d_t \approx 0.03$ eV, taking $\gamma_0 = 3$ eV, one arrives at $d_t \approx 30$ nm as a „natural limit" for the signature of quantum-size effects in a single nanotube layer. G. I. Márk remarked that impurities can drastically alter the transport properties of a nanotube. A. A. Lucas stated that „nanoscopic structure" literally means an object with dimension up to 50 nm, while objects larger than 100 nm in diameter are "microscopic structures".

P. Bernier suggested that the third dimension should also be taken into account in the classification; thus objects of one to a few microns in length should be called nanotubes and structures with lengths of several tens of microns should be called filaments. L. P. Biró objected that the diameter has a stronger influence on physical properties than the length. It is thus more important to differentiate nanotubes and filaments according to

L.P. Biró et al. (eds.), Carbon Filaments and Nanotubes: Common Origins, Differing Applications, 343–345.
© 2001 *Kluwer Academic Publishers. Printed in the Netherlands.*

their diameters. G. G. Tibbetts and M. Endo maintained that the aspect ratio and not length is the truly fundamental parameter for applications such as reinforcement.

The Chairman suggested that the roundtable consider whether growth conditions might distinguish between nanotubes and filaments. Typical growth conditions are compared in the following Table:

Table. Typical growth conditions for VGCFs and carbon nanotubes

	Filaments	Nanotubes
Temperature	More than 500 °C	More than 950 °C
Gas	CO/CO_2, HC	C, HC, CO/CO_2
Catalyst	Fe, Ni, Co	None, Fe, Ni, Co
Time	2-20 min	---
Growth rate	1 mm/min	0.3 mm/min

L. P. Biró recalled that the growth conditions for producing single-wall (SWNT) or multiwall nanotubes (MWNT) are not the same, and the rates at which these two systems grow are different. For instance, the conditions used in Namur are $T = 700$ °C with C_2H_2 or C_2H_4 gases for MWNTs; $T = 1000$ °C and CH_4 gas for SWNTs. P. Bernier reported that the HIPCO process, based on the disproportionation of CO, can be carried out as low as 600 °C, but best yields are achieved at 1000 °C. L. P. Biró remarked that there is not a drastic difference between the formation temperatures of carbon filaments and carbon nanotubes. M. Endo pointed out that the ultimate goal is to make carbon filaments and nanotubes at *room* temperature with innovative chemical techniques in order to save energy and reduce the production cost. R. Ehrlich added that the formation of SWNTs or MWNTs has been achieved at 300 °C by the decomposition of fullerenes in the presence of metal catalysts. Therefore, low temperature (possibly room temperature) formation should be possible.

The discussion then ranged to the issue of whether growth with catalysts or without might distinguish between NT's and filaments in a more fundamental way. G. Tibbetts noted that much of the early VGCF growth was performed by researchers who did not understand that catalytic particles were necessary for such growth, and that frequently the catalyst particles cannot be found at either end of a filament after cessation of growth. So this is a difficult experimental issue, and we might easily be confused about it. One example is that researchers believe that SWNT's are more easily grown in a system with catalyst containing electrodes, while MWNT's are better grown with no catalyst. However, the MWNT's are really formed by the very few SWNT's grown in the discharge from impurity particles, and they are substantially thickened because there are few competitors for the active carbon atoms. M. Endo responded negatively to this, citing the high purity of benzene he uses and maintaining that he can grow substantial numbers of NT's with no impurity or catalyst, even though yield improves with a sulfur catalyst.

M. Terrones reported that growth rates with carbon arc discharge are up to 5 micron/s. Even if the arc is maintained for a longer period of time, the nanotubes do not increase in length (a 20 min arc yields 10 micron long MWNTs). L. P. Biró added that high-energy irradiation with heavy ions also leads to extremely fast growth rates for carbon nanotubes. Ph. Lambin pointed out that, in the catalytic CVD process used in Namur, the growth rate for SWNTs is of the order of 5 microns per *minute*. L. P. Biró commented that the growth rates may be related to the rate of the carbon generation in the system; therefore, growth is controlled by the availability of active C atoms.   J. L. Figueiredo added that in the case of filaments the growth rates are also related to the impinging of the carbon-containing molecules to the surface of the catalyst, and that can be highly temperature dependent. P. Bernier noted that that filaments need more carbon than SWNTs due to their much larger diameter, therefore the growth velocity should be different. For instance, ca. 20 atoms are needed to elongate a SWNT by one carbon atom, yet several thousands are required for the elongation of a thicker filament. M. Endo added that the growth rate of carbon nanotubes may be extremely dependent on the process (e.g. CVD, catalysis or carbon arc discharge) as different temperatures are involved.

At the conclusion of a long discussion on a warm afternoon, the group began to wax philosophical. G. G. Tibbetts concluded with a surmise that two such similar structures as filaments and NT's , formed in related ways, may really be identical creatures at the NT core level. "If it walks like a duck, if it swims like a duck, and if it quacks like a duck, then it MUST BE A DUCK!", to which M. Endo added "In conclusion, we see what we want to see!".

ROUND TABLE 2

"Properties of Imperfect Carbon Structures"

Notes taken by J. M. Nugent and C. Neamtu

Introduction by the chairman: P.M. Ajayan

The discussion starting point was to define imperfections and characterize disorder:
- Definition of imperfections
  - Structural disorder
  - Degree of graphitization
  - Organization of graphite planes
- Charaterization of disorder
  - Local vs. global disorder
  - X-ray, Raman, TEM/SEM, STM/AFM
- Origins of disorder
  - Graphitization mechanism
- Properties
  - Mechanical
  - Electrical
  - Thermal
  - Electrochemical

Finally, we should consider if we *really want* perfect carbon nanotubes or fibers.

Summary of the discussion:

In general, the definition of what is a defect in a carbon fiber or nanotubes was discussed without a clear conclusion. It was decided that the application of the nanotube or fiber will dictate what should be regarded as a defect. For electrical applications of nanotubes, defects drastically alter the characteristics. For the case of VGCF composite mechanical properties, G. Tibbetts pointed out that a highly graphitized fiber might have been expected to give better tensile strength, yet the opposite was observed

This led to a discussion regarding plasma treatment to control the surface structure and chemistry of nanotubes and fibers for adhesion to a polymer matrix. A poster describing the use of Ar/NH_3 plasma treatment of nanotubes to attach functional groups at 1 out of every 1000 carbon atom sites was mentioned. The consensus of the participants was that the addition of functional groups to nanotubes is primarily interesting for mechanical reinforcement. P. M. Ajayan pointed out that it is not known whether the addition of functional groups will affect the electrical properties favorably or unfavorably. For enhancing electrical conductivity using carbon fibers, it is not so important to have a perfect structure. G. Tibbetts and C. Bernardo also pointed out that in making composites with a combination of desirable properties (ie. electrical conductivity and mechanical strength), it is often necessary to accept a compromise between the structural perfection of the fibers used and the desired mechanical strength. Z. Kónya pointed out that functional groups can also be added through oxidative treatments. S. Bonnamy pointed out that plasma treatment is used to enhance the mechanical properties in PAN based fiber composites. A discussion developed about

347

L.P. Biró et al. (eds.), Carbon Filaments and Nanotubes: Common Origins, Differing Applications, 347–349.
© 2001 *Kluwer Academic Publishers. Printed in the Netherlands.*

nanotubes with structures other than functional groups. N. Neves asked about making a branched nanotube to improve the mechanical interlocking and enhance the mechanical properties of a composite material. P. M. Ajayan and M. Terrones noted that this type of structure has been observed by M. Endo. A discussion then centered about defining what a nanocomposite is. P. M. Ajayan replied that the definition of a nanocomposite is that one phase of a material has at least one dimension on the nanoscale. Ph. Lambin pointed out that it is difficult to define and determine the electrical properties of a nanotube in a composite material.

M. Terrones showed a slide of a Haeckelite nanotube the structure of which incorporates many pentagons and heptagons, which are considered as defects in a normal, graphite-like nanotube. He pointed out that the mechanical properties would be the same for both perfect and defected materials, but that the electrical properties would differ. The problem is that we don't know how to make these tubes. "So you can tailor the properties by introducing *things*." N. Grobert noted that the term defects can have a negative connotation when in fact they can be very useful for certain applications. B, N, and S were discussed and the many questions about their effects on the properties of both nanotubes and fibers were raised. G. Tibbetts responded for the fiber group by saying that proper fiber function seems to be unhampered by these impurities.

P. M. Ajayan pointed out that there has to be examination of defects at the atomic as well as the macroscopic scale. While atomic defects are interesting to the nanotube community in terms of electrical properties, atomic defects may not be important to the properties of fibers at all. P. M. Ajayan asked the fiber experts many times about the classification and characterization of defect structures in carbon fibers. P. M. Ajayan also questioned the fiber community about their definition of a perfect fiber. S. Bonnamy, C. Bernardo and G. Tibbetts stated that fibers are not perfect and that the properties important for practical applications do not depend on attaining a perfect structure. S. Bonnamy pointed out that the term "degree of graphitization" is not relevant to carbon nanotubes and that for VGCF the graphitization occurs after high temperature thermal treatment of turbostratic carbon.

Both G. Tibbetts and I. Pócsik pointed out that the fiber industry approaches applications from an engineering standpoint. When developing an application, one defines what properties are needed and then optimizes the material to fit that criteron. This approach seems to be much different from that of the nanotube community, which tries to understand each structure and then find applications for it. Terrones pointed out that nanotubes would probably have been identified in the 1950's if TEM technology were developed to the present standard, and reminded the listeners of M. Endo's comment of yesterday about microscopy, "You see what you want to see".

A debate originated on the usefulness of global as opposed to local analysis techniques for defecting defects. Raman, TEM, AFM and STM were discussed and the limitations were pointed out. Raman is not really useful for detecting defect structures in nanotubes according to E. Obraztsova. For many other techniques discussed, only one tube can be examined at a time. M. Lake listed some analysis techniques for fibers: X-ray, ESCA, SIMS, BET, and inverse gas chromatography are all used to understand the surface chemistry and structure. The general consensus is that to fully understand a structure, a series of techniques is necessary.

Ph. Lambin raised the point that there is a mathematical model for a perfect nanotube and that there is no such model for a fiber. Without a model, it is difficult to discuss defects. J. Charlier disagreed that there was no model for fibers, and cited M.

Dresselhaus, who modeled carbon fibers even though perhaps not on the atomic scale. He also commented that defects can be defined by the scale of the model used for a specific system.

Key Points of Discussion:

1) How are defects defined for both carbon nanotubes and fibers and what is perfection?
2) How do we characterize the defects?
3) Defect structures can have desirable properties for one application but detrimental properties for another application.
4) There is a difference between the engineering approach in fiber based composites and the scientific point of view in nanotube studies.
5) It really may not be necessary to understand defect structure, depending on what applications are envisaged.
6) Some differences in studying fibers and NT's may come from differences in the analytical capability available at similar stages in the development of these two materials, as well as the relative sizes of the systems.

ROUND TABLE 3

"Likely Applications: Near Term and Long Term"

Notes taken by F. W. J. van Hattum

Introduction by the chairman: M. L. Lake

The Chairman cited lessons learned from large US government investments in R&D programs:

- In general, research investments create opportunity windows, but only for a limited time.
- Even if the original research goals are not reached, industrial spin-offs can still make the research successful.

Keeping these lessons in mind, the Chairman suggested as a discussion framework the creation of a list of opportunities and barriers for near term and long term applications that would have the economic potential to create viable new industries.

Summary of the discussion:

C. A. Bernardo proposed that a sharp division between nanotubes (NTs) and filaments (NFs) may not be needed from the point of view of applications.

D. D. L. Chung pointed out that electromagnetic and electrochemical applications can be considered as near term applications for NFs. Some applications require a thermal conductivity of polymer matrix composites of only 11 W/mK (easily attainable with ceramic particles as fillers). For thermal pastes it is not so much the filler thermal conductivity but the ability of the filler particles to touch each other in order to supply good thermal contact that dominates.

The Chairman proposed that the roundtable focus its discussion on applications that can create new industries.

Ph. Lambin noted that field emission devices seem to be promising, although it is not yet known if NTs will be the best candidate. However, the cost of NTs are currently the limiting factors for application in field emission devices. P. M. Ajayan added that MWNTs as produced by arc discharge may cost $60-100/g. M. Terrones made the observation that in Korea and Japan huge amounts of money are being invested in research on application of NTs in field-emission devices. Samsung Korea has already produced with this technology a large Flat Panel Display (FPD), and displays for hand-held cameras. The electronics industry in Japan requires large quantities of NTs for several potential applications, so producing sufficient quantities of NTs is really a fundamental issue. Similar applications could be expected for NFs with NTs at the tip as well. J.-C. Charlier mentioned that improvements might be expected after optimizing NTs for specific purposes; for instance, boron atoms on an NT tip could enhanced field emission.

The production of nanotube based FPDs may prove to be a big market, according to L. P. Biró, which again points to the need for large-scale production of NTs. Growing NTs catalytically seems to be the most promising method, because we

351

L.P. Biró et al. (eds.), Carbon Filaments and Nanotubes: Common Origins, Differing Applications, 351–354.
© 2001 *Kluwer Academic Publishers. Printed in the Netherlands.*

can build on the industrial experience accumulated in NF production. Flat panel displays can be considered to be a near term application of NTs. Recently the feasibility of growing well oriented, regular arrays, 'forests' of NTs on any surface has been demonstrated. Growth on glass would allow the use of cheap and flat support material for large area FPDs. These 'forests' could also be used in water or food purification.

P. M. Ajayan pointed out that the symbiosis between bio- and nanotechnology will increase. NTs will allow new structures and manipulation on the nanoscale, meaning that characterization of NTs will certainly become an issue. The catalytic process is now the most promising large-scale production method for NTs; moreover, those using the arc discharge method should focus on industrially available arcs for the sake of practicality.

The Chairman asked the roundtable to consider electronic and electrochemical applications:

Electrochemical applications are promising for both NTs and NFs in batteries, capacitors, and fuel cells, said D. D. L. Chung. Careful comparison of these novel materials with traditional carbons in electrochemical systems should focus on the *complete* cell system. Application of carbon filaments, including Ni-coated filaments, in conducting concrete for electromagnetic shielding is 'ready to go'. Near term applications also exist for mesoporous carbon in purification, where activated carbon filaments show unique properties.

Let us now focus on long-term applications and barriers, said the Chairman.

NTs are promising in nanoelectronics said L.P. Biró. However, absence of large-scale production limits their opportunities. Two approaches might be envisaged: 'On the spot' growth of NTs which would dramatically reduce the problems arising in contacting each NT, or large-scale production and NT selection before the fabrication of components. J.-Ch. Charlier remarked that much more research will be necessary on this issue, as it is currently impossible to select NTs for desirable properties. For nanoelectronics, nanowires and junctions seem better suited than NTs. The chemistry of NTs might not allow us to functionalise SWNTs for specific properties in the immediate future. In the longer term, application of these SWNTs as sensors can be considered, according to M. Terrones. For application of NTs as semiconducting/metallic tubes, precise chirality control is required which is not currently possible. In Ph. Lambin's opinion, this would require controlled growth, which has so far been little investigated due to lack of analytical tools to study growth *in situ* at the atomic level. L. P. Biró noted that perhaps we can utilize the diagnostic methods used in plasma studies, especially for the arc-discharge technique.

P. M. Ajayan pointed out that possible long term applications for NTs can be sought in the field of biotechnology and biochemistry, where there is a huge need for new technology, especially for molecular recognition. The use of nanotools might even be a near term application. H. Kanzow suggested that the NT alternator (electromechanical) devices as studied in Montpellier might a near term application, although cost would again be a barrier. He affirmed that this applications has been tried unsuccessfully with traditional carbon fibers.

M. Terrones remarked that in general results from experiments on NTs seem to change from application to application and sometimes from sample to sample, so exhaustive study and testing will be required.

The Chairman asked for suggestions for structural applications

N. Grobert commented that so far during this institute the only structural applications discussed have utilized matrix materials. Could NFs/NTs possibly be used without matrix materials?

M.L. Lake responded that his company has investigated several promising applications that used fibers in paper form.

C. A. Bernardo pointed out that research on NF composites has so far been mainly focussed on attaining good mechanical properties. The focus is now shifting to electrical and thermal properties. Future research will focus on preform threads comprised of NFs and polymer (with specifically tailored properties) to incorporate in structures. D.D.L. Chung suggested that a near term structural application could be the use of NFs as an intermediate layer between traditional fiber plies for increased damping.

P.M. Ajayan asked if smaller reinforcements would necessarily be better. M.L. Lake commented that in composite theory strength is not a function of fiber diameter; generally, strength is simply dominated by defects. F.W.J. van Hattum commented that micromechanical modeling indicates that in many regimes of length/diameter the matrix material will be the limiting tensile strength.

Collaboration and conclusion

P. M. Ajayan suggested that it would be helpful to organize possible collaborations.

C.A. Bernardo volunteered that 10-100 g of NTs would be sufficient to use in rheological experiments to obtain valuable information on the use of NTs as reinforcement in polymer matrix composites. Ph. Lambin suggested using a mix of VGCFs and NTs for applications to composites.

M. L. Lake commented that cost, growth control and characterization of the materials seem to be common concerns among many groups. M. Terrones suggested performing tests and research in different labs on the same filaments and NTs to increase insight and make optimum use of different expertise .

354

Table 1: Near and Long Term opportunities and barriers for NT/NF applications

	Opportunities	*Barriers*
Near Term	Field Emission devices (NT&NF)	Cost Production
	EMI-shielding (EMI gaskets) (NF)	-
	Electrochemical applications (NT&NF)	Cost Research
	Electromagnetic concrete (NF)	-
	Fluid purification (NF)	Cost
	Damping in lightweight structural polymer-matrix composites (NF)	-
	Lithium-ion batteries (NF)	Existing vapor-grown carbon fibers
Medium and Long Term	Composite structures made with NF/polymer fibers (NF)	Develop intermediate technology
Long Term	Nanotechnology tools (NT)	Characterization Interface with external world
	Biotechnological applications (NT&NF)	Research
	Nanoelectronics (NT)	Growth control & assembly
	Sensors (NT)	Cost

SUBJECT INDEX

A

activated carbon, 41, 86, 122, 282, 283, 284, 352

activated carbon fibers, 122, 282

activation, 6, 112, 115, 122, 124, 125, 130, 283

active sites for absorption, 155

additive in the electrode, 58

adjacent coaxial tubes, 149

AFM, 101, 102, 121, 122, 123, 124, 129, 131, 174, 175, 255, 256, 257, 258, 259, 260, 261, 262, 266, 316, 318, 319, 347, 348

aligned nanotubes, 172, 173, 321

alignment, 104, 172, 188, 191, 192, 193, 194, 279, 281, 289, 292, 293, 294, 295, 296, 297, 299, 302, 334

alloy phase generation, 94

alumina, 2, 65, 70, 86, 88, 91, 92, 94, 258, 326

alumina substrate, 65, 70

amorphous, 58, 65, 79, 86, 90, 92, 94, 95, 98, 101, 150, 160, 166, 174, 280, 311

amorphous carbon, 58, 79, 86, 90, 92, 94, 95, 98, 101, 174

amorphous-like tip, 151, 160, 165, 167

anisotropy, 8, 236, 238

anti-Stokes, 37, 38, 39

apparent diameter, 223

apparent height, 225, 261

application as polymer fillers, 104

Applied Sciences, Inc., 9, 196, 291, 299, 313, 341

arc discharge, 41, 75, 76, 77, 78, 80, 81, 102, 134, 150, 167, 172, 174, 180, 344, 345, 351, 352

arc discharge method, 77, 78, 102, 134, 180, 352

armchair, 12, 13, 14, 21, 22, 23, 24, 26, 33, 34, 35, 44, 100, 150, 151, 152, 153, 154, 157, 159, 160, 163, 165, 166, 167, 168, 173, 236, 237, 238, 241, 266, 272

asymmetry, 31, 143, 225, 226, 238, 270, 273

asymmetry in the tunneling current, 225

atomic force microscopy, 123, 255

atomic helices, 201

atomic rearrangement, 153

atomic resolution, 29, 60, 219, 220, 221, 225, 226, 227, 233

atomic scattering factor, 201

atomic structure, 197, 219, 266, 269, 270

axial period, 206

B

B and N profiles, 141

B doping, 150, 151, 156, 158, 168

B-C pair, 159

BCN, 171, 181

B-C-N nanotubes, 134, 144

$B_xC_yN_z$, 171, 180

B-doped carbon nanotubes, 158

bending, 55, 158, 166, 257, 258, 273

bending of the cantilever, 257

Bessel functions, 209

BN - C nanotubes, 133

BN MWNTs, 167, 168

BN nanotube, 141, 150, 151, 159, 160, 161, 162, 163, 164, 165, 166, 167, 168

BN SWNTs, 163, 165, 166, 168

bond length reduction, 161

boron nitride SWNTs, 150

Bragg circles, 212

Brillouin zone, 14, 17, 18, 20, 21, 25, 30, 151

broadening, 224, 225

buckling, 161

bulk density, 8, 188

bundles of NT, 208

burn-off, 123, 125, 127, 128, 129, 130

C

cantilever, 129, 256, 257, 319

355

H

I

J

K

L

364